THE ARRL ANTENNA COMPENDIUM VOL 4

Editor
R. Dean Straw, N6BV

Assistant Editor
Paul M. Danzer, N1II

Production
Shelly Bloom, WB1ENT
Dan Wolfgang
Steffie Nelson, KA1IFB

Technical Illustrations
Dave Pingree, N1NAS
Joe Costa

Cover Design
Sue Fagan

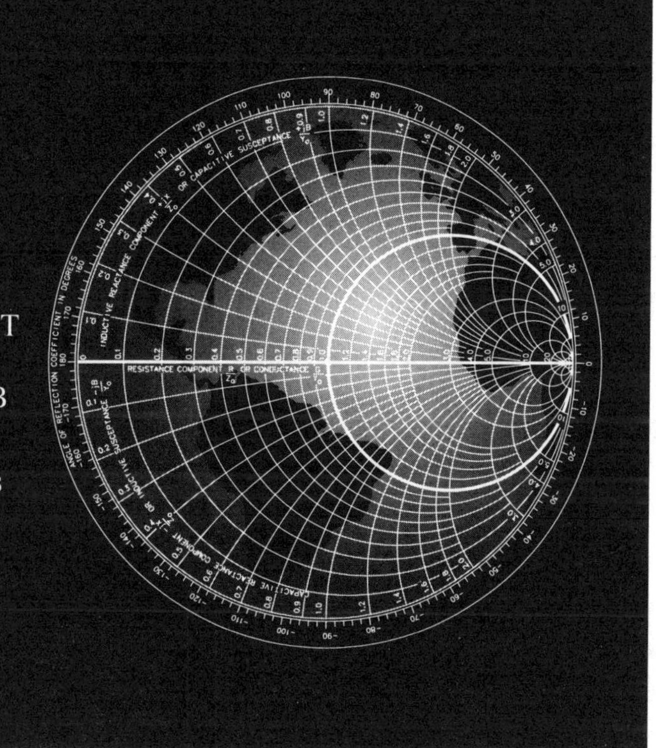

Printed in USA

First Edition
Second Printing, 1996

ISBN: 0-87259-491-2

Foreword

Hams love antennas! The ARRL receives far more antenna manuscripts than any other sort, far more than can possibly be published in *QST*. Almost ten years ago, the first volume in the *ARRL Antenna Compendium* series went to press, providing a great vehicle for many worthy articles that might otherwise not see the light of day.

This is the fourth volume in this popular series. It contains 38 previously unpublished articles, covering a wide range of antenna-related topics—all the way from math-intensive, heavyweight discussions like Frank Witt's article about the scientific optimization of the 80-meter dipole, to fun antennas for specific purposes, such as Ricky Bibby's balloon-supported Field Day loop.

Are you fascinated with 10-meter longpath propagation? So is Carl Leutzelschwab. Check out his revealing article in this volume. Are you heavily into computer modeling of antennas? So are a lot of antenna specialists, like Brian Beezley, Al Christman, Jim Breakall and Jack Belrose. They all have contributed to this volume.

Are you an HF mobile person, wondering about the relative efficiency of different loading schemes for mobile whips? Three articles in this volume deal with this very subject, in considerable detail. Or are you a low-frequency aficionado? (You might just as well be, if you're going to be able to deal with the ups and downs of the sunspot cycle!) Seven articles in Volume 4 are devoted to 80 and 160-meter skyhooks, some of them pretty impressive. Take Lloyd Koontz's 80-meter Yagi with a 338-foot boom, for example. Wouldn't we all love to have a monster skywire like that? Hello, Japan!

For the first time in this series, we've bundled with the book itself a disk containing the source data used to model many of the antennas. Hams interested in the details of antenna modeling can gain a lot of insight into the process by looking over source data created by experts. By bundling the data with the book, the League can provide the reader with valuable information, take advantage of volume savings, and actually save by *not* trying to offer data disks separately, as was done for earlier volumes.

The data files are used with several different commercial modeling programs (*MN*, *AO*, or *NEC/Wires* by Brian Beezley, *ELNEC* by Roy Lewallen, or *NEC2*.) Please remember that we do not provide the modeling programs themselves, only the data that work in them!

In short, there's something for virtually every antenna enthusiast in Volume 4 of *The ARRL Antenna Compendium*! Perhaps you are inspired to write an antenna article of your own? We'd love to see it, as we prepare next for Volume 5.

David Sumner, K1ZZ
Executive Vice President
Newington, Connecticut
January 1995

Instructions for Accompanying Diskette

The diskette bundled in the back of *The ARRL Antenna Compendium, Vol. 4* includes numerous data files created by the authors to analyze their antennas, using commercially available antenna modeling software such as *NEC2*, *NEC/Wires*, *ELNEC*, *MN* and *AO*.[1] [**Note**: the ARRL does not include the modeling software itself on the diskette, only data for these programs!] Several authors also wrote special analysis programs for their articles. Executable versions of these are also on the diskette, together with source code, when applicable.

The diskette data is organized into 17 separate subdirectories, each named using the author's amateur call sign. For example, the data corresponding to the article by Michael Orr, AA2PE, is found in the \AA2PE subdirectory, while the two articles by Peter Dodd, G3LDO, refer to disk files found in the \G3LDO subdirectory. The *MOBILE.EXE* program by Leon Braskamp, AA6GL, is in the \AA6GL subdirectory.

With the exception of the executable programs, it is unlikely that you will want to transfer the entire diskette to your hard disk. However, we do recommend that you copy the original diskette and store it safely as a backup, should you accidentally overwrite a file on the working copy. When you wish to examine or use a particular data file, get into the appropriate subdirectory from the DOS prompt using the "CD" (Change Directory) DOS command. The files should be examined using a text word processing program, since each file is an ASCII data file.

Each antenna data file has a distinct filename extension corresponding to the antenna-analysis program in which it is used. The filename extensions on the disk are:

*.NEC	—	used for the *NEC2* program
*.ANT	—	used for the *MN*, *NEC/Wires* or *AO* programs by K6STI
*.EN	—	used for the *ELNEC* program by W7EL
*.BAS	—	BASIC source code
*.PAS	—	PASCAL source code
*.EXE	—	executable file

Even if you are an experienced antenna modeler, you will gain some insight into how the experts work by examining their data files. Some very interesting techniques are displayed in a number of the data files, and it certainly beats typing in the data manually when you wish to see if you can possibly improve or "tweak" a design any further!

[1] The *NEC2* program is available from the Applied Computational Electromagnetics Society, c/o Dr. Richard W. Adler, Code 62AB, Naval Postgraduate School, Monterey, CA 93943.

NEC/Wires, *MN*, and *AO* are available from Brian Beezley, K6STI, 3532 Linda Vista Dr, San Marcos, CA 92069, 619-599-4962.

ELNEC is available from Roy Lewallen, W7EL, PO Box 6658, Beaverton, OR 97007.

Contents

80 and 160-meter Antennas

Antenna Modeling

Mobile Antennas

Multiband Antennas

Portable/Temporary Antennas

Propagation

Receiving Antennas

Rotators and Measurements

Transmatches and Transformers

Transmission-Line Measurements and Accessories

VHF/UHF Antennas

About the American Radio Relay League

The seed for Amateur Radio was planted in the 1890s, when Guglielmo Marconi began his experiments in wireless telegraphy. Soon he was joined by dozens, then hundreds, of others who were enthusiastic about sending and receiving messages through the air—some with a commercial interest, but others solely out of a love for this new communications medium. The United States government began licensing Amateur Radio operators in 1912.

By 1914, there were thousands of Amateur Radio operators—hams—in the United States. Hiram Percy Maxim, a leading Hartford, Connecticut, inventor and industrialist, saw the need for an organization to band together this fledgling group of radio experimenters. In May 1914 he founded the American Radio Relay League (ARRL) to meet that need.

Today ARRL, with more than 170,000 members, is the largest organization of radio amateurs in the United States. The League is a not-for-profit organization that:

- promotes interest in Amateur Radio communications and experimentation
- represents US radio amateurs in legislative matters, and
- maintains fraternalism and a high standard of conduct among Amateur Radio operators.

At League headquarters in the Hartford suburb of Newington, the staff helps serve the needs of members. ARRL is also International Secretariat for the International Amateur Radio Union, which is made up of similar societies in more than 100 countries around the world.

ARRL publishes the monthly journal *QST*, as well as newsletters and many publications covering all aspects of Amateur Radio. Its headquarters station, W1AW, transmits bulletins of interest to radio amateurs and Morse code practice sessions. The League also coordinates an extensive field organization, which includes volunteers who provide technical information for radio amateurs and public-service activities. ARRL also represents US amateurs with the Federal Communications Commission and other government agencies in the US and abroad.

Membership in ARRL means much more than receiving *QST* each month. In addition to the services already described, ARRL offers membership services on a personal level, such as the ARRL Volunteer Examiner Coordinator Program and a QSL bureau.

Full ARRL membership (available only to licensed radio amateurs) gives you a voice in how the affairs of the organization are governed. League policy is set by a Board of Directors (one from each of 15 Divisions). Each year, half of the ARRL Board of Directors stands for election by the full members they represent. The day-to-day operation of ARRL HQ is managed by an Executive Vice President and a Chief Financial Officer.

No matter what aspect of Amateur Radio attracts you, ARRL membership is relevant and important. There would be no Amateur Radio as we know it today were it not for the ARRL. We would be happy to welcome you as a member! (An Amateur Radio license is not required for Associate Membership.) For more information about ARRL and answers to any questions you may have about Amateur Radio, write or call:

ARRL Educational Activities Dept
225 Main Street
Newington CT 06111-1494
(203) 666-1541
Prospective new amateurs call:
800-32-NEW HAM (800-326-3942)

80 and 160-Meter Antennas

The Slant-Wire Special

By Al Christman, KB8I
c/o Grove City College
100 Campus Drive
Grove City, PA 16127-2104

KB8I thoroughly analyzes a 160-meter directional array created by adding slant-wire reflectors to a grounded vertical monopole.

Introduction

This article shows how to add one or more "slant-wires" to a conventional ground-mounted vertical-monopole antenna to produce a very simple directional array with 5 dBi of forward gain and a front-to-back ratio of 19 dB. If desired, the original omnidirectional radiation pattern can also be restored by simply "floating" the slant-wire(s).

Background

Some time ago, the FCC published and distributed a Notice of Inquiry (NOI) in order to determine what portions of their rules for directional AM-broadcast antenna arrays should be changed in order to better meet the needs of the industry. One of the respondents to this NOI quoted an interesting paper presented at the 1987 Broadcast Engineering Conference of the National Association of Broadcasters.[1] I spoke to the author of this paper, and realized that the concept he described could easily be applied to the ground-mounted vertical-monopole antennas hams often use for "low-band" communications. As a result, I performed some computer-modeling studies utilizing *ELNEC* version 3.03,[2] and the outcome is presented in this article.

Start with a Tower

In practice, the slant-wire concept is applied to a single preexisting vertical monopole antenna, which is generally made of tower sections. Normally the tower is ground-mounted and fed across a base insulator, with a ground system composed of many radials. In AM broadcast usage, there are often 120 buried radials, and the length of each is usually 0.25 λ or more. Please note that in this article I have not tried to analyze the "slant-wire" system when utilized in combination with elevated radials.

Since the idea seems to have originated within the AM broadcast community, I decided to model the slant-wire antenna on the 160-meter amateur band, at a frequency of 1.84 MHz. The basic design used in my studies consists of a 130-foot vertical radiator made of aluminum and having an equiva-

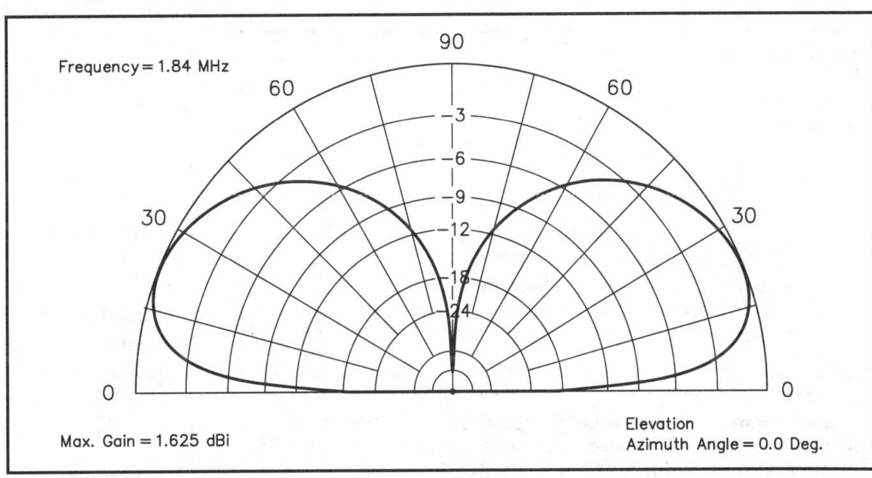

Fig 1—Elevation-plane radiation pattern for a 130-foot base-insulated tower operating at 1.840 MHz. An extensive radial system functions as a ground for this vertical monopole.

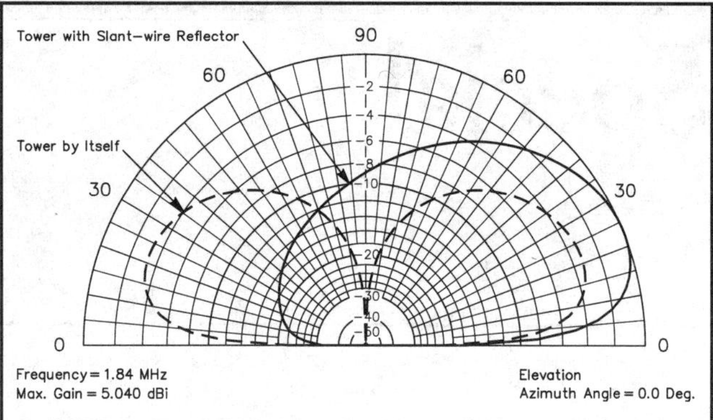

Fig 2—Elevation-plane radiation pattern for the 130-foot tower when a single slant-wire reflector is added, compared to that of the tower in Fig 1 by itself. The slant-wire is 133.5 feet long, and its grounded lower end is located 90 feet away from the base of the tower, opposite to direction of main lobe.

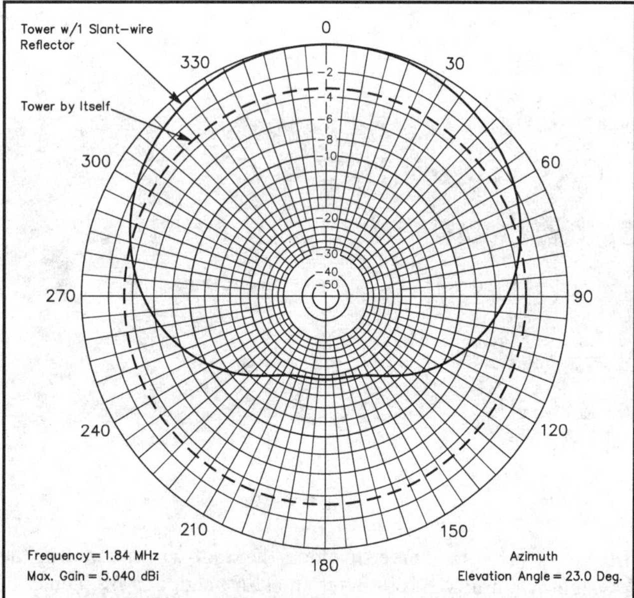

Fig 3—Azimuthal-plane radiation pattern at elevation angle of 23° for the 130-foot tower with slant-wire reflector. Again, the reference antenna is the tower by itself.

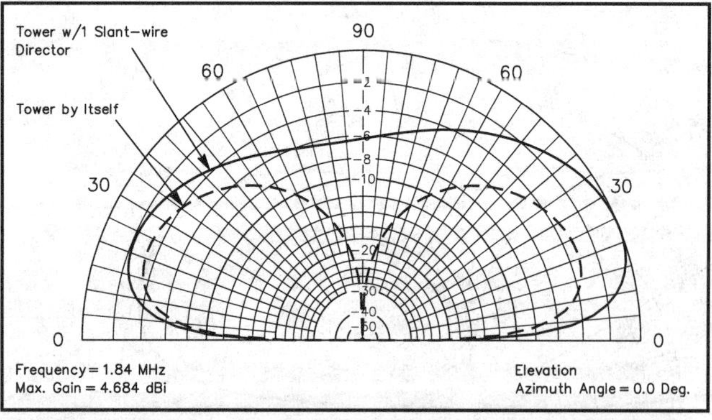

Fig 4—Elevation-plane radiation pattern for the 130-foot tower with single slant-wire, retuned now as a director, together with the reference tower by itself. The slant-wire is 133.5 feet long, and its grounded lower end, which is loaded by a capacitive reactance of 40 Ω, is located 90 feet away from the base of the tower.

lent diameter of 8 inches. The ground constants used in the *ELNEC* program are 13 for the relative permittivity, and 0.005 Siemens per meter for the conductivity. The elevation-plane radiation pattern, shown in **Fig 1**, indicates a maximum gain of 1.625 dBi, at a take-off angle of 22°. The calculated input impedance is 37.36 + *j* 6.46 Ω, which indicates that the tower is slightly long electrically (the reactive component is inductive), even though its physical height is somewhat less than λ/4 (133.65 feet at 1.84 MHz).

Add One Slant-Wire

The first step in the design of this array is to choose the proper location and length of the slant-wire so that it will act as a parasitic element. Initially, the base of the slant-wire was placed at an arbitrary distance from the vertical radiator, and grounded directly to earth. The #12 AWG slant-wire is oriented so that its upper end (if it were long enough) would be tied directly to the top of the tower. After trying several different combinations of length and spacing, I found that a wire length of 133.5 feet gave good results as a reflector when the base of the slant-wire was located 90 feet away from the tower.

With the lower end of this slant-wire solidly grounded to earth, *ELNEC* predicts a maximum forward gain of 5.040 dBi at a take-off angle of 23°, and a front-to-back ratio of 19.038 dB. The feedpoint impedance at the base of the tower is 36.09 + *j* 55.35 Ω. The elevation-plane radiation pattern is displayed in **Fig 2** and the corresponding azimuthal-plane radiation pattern in **Fig 3** reveals a half-power beamwidth of 146°. Note that the slant-wire, like the tower itself, requires a high-quality RF ground connection to work well. For reference, the response of the tower by itself is overlaid on both the elevation and azimuth patterns.

To restore the conventional omnidirectional pattern normally expected for a single vertical monopole antenna, it is only necessary to lift the ground connection at the base of the slant-wire, so that it "floats" with respect to earth. The presence of the open-circuited slant-wire has only a very minor effect on the radiation pattern. The input impedance for the antenna in the omnidirectional mode is 35.98 + *j* 6.94 Ω. With a drive power of 1500 W into the base of the tower, the current at the bottom of the short-circuited slant wire is 8.53 A, and the potential across the base of the slant-wire (when it is open-circuited) is about 236 V.

Further modeling was performed with *ELNEC* to determine if the reflector could be made to function as a director by inserting an appropriate series reactance at the base of the grounded slant-wire. I found that a capacitive reactance of 40 Ω (2162 pF at 1840 kHz) would provide a maximum gain of 4.684 dBi at a take-off angle of 22°, but the front-to-back ratio was a disappointing 2.5 dB! The elevation-plane radiation pattern for this director configuration is shown in **Fig 4**—the lack of rejection off the back of the main lobe is readily apparent. These results are somewhat discouraging—it appears that the length and location of the slant-wire are nearly optimum for reflector operation, but it may not be possible to retune the same wire to perform effectively as a director.

Fortunately, the solution to this problem is relatively easy—multiple reflector slant-wires (two to four) can be installed, spaced at equal intervals around the tower. When these slant-wires are open-circuited at their bases, they essentially disappear from the antenna array. On the other hand, a slant-wire whose base is grounded functions as a parasitic reflector, and the direction of maximum gain is shifted accordingly. If omnidirectional radiation is desired, then all slant-wires should be open-circuited at their lower ends, so that the system acts like an isolated vertical monopole.

Two Slant-Wires

Fig 5 illustrates the elevation-plane radiation pattern for a system utilizing two slant-wires located on opposite sides of the tower,

180° apart, when only the wire to the rear of the main lobe is grounded. Each wire is 133.5 feet long, and the base of each one is spaced 90 feet horizontally from the bottom of the tower. The gain is just over 5 dBi, and the input impedance is about 36.5 + *j* 53.9 Ω. The front-to-back ratio is better than 19 dB, with a half-power beamwidth of 144°, as shown by **Fig 6**. This pair of drawings emphasizes the amount of rejection of unwanted signals that can be achieved with this array, simply by grounding the appropriate reflector. For an input power of 1500 W, the maximum potential at the base of the open-circuited slant-wire is 277.2 V, and 8.4 A of current flows at the base of the grounded parasitic element.

Again, omnidirectional radiation from this array is achieved by disconnecting the bases of both slant-wires from ground, so that each one is detuned or floating. At the legal power limit in the omnidirectional mode, the potential at the base of each ungrounded slant-wire is 232.7 V. The tower's input impedance changes very little from the one-slant-wire case.

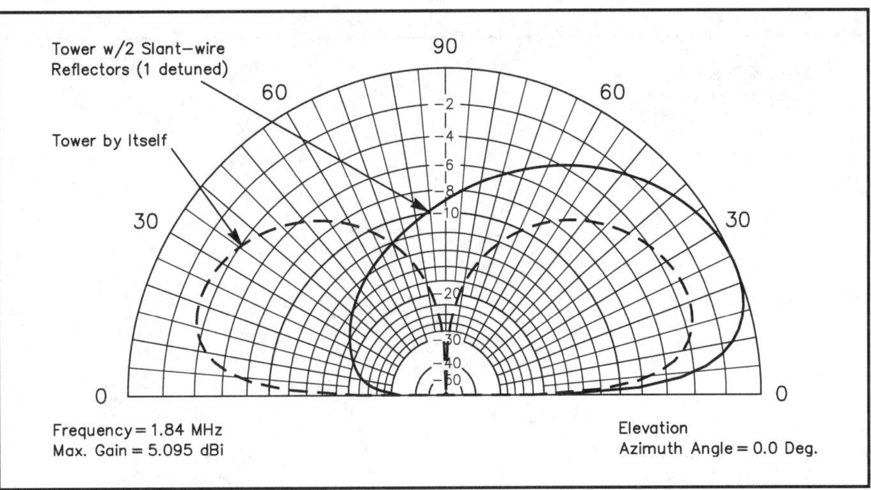

Fig 5—Elevation-plane radiation pattern for the 130-foot tower with two slant-wire reflectors spaced 180° apart. The slant wire to the rear is grounded at its base and acts as a reflector, while the wire in the direction of the main lobe is open-circuited at the ground connection and is effectively detuned.

Three Slant-Wires

In this configuration there are three slant-wires, spaced equally at 120° intervals around the tower. As usual, each is 133.5 feet long, and the lower ends are all 90 feet away from the tower base. The principal elevation- and azimuth-plane radiation patterns are shown in **Fig 7** and **Fig 8** respectively. The peak forward gain is just over 5 dBi, the front-to-back ratio is 19+ dB, and the feedpoint impedance is 36.94 + *j* 52.98 Ω.

We note here that the size and shape of the major radiation lobe remain almost constant whether we have one, two, or more slant-wires in the system, provided that only one wire at a time is active as a reflector. The main advantage of using additional slant-wires is the ability to rotate the main lobe around the compass in smaller intervals. This enables us to point the front of the beam in a desired direction, or to aim the null at an offending station. Maximum values for the open- and short-circuited slant-wires, under full-power operating conditions, are 254.5 V and 8.34 A respectively.

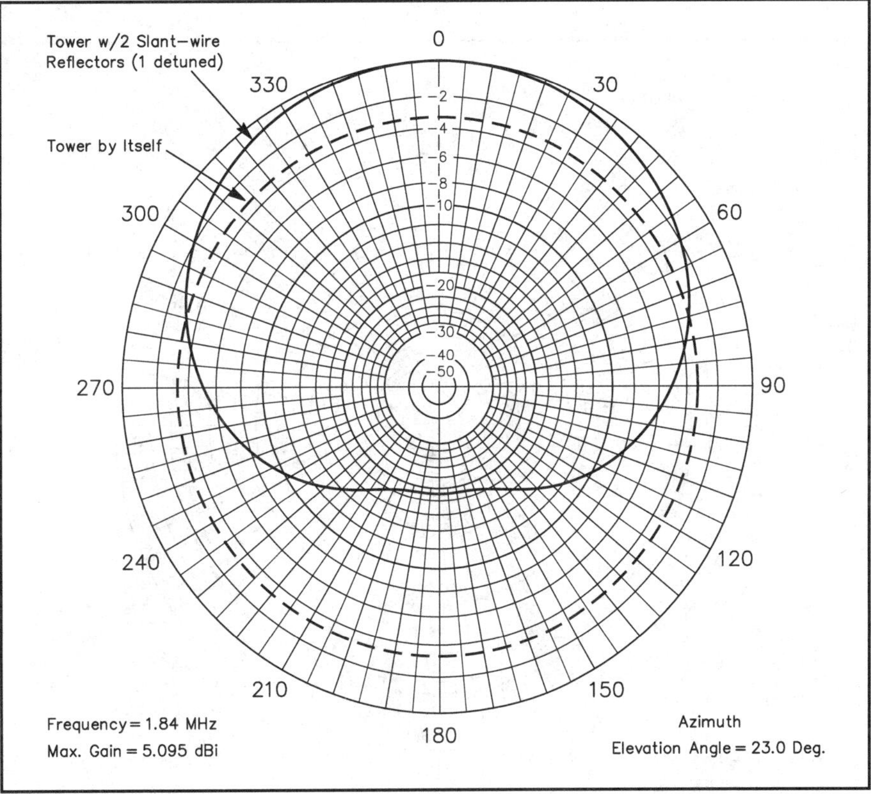

Fig 6—Azimuthal-plane radiation pattern for the 130-foot tower in Fig 5, with two reflectors, one detuned, compared to the tower by itself.

With all three of the slant-wires isolated from ground, the result is the normal omnidirectional pattern. There is only a tiny bit of noncircularity in the radiation pattern, amounting to about 0.01 dB. The maximum voltage appearing at the base of each slant-wire is 230.4 V for an input power of 1500 W to the tower. As usual, the input impedance at the base of the tower changes to that of the tower by itself when switching from the directional to the omnidirectional mode, and is now 33.5 + *j* 7.9 Ω.

In an attempt to achieve more gain and a higher front-to-back ratio, I modeled the three-slant-wire array with two of the slant-wires grounded, hoping that an additional reflector might improve overall performance. The results, shown in **Fig 9** and **Fig 10**, were definitely mixed. Using two active reflectors provides a bit more gain (about a quarter of a dB) and a narrower beamwidth in the forward direction, but the front-to-back ratio deteriorates badly, falling by about 12 dB, or two S-units! Also, the real part of the input impedance has decreased from 30+ Ω to less than 8 Ω, which is very bad news indeed.

Four Slant-Wires

The final version of this antenna has four slant-wires, which are spaced in 90° increments around the central tower. The slant-wires could be directed toward the ordinal

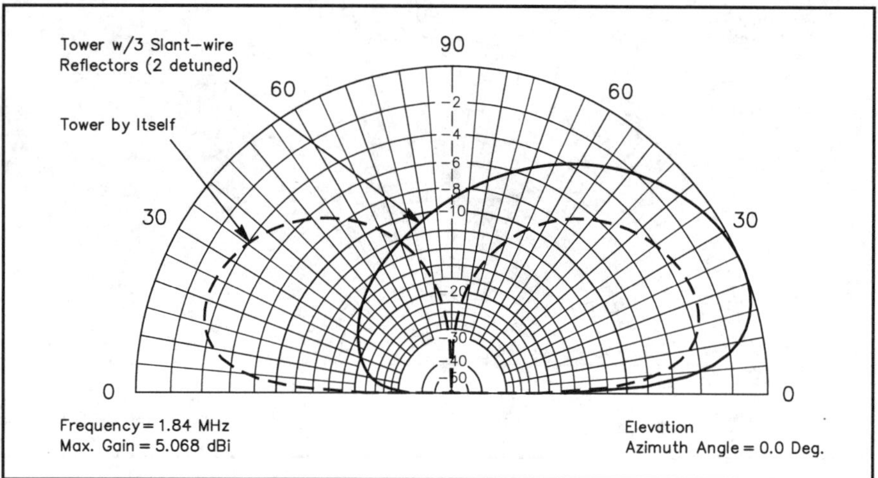

Fig 7—Elevation-plane radiation pattern for the 130-foot tower with three slant-wires spaced 120° apart. The slant-wire directly to the rear of the main lobe is grounded at its base and acts as a reflector, while the other two are open-circuited and are effectively detuned.

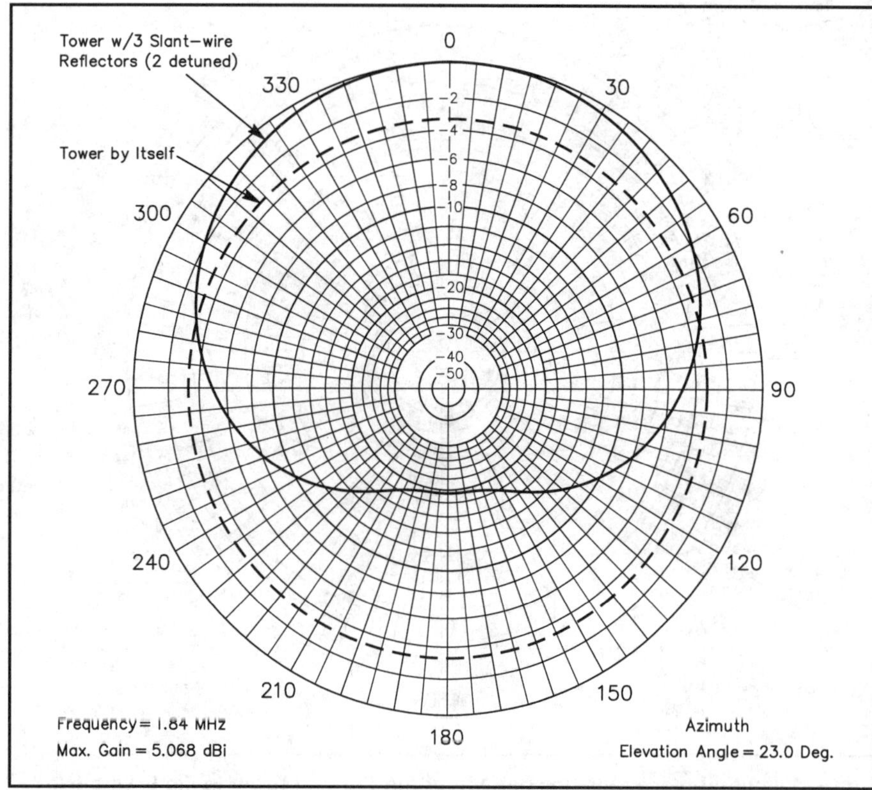

Fig 8—Azimuthal-plane radiation pattern for the system in Fig 7, with three reflectors, two detuned. The pattern for the tower by itself is shown for reference.

points of the compass (N-S-E-W), or perhaps NE-SE-SW-NW, as is commonly done here in the northeastern part of the USA. **Fig 11** and **Fig 12** display the principal elevation- and azimuthal-plane radiation patterns for this array, assuming that only one slant-wire is grounded at a time to produce reflector action. Maximum forward gain is 5+ dBi, the front-to-back ratio is nearly 20 dB, and the feedpoint impedance is 37.28 + j 52.11 Ω. The maximum short-circuit current and open-circuit voltage at the bases of the slant-wires, during full-power operation, are 8.3 A and 274 V respectively.

Again, floating all four slant-wires above ground results in the usual omnidirectional pattern. The input impedance in this case is 32.38 + j 8.39 Ω. The potential between the lower ends of the slant-wires and ground is 228.4 V at 1500-W input to the tower.

An investigation was conducted to determine what would happen if more than one of the slant-wires were grounded at the same time. When two slant-wires are used as reflectors, the patterns of **Fig 13** and **Fig 14** are produced. This configuration yields a slight increase in forward gain, accompanied by a large reduction in both front-to-back ratio and input resistance.

Finally, three of the slant-wires were grounded and used as reflectors, leading to the results illustrated in **Fig 15** and **Fig 16**. Surprisingly, this method produces the worst performance of all, with values for gain, front-to-back ratio, and feedpoint resistance that are all quite low.

The Four-Square

It seems logical to compare the "Slant-Wire Special" to the classic *four-square* array favored by many low-band DXers. Accordingly, an *ELNEC* model for the four-square was prepared for analysis; it is composed of four 130-foot aluminum towers, 8 inches in diameter, spaced 133.65 feet apart. Using the same ground constants as before, the elevation- and azimuthal-plane radiation patterns shown in **Fig 17** and **Fig 18** were produced, compared as usual with the response of a single tower by itself for reference. Notice that the forward gain and front-to-back ratio for the four-square are both superior to those of the slant-wire array. This is especially evident in **Fig 19**, which includes the elevation patterns for both antennas on the same graph.

However, the Slant-Wire Special occupies a smaller area and is much less expensive to construct, since it requires only one tower. The feed system is also much simpler, requiring only four relays and some control cable. (It would also be a good idea to include two matching networks at the base of the tower, one for the directional mode and one for the omnidirectional mode.) Each of the slant-wires needs a good ground system for proper operation, so an interconnection to the main tower ground system or the installation of additional radials at the base of each slant-wire will be necessary.

Operation on 80 Meters

The Slant-Wire Special can also be scaled to other bands, although ground losses have a progressively greater impact as one moves higher in frequency. Computer modeling was carried out at 3.8 MHz utilizing an aluminum tower 6 inches in diameter and 63 feet tall. Assuming the same ground constants as for 160 meters, the tower alone (over a good ground system) yields a peak gain of 0.814 dBi at a take-off angle of 24°, with an input impedance of 37.94 + j 7.93 Ω.

A good length for the slant-wires is 65.2 feet (#12 AWG), and the spacing from the tower base should be 50 feet. Using a four-wire system, the maximum forward gain is 4.128 dBi at a take-off angle of 25°, and a front-to-back ratio of 20.196 dB is predicted by *ELNEC*. The feedpoint impedance in the directional mode is 45.56 + *j* 47.96 Ω. At full input power, the peak values of the operating parameters for the slant wires are 262.9 V and 7.34 A. In the omnidirectional mode, the gain is 0.882 dBi at a 24° take-off angle; the input impedance is 33.16 + *j* 9.57 Ω, and the slant-wire open-circuit potential is 211.8 V for 1500 W of drive.

Conclusion

This article has described a simple modification to a conventional ground-mounted vertical antenna to achieve a switchable directional system that provides meaningful gain and front-to-back ratio without having to construct a full-blown driven array.

Acknowledgment

The author gratefully acknowledges the assistance of Grant Bingeman, whose original article provided the impetus for this project.

Notes and References

[1]Grant W. Bingeman, "An Economical Directional Antenna for AM Stations," 41st Annual Broadcast Engineering Conference of the National Association of Broadcasters, 1987. (Mr. Bingeman is employed by the Continental Electronics Division of Varian, Inc, Dallas TX.)

[2]*ELNEC* is available from Roy Lewallen, W7EL, PO Box 6658, Beaverton, OR 97007.

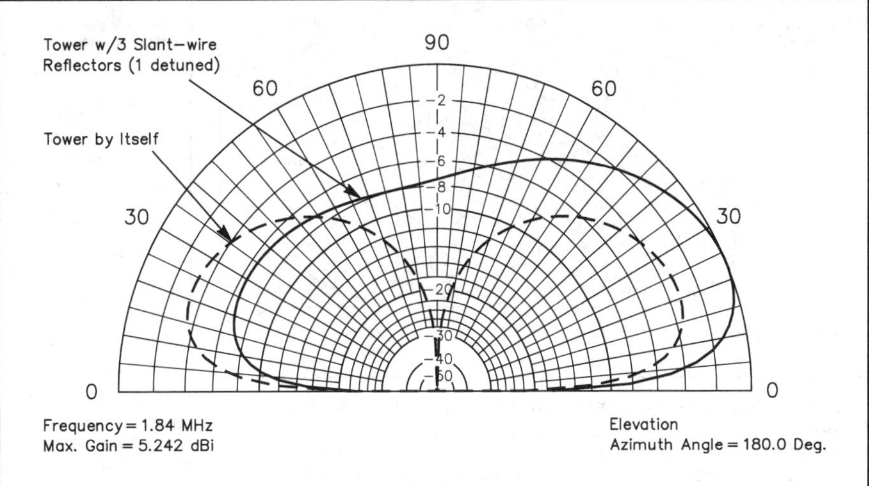

Fig 9—Elevation-plane radiation pattern for the 130-foot tower with three slant-wires spaced 120° apart, compared with the tower by itself. Here, the two slant-wires at the sides are grounded at their bases and act as reflectors, while the remaining one in-line with the direction of the main lobe is open-circuited and effectively detuned.

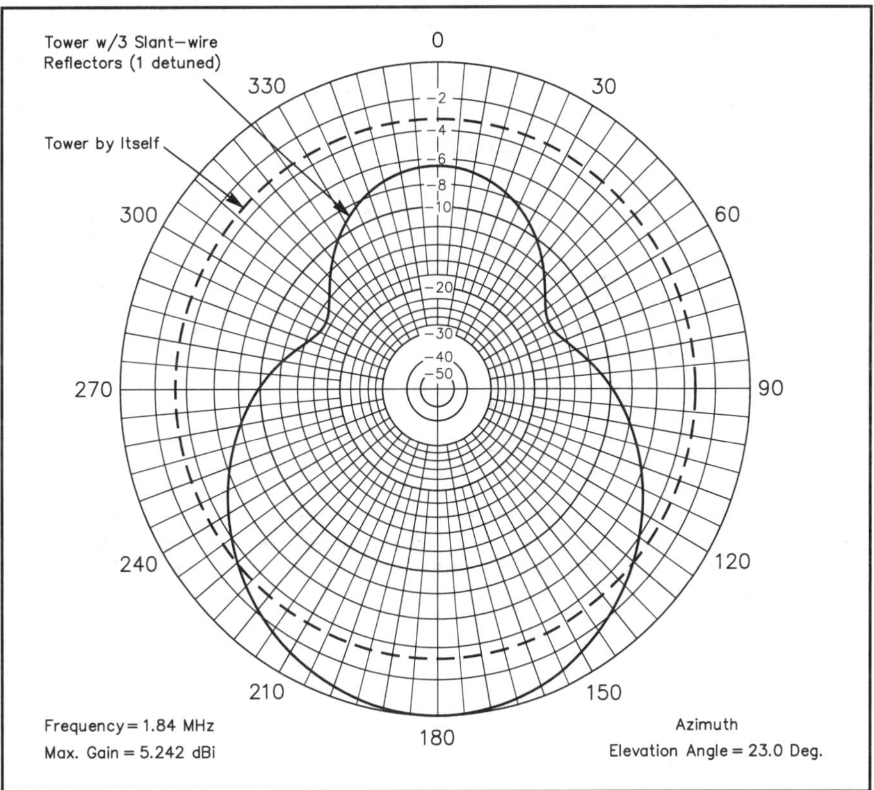

Fig 10—Azimuthal-plane radiation pattern for the system in Fig 9. The main lobe is reversed compared to the pattern in Fig 8 (where the reflector is opposite the direction of the main lobe), but the F/B ratio has deteriorated.

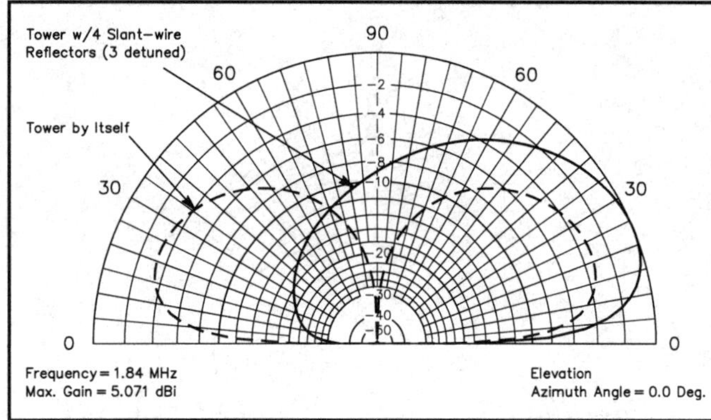

Fig 11—Elevation-plane radiation pattern for the 130-foot tower with four slant-wires spaced 90° apart, compared as usual to the tower by itself. One slant-wire is grounded at its base and acts as a reflector, while the other three are open-circuited and are effectively detuned.

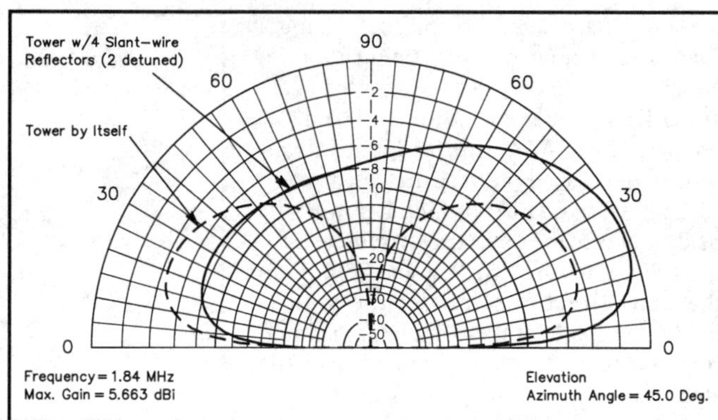

Fig 13—Elevation-plane radiation pattern for the 130-foot tower with four slant-wires spaced 90° apart. Here, two slant-wires are grounded at their bases and act as reflectors, while the remaining two are open-circuited. The forward gain is increased very slightly over that in Fig 11 (where a single active reflector is used), but the F/B has deteriorated badly.

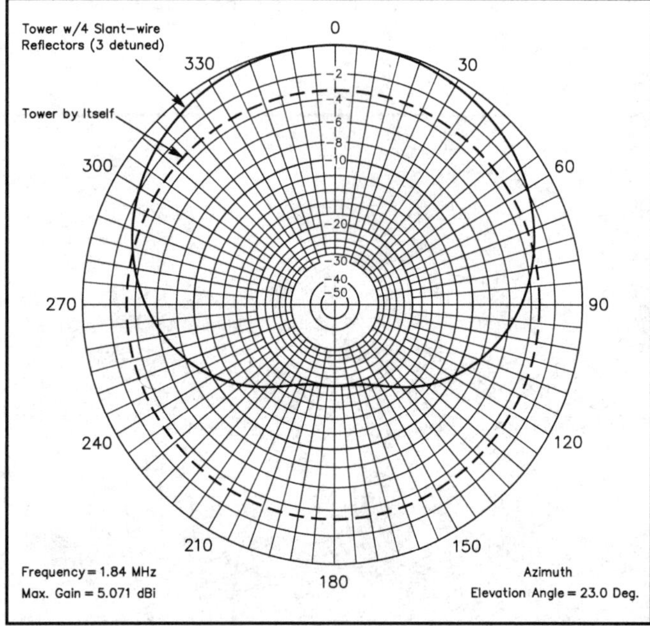

Fig 12—Azimuthal-plane radiation pattern for the system in Fig 11. When detuned, the effect of the other three slant-wire reflectors is minimal.

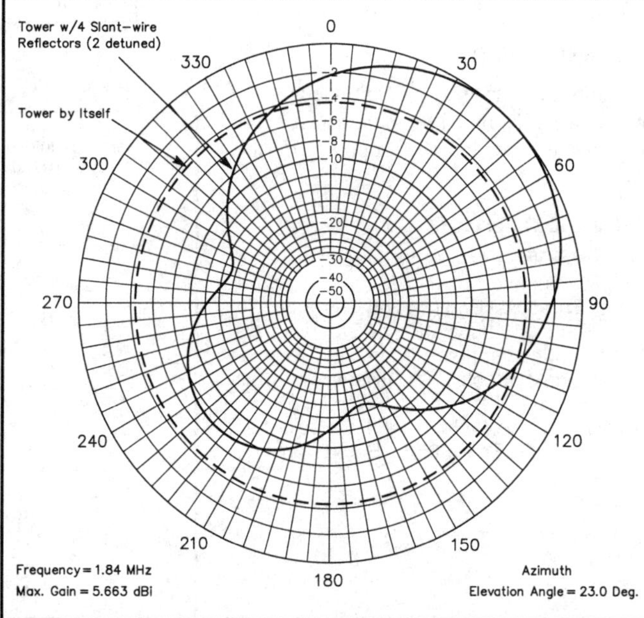

Fig 14—Azimuthal-plane radiation pattern for the system in Fig 13. The F/B has deteriorated to about 7 dB.

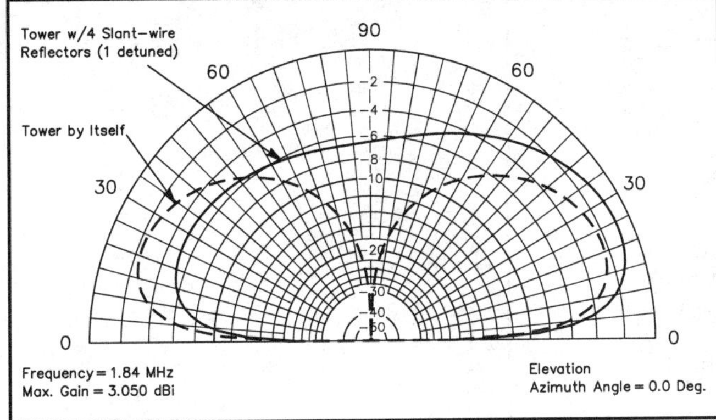

Fig 15—Elevation-plane radiation pattern for the 130-foot tower with four slant-wires spaced 90° apart. Here, three slant-wires are grounded at their bases and act as reflectors, while the remaining one is open-circuited and effectively detuned. Now the gain falls off compared to Fig 11 (where a single active reflector is used), and the F/B deteriorates further.

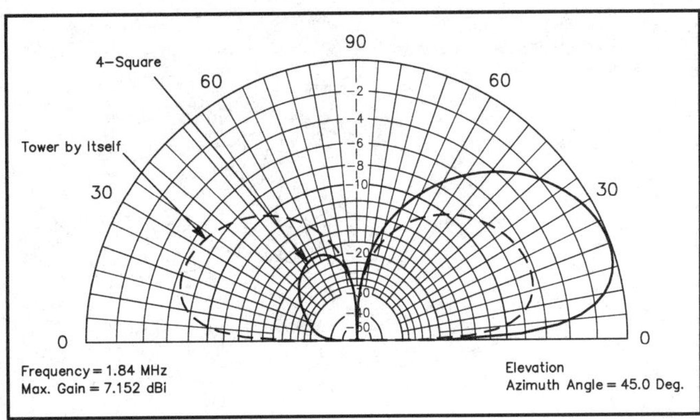

Fig 17—Elevation-plane radiation pattern for the classic "four-square" vertical array, compared to pattern for a 130-foot tower by itself.

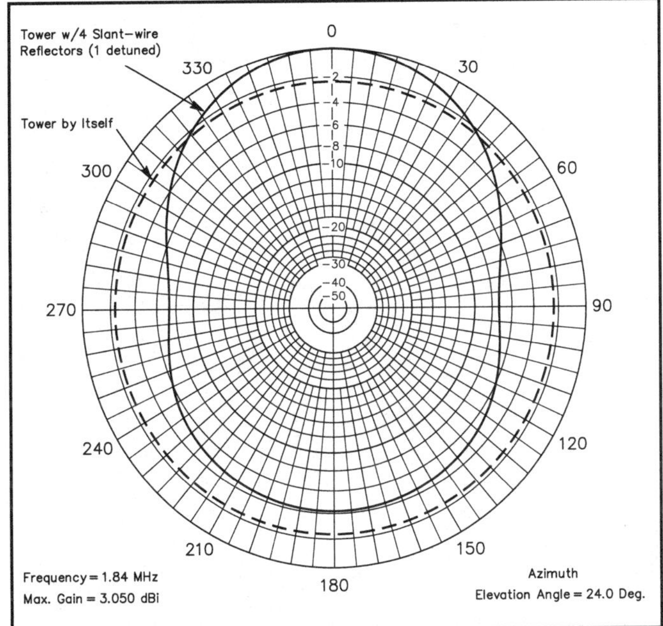

Fig 16—Azimuthal-plane radiation pattern for the system in Fig 15. The F/B is down to about 4 dB, and the gain is not much greater than that of the tower by itself.

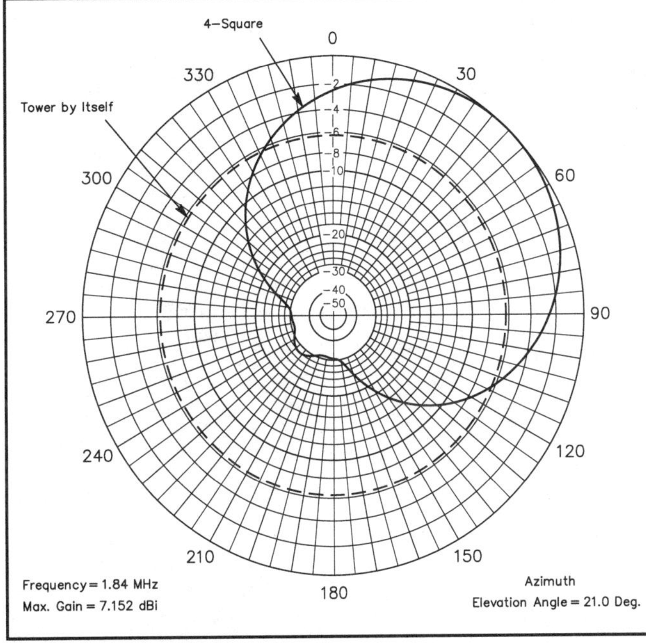

Fig 18—Azimuthal-plane radiation pattern for the classic "four-square" vertical array, compared to the tower by itself.

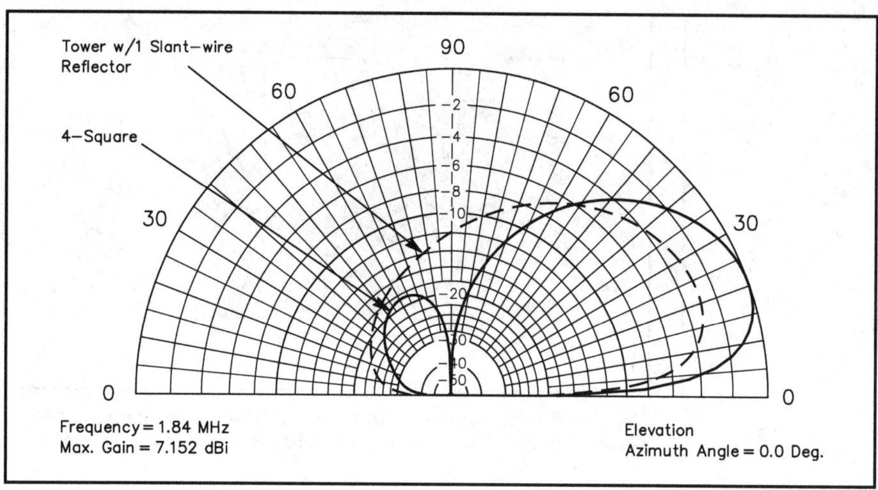

Fig 19—A comparison of the elevation-plane radiation patterns for the four-square and a single active-reflector slant-wire system. The four-square has almost 2 dB more forward gain, and a slightly superior F/B, but it takes a lot more resources compared to the Slant-Wire Special.

The 160-Meter Sloper System at K3LR

By Al Christman, KB8I
1101 W Main Street
Grove City, PA
16127

Tim Duffy, K3LR
102 Elliot Road
West Middlesex, PA
16159

Jim Breakall, WA3FET
308 Lutz Lane
Port Matilda, PA
16870

The design process that created the monster 160-meter array at K3LR is shown here in detail.

Introduction

The August 1994 issue of *QST* published a shortened version of this material. Here, the authors go through the details behind the design process of this gain array for 160 meters.

During the spring of 1992, we began the development work on a 160-meter antenna system composed of half-wave slopers to be installed on Tim Duffy's 190-foot tower. Our goal was to build an array that would provide forward gain toward any one of several switch-selectable compass points, but we also wanted to have the ability to choose an omni-directional mode as well.

Good rejection of high-angle signals was an important requirement, as was overall efficiency. The first part of this article is a review of the theoretical design process, and the second part describes the construction, testing, and operation of the actual array.

Background

Perhaps the most well-known directional antenna using slopers is the 40-meter array of Dave Pietraszewski, K1WA.[1] His design (see **Fig 1**) uses five identical half-wave sloping dipoles spaced uniformly around a tall support mast, which should be high enough that the dipoles descend toward the ground at an angle of 60° below horizontal. All five radiators are fed with equal lengths (slightly over 135°) of coax. Only one element is driven at a time, and the other four open-circuited feeders function as loading inductors so that all of the passive elements act as reflectors.

Because of height constraints at K3LR, an angle of only about 45° below the horizontal can be achieved, rather than Dave's desired 60° value. If a K1WA-style array (with 135° long feeders) was installed at K3LR's QTH, *ELNEC*[2] predicted that the principal-

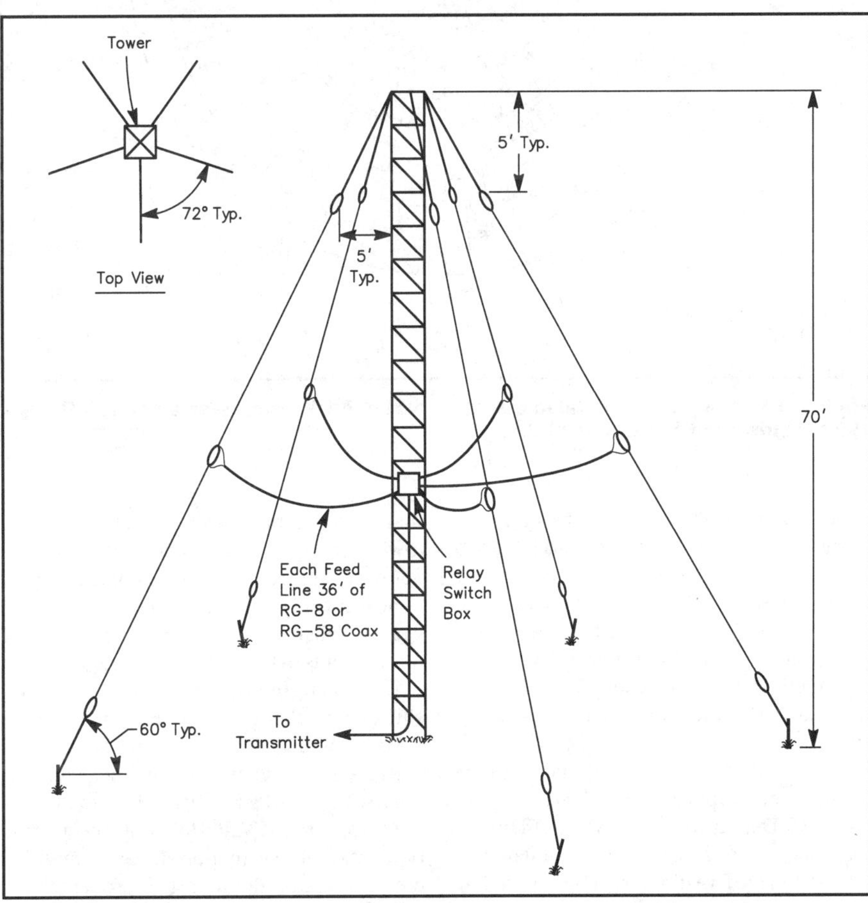

Fig 1—The K1WA Sloper System uses five identical λ/2 sloping dipoles spaced uniformly around a tall mast. Each feeder has an electrical length of about 135°.

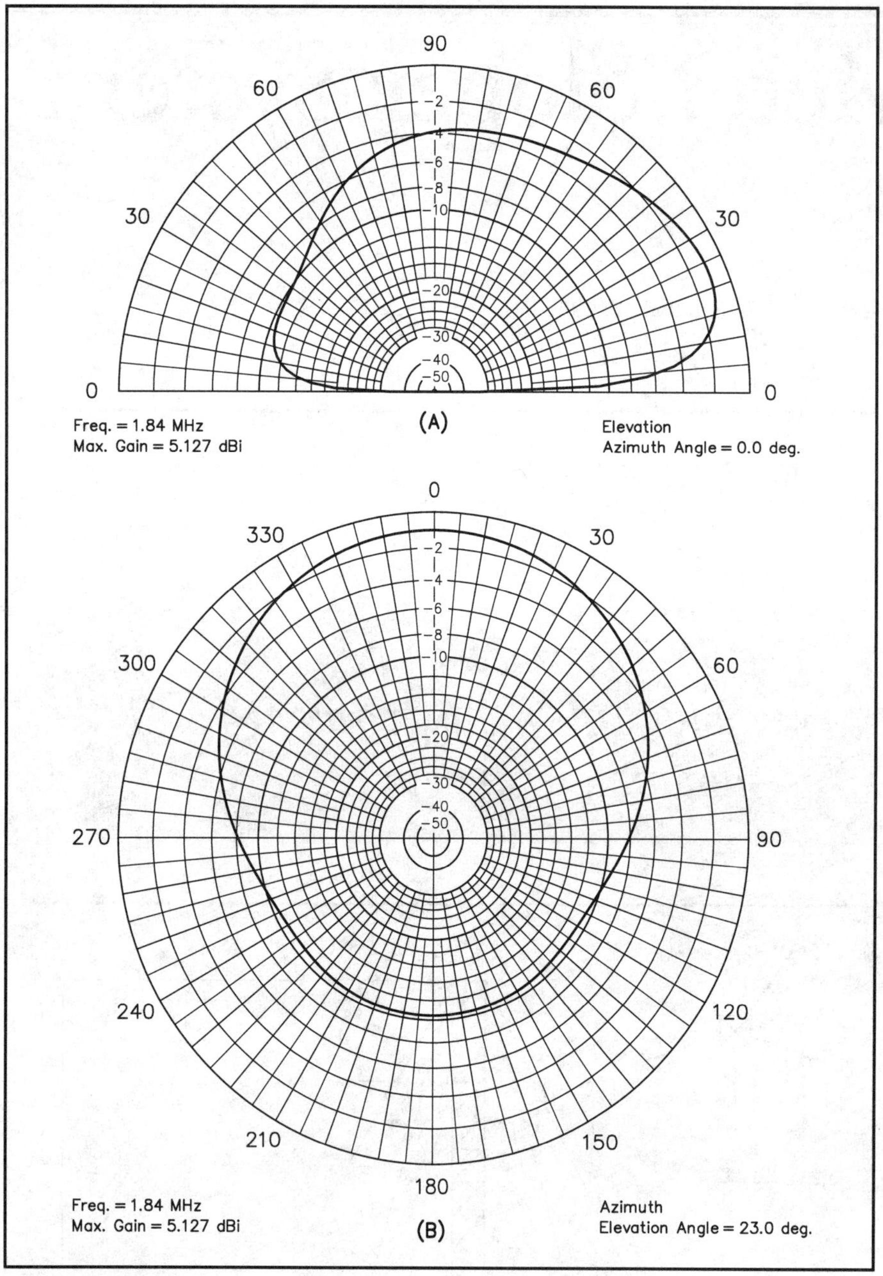

Fig 2—At A, computer-generated elevation plot of K1WA sloper array on K3LR's 190-foot tower; at B, azimuth plot.

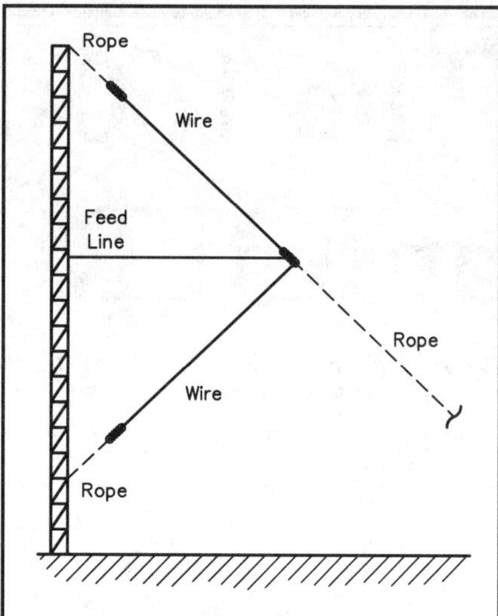

Fig 3—The K8UR sloper array uses four identical λ/2 sloping dipoles spaced uniformly around a tall mast. The lower half of each dipole is pulled in toward the tower.

plane radiation patterns would be similar to those shown in **Fig 2**. This antenna has good low-angle gain, but the rear lobe is very large, and there is far too much radiation at high angles. It is possible to increase the rejection off the back of the array by making slight adjustments in the length of the phasing lines, but the improvement is rather small.

A second type of directional antenna using slopers appeared in an article published by Dennis Mitchell, K8UR.[3] This directional antenna uses only four elements, rather than the five dipoles recommended by K1WA. In addition, Dennis "pulled in" the lower half of each radiator (see **Fig 3**) so that it slopes back toward the base of the tower. Changing the geometry in this manner produces a signal whose polarization is almost entirely vertical, depending upon the exact disposition of the wires. Finally, all of the elements in the K8UR antenna are driven with equal-amplitude currents phase-shifted by either 0, 90, or 180°, just like the classic "4-square" phased-vertical array.

Fig 4 shows what would happen, according to *ELNEC*, if a K8UR-style antenna was hung from the K3LR 190-foot tower. This type of sloper system produces a small increase in gain, and a vast improvement in front-to-back ratio, when compared to the K1WA array. The big disadvantage of the K8UR antenna is the complexity of the phasing and matching networks, which are essentially identical to those of the "4-Square."

Design

The initial design of the K3LR array combines some of the best features from both the K1WA and K8UR antennas. The physical appearance of this particular system is the same as that of K8UR, consisting of four identical "bent" dipoles spaced at 90° intervals around the tower. **Fig 5** shows a portion of this array, including all of the design dimensions. When used in the directional mode, however, this antenna is electrically similar to that of K1WA, because only one element is driven at a time. The remaining three dipoles are inductively loaded by open-circuiting the far end of each individual feeder, so that all three of them act as parasitic reflectors. Only four elements are used (rather than five), because modeling with *ELNEC* indicated that there were no performance advantages to be gained by using the extra dipole. **Fig 6** illustrates the principal-plane radiation patterns for this early-stage K3LR model in the directional mode.

The elevation-plane patterns for both the K1WA and K3LR antennas are given in **Fig 7**, which clearly illustrates the superiority of the K3LR design. **Fig 8** is a similar drawing comparing the K3LR and K8UR arrays. It reveals that the K3LR system has slightly less forward gain than the K8UR system. However, the K3LR sloper design provides better rejection "off the back of the beam" at virtually all take-off angles—we felt this was a distinct advantage.

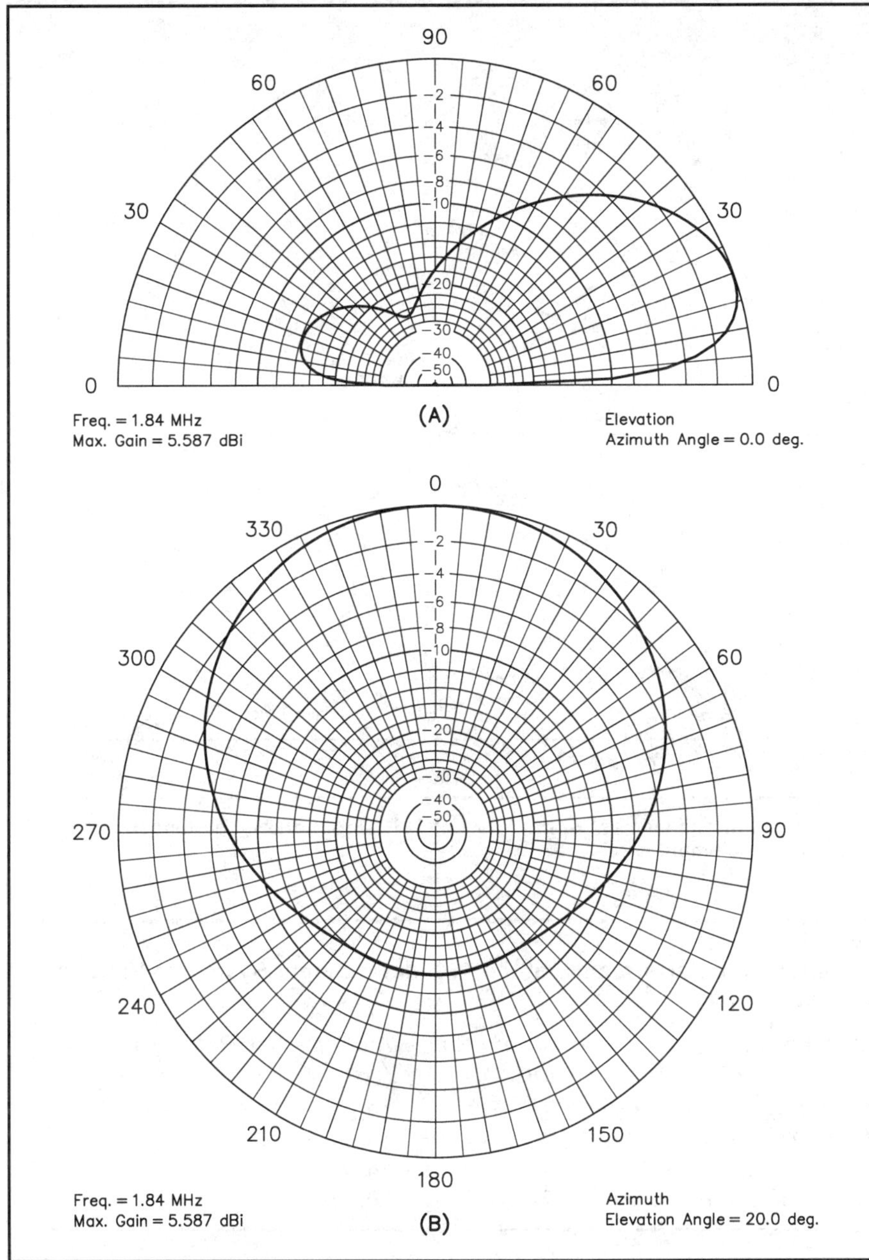

Fig 4—At A, elevation plot of K8UR sloper array on K3LR's 190-foot tower; at B, azimuth plot.

Experimentation with *ELNEC* on the K3LR basic design indicated that the best combination of gain and front-to-back ratio would occur when the parasitic elements were loaded with about 100 Ω of inductive reactance. To produce this amount of loading, an electrical length of 153.45° of open-circuited 50-Ω line was needed at the center of each element. Because of the long cable lengths, we chose to use RG-8X for the feeders (instead of RG-213) to reduce the suspended weight. With a velocity factor of 78%, this works out to be 177.74 feet of actual transmission line, assuming a design frequency of 1.840 MHz.

At this point in the analysis, we decided to consider the effects on system performance caused by losses in the RG-8X feeders. A phone call to the Belden Corporation yielded all of the electrical parameters for type 9258 cable. The cable attenuation is roughly 0.61 dB per 100 feet at 1.840 MHz. In addition, calculations revealed that the actual characteristic impedance of the transmission line was not exactly 50 Ω resistive, but closer to 49.78 + *j* 1.82 Ω.[4]

Because of line losses, the input impedance at the ends of the open-circuited feeders was no longer purely inductive (*j* 100) but had become approximately 25.76 + *j* 93.45 Ω. When this corrected value was substituted into *ELNEC*, the resulting radiation patterns were rather surprising. **Fig 9** shows that there is a small reduction in forward gain, but a dramatic improvement in front-to-back ratio! **Fig 10** is a "before and after" comparison showing how the patterns changed shape when cable attenuation was taken into account.

To obtain omni-directional radiation from the K3LR antenna, it was decided to simply feed all four of the slopers together in parallel. The elevation and azimuthal-plane patterns for such an arrangement are given in **Fig 11**. Notice that the azimuthal pattern is almost perfectly circular. In the omni-directional mode, *ELNEC* predicts that the initial-design K3LR array would perform very similarly to a single quarter-wave vertical monopole mounted over an excellent ground system.

Although the directivity of this initial design was good, we wanted more gain, so a decision was made to add four elevated radials to the *ELNEC* model. These radials would be horizontal, and mounted at a height of about 10 to 15 feet above the ground. In terms of azimuth, each radial is positioned midway between two adjacent slopers. The length of each radial was selected as λ/4 at 1.840 MHz, and all four were to have their inner ends connected directly to the tower.

Since these radials would be very close to the earth (in terms of wavelength) we realized that *ELNEC* might not give us the kind of accuracy we wanted. Therefore, we decided to utilize *NEC*, the full-blown "*Numerical Electromagnetics Code—M*ethod of Moments," which was developed at Lawrence Livermore National Laboratory. *MININEC* and its derivatives (such as *ELNEC* and *MN*) are related to *NEC*, but are much less complicated and have fewer features. In order to be really precise, we felt that it would be necessary to include all of the "smaller" antennas, which were already mounted on the K3LR 190-foot tower. So, two full-size 3-element 40-meter beams (at 100 and 190 feet), a 6-element 10-meter beam (at 198 feet), and a 3-element 20-meter beam (at 60 feet) were added to the NEC computer model, along with the tower, the four slopers and the four elevated radials.

Fig 12 shows the model for the complete K3LR 190-foot tower, including the 160-meter sloper system and the four HF beams. The principal elevation- and azimuthal-plane radiation patterns for this array are shown in **Fig 13**. *NEC* indicates a maximum gain of 3.74 dBi at a take-off angle of 20°, and a front-to-back ratio of 19.10 dB. The driving-point impedance predicted by the computer (for any one of the four elements) is 80.4 + *j* 65.2 Ω at a frequency of 1.840 MHz. **Table 1** shows the

NEC-generated values for the forward gain and front-to-back ratio when the top-band antenna is mounted on the 190-foot tower both with or without the HF beams, and with or without the four elevated horizontal radials.

Since the HF beams were definitely going to remain on the tower, we decided to build the array *with* the four elevated radials in order to take advantage of the slight bit of extra gain and the improved front-to-back ratio promised by *NEC*.

At this point it was time to collect all of the pieces and build the entire system. After this task was completed, adjustments and on-the-air testing could begin.

Construction

The four dipoles were constructed from standard #14 black insulated electrical wire. Although the insulation on the wire was not taken into account explicitly during the computer-modeling process, it will lessen the effects of weather on the wire. We expected somewhat different element lengths because of this change. Budwig center insulators were used because they provide a convenient termination to a UHF female connector and are weatherproof. The end insulators are ribbed units that are six inches long.

All of the elements are fed with ferrite-bead current baluns. These baluns are made from 14 inches of RG-142 Teflon coaxial cable, with fifty FB-73-2401 beads slipped over the top, and a male UHF Teflon connector on either end. The 14-inch section of coax was taken into account when determining the correct length for the individual transmission lines.

Belden RG-8X was chosen for the sloper feed lines. It was selected because it is high quality, can handle 1500 W at 1.8 MHz and is lightweight. Male Teflon UHF connectors were installed on both ends of each feed line. We calculated a length of 176.212 feet for the RG-8X when connected to 14 inches of RG-142 to obtain the 153.45° feeders. Final lengths were verified with an HP-8752 network analyzer. The calculated lengths using the manufacturer's velocity factor and the electrical length as measured on the network analyzer were within 3 inches for all cables. As 3 inches represents a very small frequency shift at 1.8 MHz, the antenna builder can be confident when using good quality cable that our results should be repeatable.

We used a fiberglass Hoffman box to house the relay switching hardware. The fiberglass enclosure was chosen so that the coax connectors would be insulated from each other without having to use special hardware. Four double-pole double-throw relays were used for driven-element selection. The relays are manufactured by Deltrol and are model RNF-100-DP. They have a 10 A, 5 kV contact rating and were purchased from Surplus Sales of Nebraska. These relays have a minimal impact on circuit losses and have high isolation. Five UHF chassis-mount female connectors were installed on top of the fiberglass enclosure and the relays were mounted beside each antenna connector. These relays have 12-V coils.

It is important to note that a conventional antenna switch will not work for this project. Since the unused elements must be open-circuited, both the center conductor and the shield must float to allow the coaxial feeder to act as a load. Typical commercial antenna switches float only the center conductor.

Erection

The elements are attached to the tower with lengths of black "well rope," which is very good for outdoor installations. Since its primary use is to retrieve submerged pumps, this type of rope withstands weather very well. Four pulleys were installed on the tower at the 190-foot level. The slopers are pulled up with the ropes, and are adjusted so that the outer ends of the wire elements are about 3 to 5 feet away from the tower.

Well rope is also tied to each center insulator and is used to pull the elements away from the tower. These ropes are anchored to wooden stakes that are oriented so that the elements are aimed in compass directions of 45°, 135°, 225°, and 315°.

Care was taken to make sure that the feed lines to each element came away from the dipoles as close to horizontal as possible, to minimize interaction with the radiating ele-

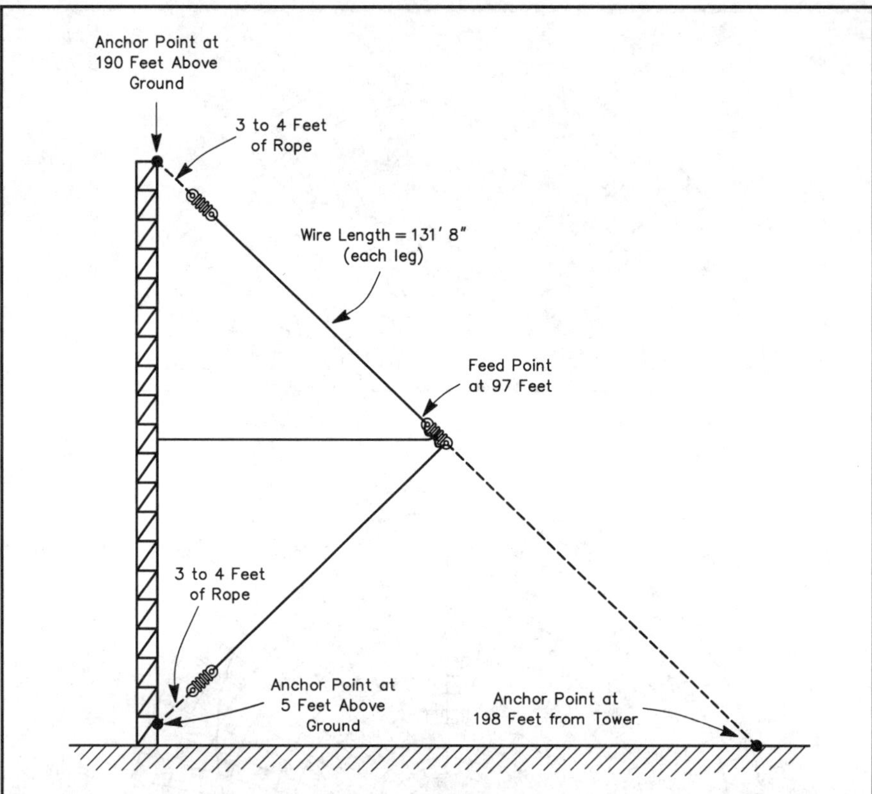

Fig 5—Detail of one element of the four in K3LR sloper array, as compensated for insulation on element wires.

Table 1
NEC-predicted Performance of the 160-Meter Sloper System in Various Configurations

Description	*Forward Gain (dBi)*	*Front-to-back Ratio (dB)*
No HF Beams, No Radials	2.88	18.59
No HF Beams, 4 Radials	3.85	15.59
4 HF Beams, No Radials	3.27	14.42
4 HF Beams, 4 Radials	3.74	19.10

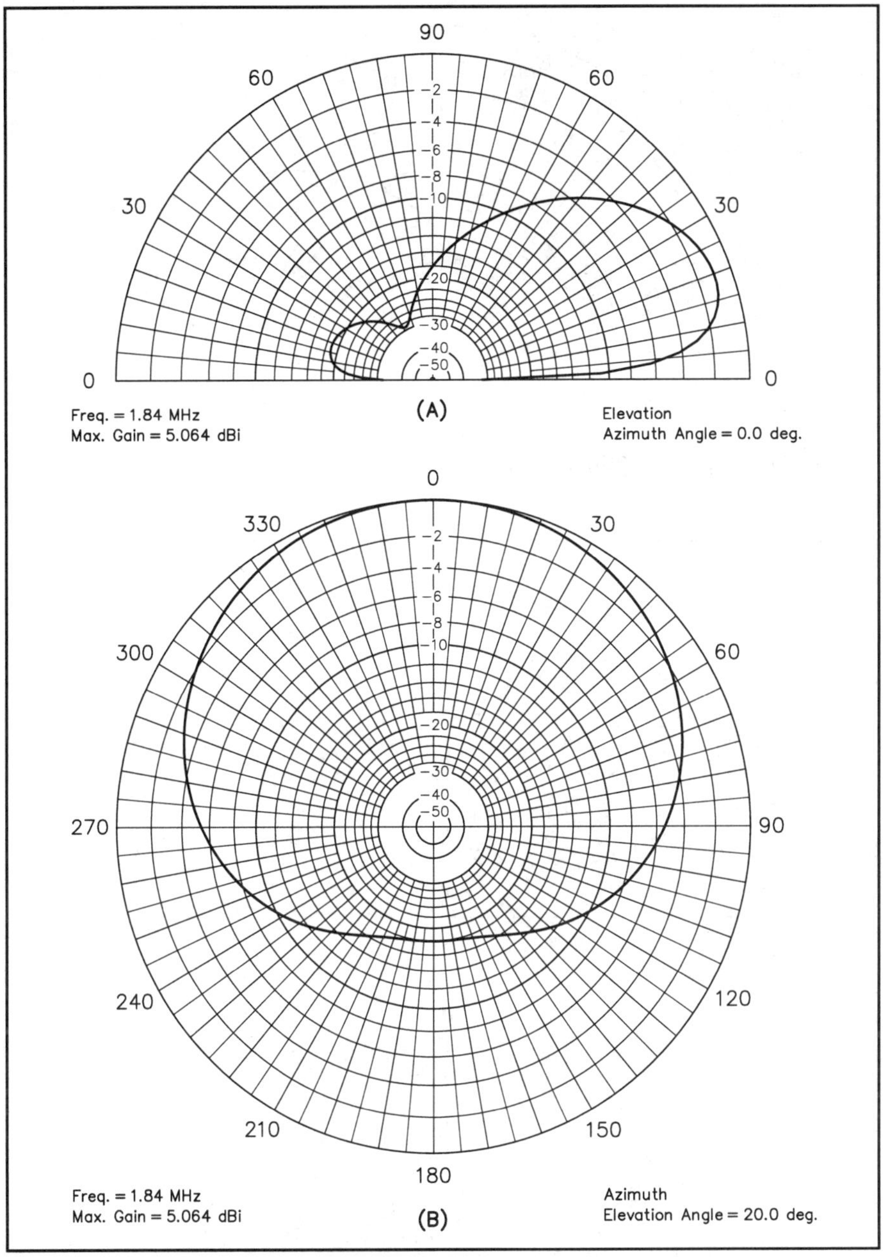

(A)

Freq. = 1.84 MHz
Max. Gain = 5.064 dBi

Elevation
Azimuth Angle = 0.0 deg.

(B)

Freq. = 1.84 MHz
Max. Gain = 5.064 dBi

Azimuth
Elevation Angle = 20.0 deg.

Fig 6—At A, elevation plot of K3LR sloper array, early stage; at B, azimuth plot.

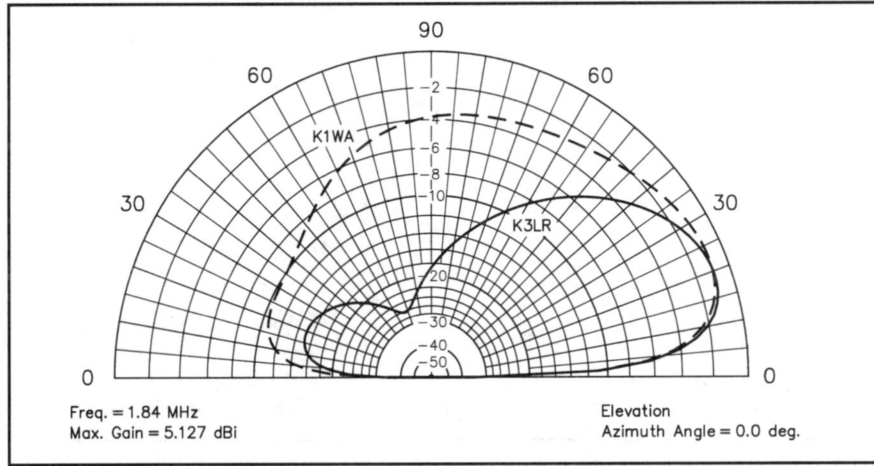

Fig 7—Comparison of K1WA and K3LR slopers.

ments. These lines are attached to the tower at the 90-foot level, and are taped to the tower from there to the ground. The 176-foot feed line lengths allow the switch box to be conveniently mounted on the ground at the base of the tower.

Four 15-foot treated 4×4-inch poles were installed at 0°, 90°, 180°, and 270° to hold up the 133-foot elevated radials. These radials are made of #14 insulated wire and are attached to the tower at a height of 13 feet.

Testing

As initially constructed (using 130 feet, 4 inch wires) the radiators were found to be a little bit short, since minimum SWR occurred at about 1.86 MHz. Each element was then lengthened by 1.3 feet to bring the resonant frequency down to 1.84 MHz. At that point, the SWR in the directional mode was so low that no impedance-matching was required. The bandwidth of the system is similar to that of a single conventional dipole, and the SWR is 1.7:1 or less for excursions of ±40 kHz from the center frequency of 1.840 MHz.

The array seemed to perform somewhat erratically at first, but it became much more stable after the "common point" was grounded. This was done by running a short heavy wire from the chassis-mounted coax connector for the main feed line (at the Hoffman enclosure) to the base of the tower, which is set in cement and grounded via three 8-foot rods. In this manner, the outer shield of the main coaxial feeder was tied directly to a good earth ground immediately adjacent to the relay switch box.

The antenna system was tested both with and without the four elevated radials, and the front-to-back ratio appeared to be around 20 dB in either case. The difference in front-to-back ratio that had been predicted by *NEC* was not noticed, either because it was too small to be discerned, or because other environmental factors in the "real world" were involved. The array seemed to play slightly better with the radials (perhaps due to some extra forward gain), so they were left in place.

There is a fair amount of "sag" in the system, because the wire elements and their supporting ropes are very long. As a result, the lower ends of the slopers actually overlap at the base of the tower, and the wires extend rather close to the ground. Thus, all four ends are spaced well apart in order to avoid arcing, which could occur if the wire elements accidentally come in contact with the tower or with each other. This problem could be completely avoided if the anchor-points for the far ends of the support ropes could be raised up off the ground, or moved farther away from the tower (which is not possible at the K3LR site). In addition, the array would fit somewhat better if the tower itself were slightly taller.

Operation

The antenna was used for the first time during the CQ World-Wide SSB DX Contest in October 1992. Station K3LR was multi-multi, and the new 160-meter antenna worked very well. Alan, N3BJ, and Scott, WR3G, were the operators, and both felt that the array was loud on the band. There were no problems either hearing or working any station, although the transmitting setup was a bit compromised by the use of an old amplifier that only put out 800 W. Nevertheless, they completed 124 QSOs, working 13 zones and 32 countries. In the 1993 SSB Contest the score was 102 QSOs, 12 zones, and 36 countries. All of these numbers stack up very well against other top multi-multi entries.

The new antenna was again utilized during the ARRL 160-Meter Contest in early December 1992. Station K3LR was entered in the multi-single category with WR3G, W3YQ and K3LR as operators. The final total was 1333 QSOs and 99 multipliers, a new all-time record for this category. During the contest the sloper system was compared to an inverted V (whose apex is up at 150 feet) located 750 feet away from the array. The 4-element antenna was always one S-unit better than the inverted V, as long as the station was at least 500 miles away. However, there are times when close-in stations are better on the inverted V.

During the 1993 CQ WW CW Contest, the top-band sloper system really shined! With nearly 200 QSOs in 72 countries, the array kept Tim's station close to the top on 160 meters. 104 Europeans were worked at K3LR, as compared with 111 European contacts made by K1AR. Since K3LR is in extreme Western Pennsylvania (only a half-mile from Ohio), this new antenna is the *secret weapon* which allows Tim to be competitive with the East Coast stations. [When this is published, the secret certainly will be out of the bag!—*Ed.*]

The omnidirectional mode of this array hasn't been fully tested, but initial observations show that received signals are down by one to four S-units when compared to the directional mode.

Other Bands

We are looking forward to receiving comments from hams who build or modify this antenna system. The design can be easily scaled to other frequencies, and suggestions for initial dimensions are given in **Table 2**. Of course, the precise lengths must be found by experimentation, and will vary from one location to another, because exact resonances will depend upon many electrical and environmental factors.

Interactions

K3LR has noticed some minor fluctuations in SWR readings on the lower 40-meter beam, as it is rotated, and he attributes these variations to the presence of

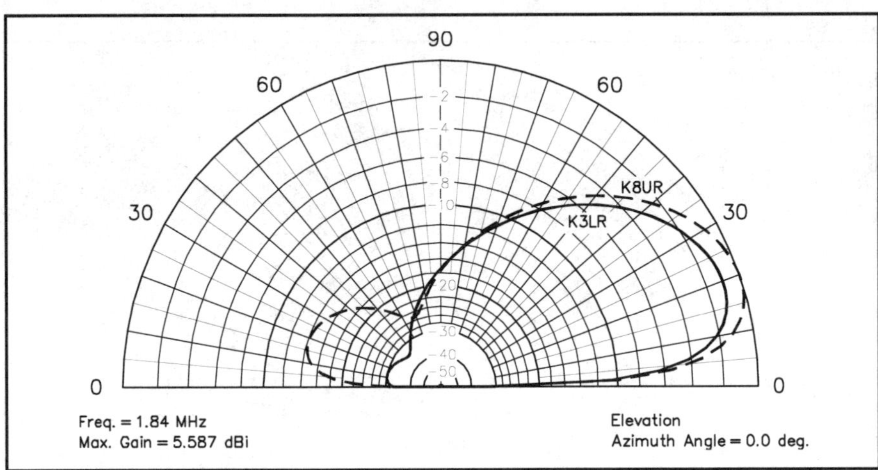

Fig 8—Comparison of K8UR and K3LR slopers.

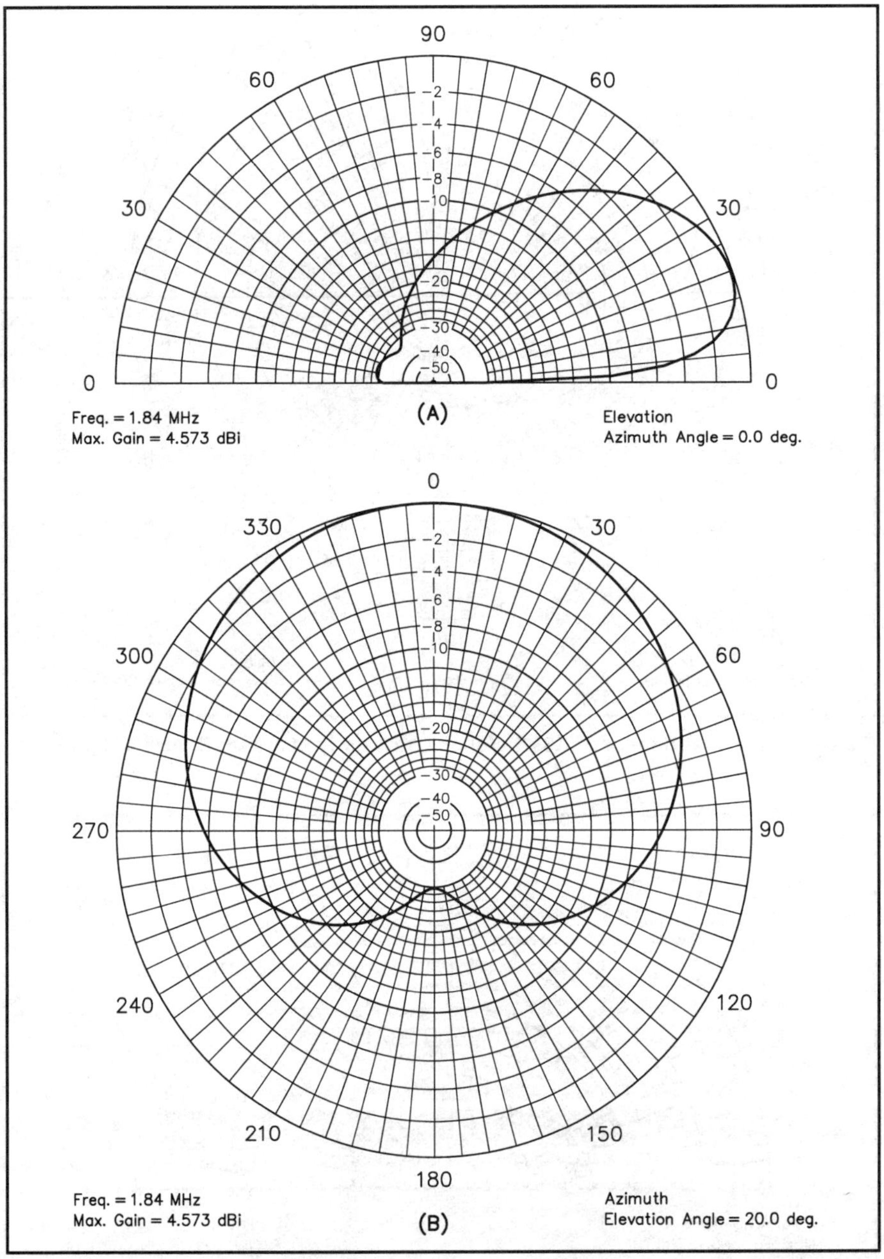

Freq. = 1.84 MHz
Max. Gain = 4.573 dBi
(A)
Elevation
Azimuth Angle = 0.0 deg.

Freq. = 1.84 MHz
Max. Gain = 4.573 dBi
(B)
Azimuth
Elevation Angle = 20.0 deg.

Fig 9—At A, elevation plot of K3LR sloper array, including feed-line losses; at B, azimuth plot.

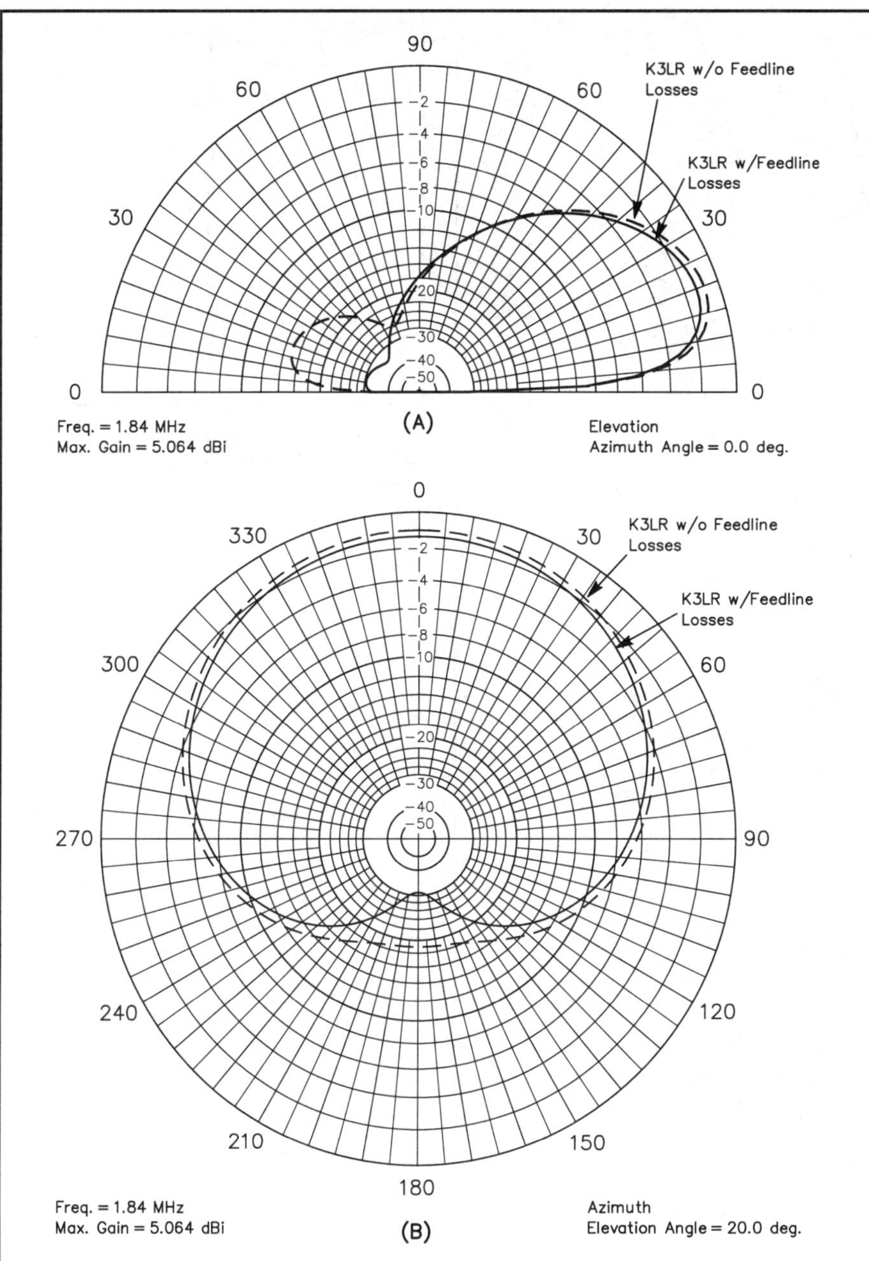

Fig 10—At A, elevation plot of K3LR sloper array, showing effect of feed-line losses compared directly with the lossless case; at B, azimuth plot.

Table 2
Approximate Dimensions (in Feet) for the Sloper System When Used on Various Amateur Bands

Freq (MHz)	*Tower Attach*	*Element Length*	*Anchor Distance*	*Feedline Length**	*Radial Length*	*Radial Height*
1.8	200	131.0	208	176.2	133.0	13
3.7	100	65.5	104	87.8	66.5	10
7.1	60	34.0	62	44.9	34.5	6

*These lengths are based upon the assumption that Belden RG-8X is used, and that a 14-inch length of RG-142 is added at the feedpoint for construction of a ferrite-bead current balun.

the 160-meter sloper system. The RG-8X feeders for the slopers are suspended at the 90-foot level, which places them only about ten feet below the 7 MHz antenna. Otherwise, there have been no discernible effects on the performance of the remaining HF beams caused by the array.

Acknowledgments

The authors would like to thank Scott Jones, WR3G, and Tim Jellison, W3YQ, for their assistance in building the antenna, and for their operational observations.

Notes and References

[1]David Pietraszewski, K1WA, "7 MHz Sloper System," *The ARRL Antenna Book*, 16th edition (Newington: ARRL, 1991), pp 4-12 to 4-14.

[2]*ELNEC* is available from Roy Lewallen, W7EL, P. O. Box 6658, Beaverton, OR 97007.

[3]Dennis C. Mitchell, K8UR, "The K8UR Low-Band Vertical Array," *CQ*, Dec 1989, pp 42-46.

[4]David K. Cheng, *Field and Wave Electromagnetics*, second edition (New York: Addison-Wesley Publishing Company, 1989), pp 440-451. The lossy transmission line formulas were taken from this book.

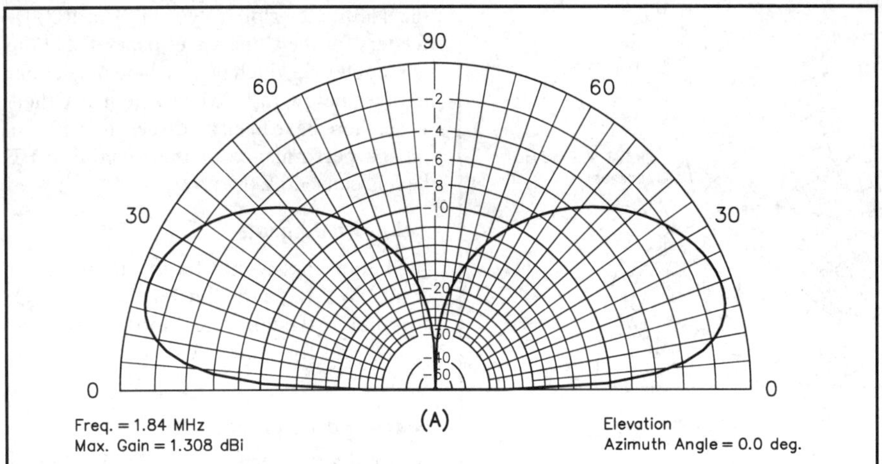

Freq. = 1.84 MHz
Max. Gain = 1.308 dBi

(A)

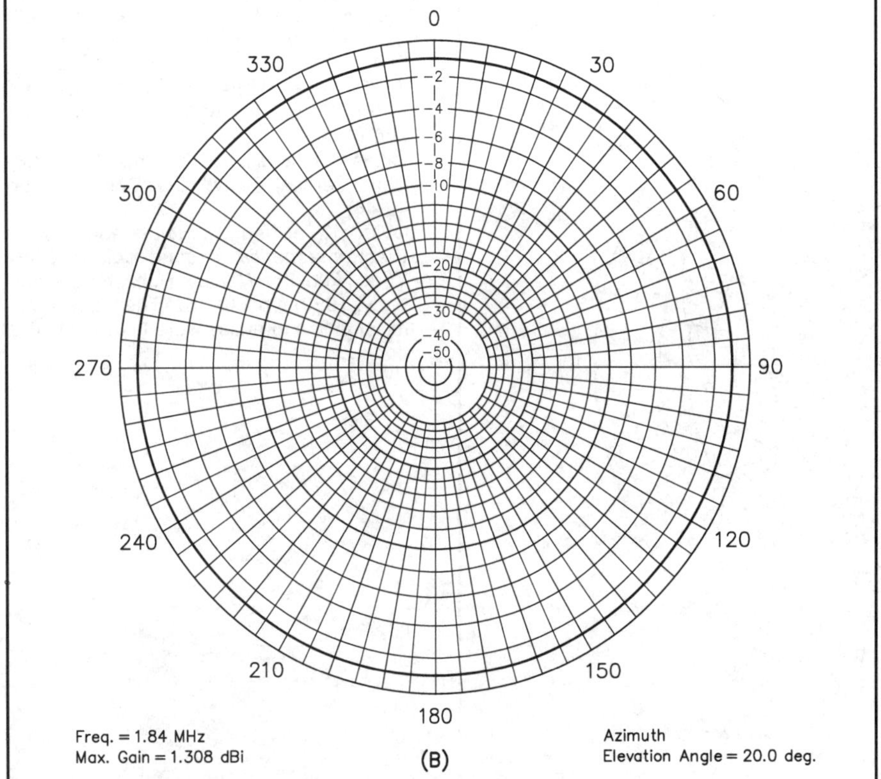

Elevation
Azimuth Angle = 0.0 deg.

Freq. = 1.84 MHz
Max. Gain = 1.308 dBi

(B)

Azimuth
Elevation Angle = 20.0 deg.

Fig 11—At A, elevation plot of K3LR sloper array, omnidirectional mode; at B, azimuth plot.

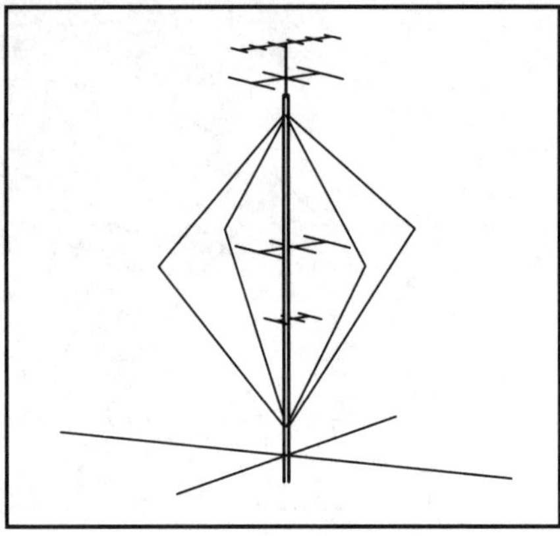

Fig 12—The 160-meter array, with radials, as mounted on K3LR's 190-foot tower.

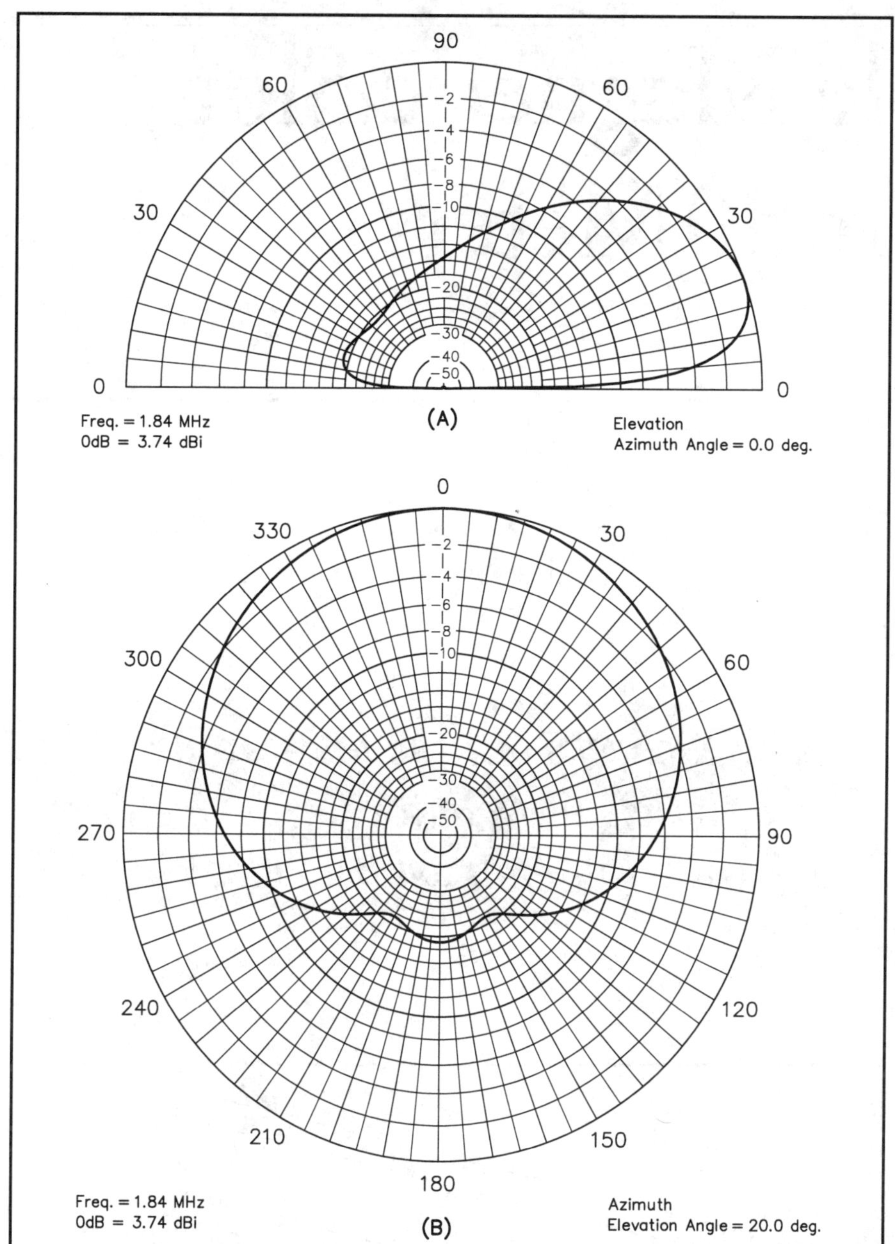

Fig 13—At A, *NEC* elevation plot of K3LR array with radials; at B, *NEC* azimuth plot.

The Double Magnetic Slot Antenna for 80 Meters

By Lew Gordon, K4VX
PO Box 105
Hannibal, MO 63401

K4VX checks out the low-profile magnetic slot radiator for 80-meter DXing. You may want to also!

In the *ARRL Antenna Compendium, Vol. 2* there is an interesting article by Russell Prack, K5RP, entitled "Magnetic Radiators—Low Profile Paired Verticals for HF." **Fig 1** shows the construction of the magnetic slot antenna. Although I have had the volume for some time, I failed to give serious consideration to Prack's design as an 80-meter DX antenna until listening to a presentation by John Brosnahan, WØUN, at the Antenna Forum at the 1993 Dayton HamVention. When I returned from Dayton I computer-modeled the magnetic slot antenna using the *NEC*[1] program.

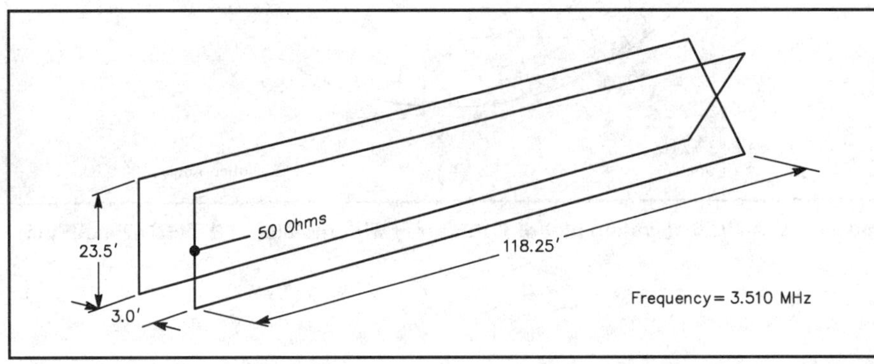

Fig 1—Diagram showing K5RP's low-profile magnetic loop for 80 meters, fed at one end.

Performance Versus Height

Although conceived to be an efficient bidirectional radiator at a very low height, I found that higher is better, as with almost all antennas. Using *NEC*,[2] I modeled K5RP's slot with the lower wires ranging from 2 to 100 feet above ground to determine the effect on performance and variation in radiation resistance.

The results shown in **Fig 2** do indeed indicate that increasing height will improve the performance, particularly at low takeoff angles. However, above 60 feet there is very little additional improvement. In fact, above 9° takeoff, the 60-foot version outperforms the 100-foot version. **Fig 3A** and **3B** show the azimuth and elevation patterns for the K5RP antenna when the bottom wire is at a height of 60 feet above ground.

Fig 4 indicates that the antenna impedance is very height-critical for impedance. It is not until a height of 30 feet that the resistive and reactive components begin to stabilize. At heights above 20 feet this antenna exhibits a large reactive component.[3]

Additional modeling also demonstrated that the separation of the horizontal wires is not critical. The results were almost identical with separations varying from 1 to 6 feet. I suspect a minimum of 3 feet should be used to prevent the two parallel wires from coming in contact in the wind.

Modeling the magnetic slot as a single loop produced an almost identical pattern as K5RP's original version; however the radiation efficiency is 81% for K5RP's version, while the single loop version dropped to 70%. This 0.6 dB difference is due to a much lower radiation resistance for the single loop.

Double Magnetic Slot

In his presentation at Dayton, WØUN suggested that a version of K5RP's magnetic slot with double the horizontal run could produce additional gain. See **Fig 5**. When modeled with *NEC*, this version,

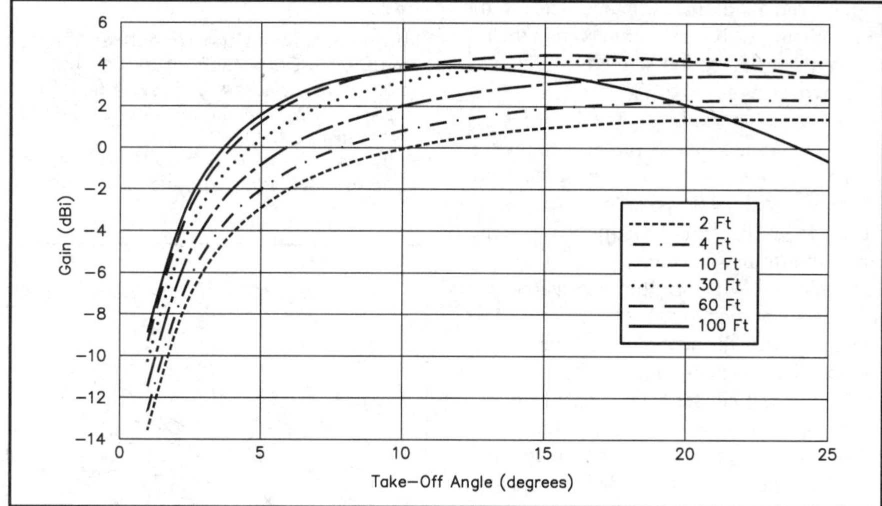

Fig 2—Performance of K5RP magnetic slot antenna versus height in feet of bottom wire over ground with conductivity of 13 mS/m and dielectric constant of 10.

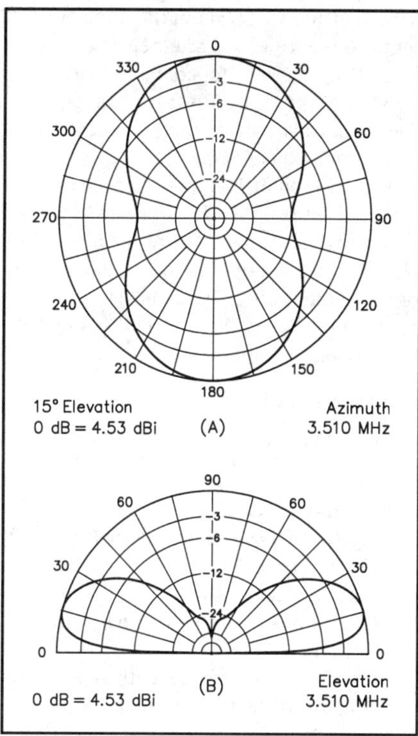

Fig 3—At A, azimuth plot for K5RP magnetic slot, where bottom wire is 60 feet over ground, at 15° elevation angle. At B, elevation plot for K5RP magnetic slot.

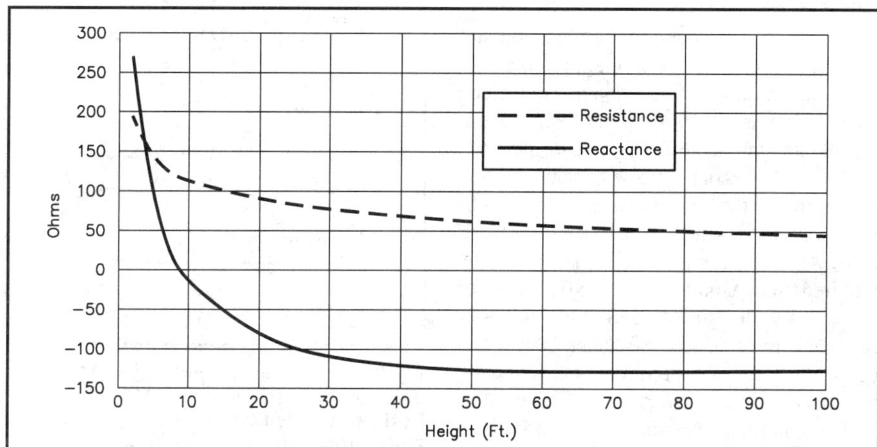

Fig 4—Impedance versus height of bottom wire over ground for K5RP magnetic slot.

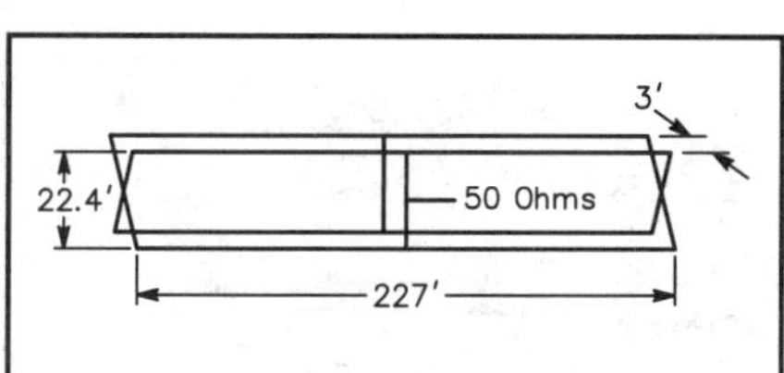

Fig 5—Diagram showing WØUN's modification of K5RP magnetic slot antenna. Overall length has been increased to 227 feet.

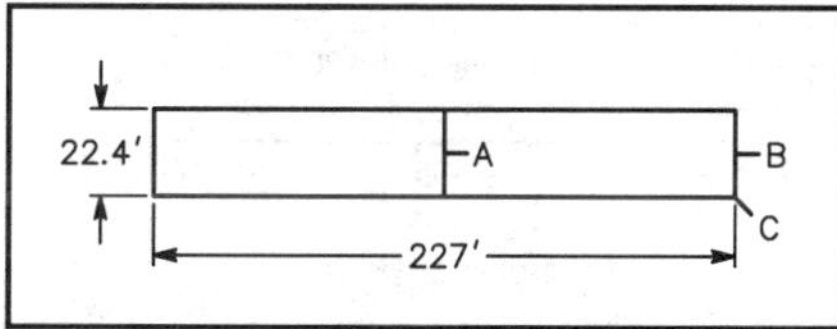

Fig 7—Diagram showing single-wire DMS, with three possible feedpoints. The most convenient is the corner feedpoint C, for mechanical considerations.

which I call the *double magnetic slot* (DMS), exhibits about 2 dB gain over the single magnetic slot. The azimuth and elevation patterns for the double-wire DMS are shown in **Fig 6A** and **6B**, compared directly with the corresponding patterns for the K5RP slot antenna.

The Single-Wire DMS

A single-wire version DMS (see **Fig 7**) fed at point A was modeled. The results indicate that a single-wire DMS is within about 0.8 dB of WØUN's double-wire DMS, but requires only one half the wire. See **Fig 8A** and 8B for the azimuth and elevation plane patterns for such a single-wire DMS.

Additional modeling with the feedpoint at B and C in Fig 7 produced almost identical patterns in either the azimuth or elevation planes. This surprising result prompted examination of the *NEC* output files for the current distribution in the vertical sections. *NEC* shows that the current is binomial-dis-

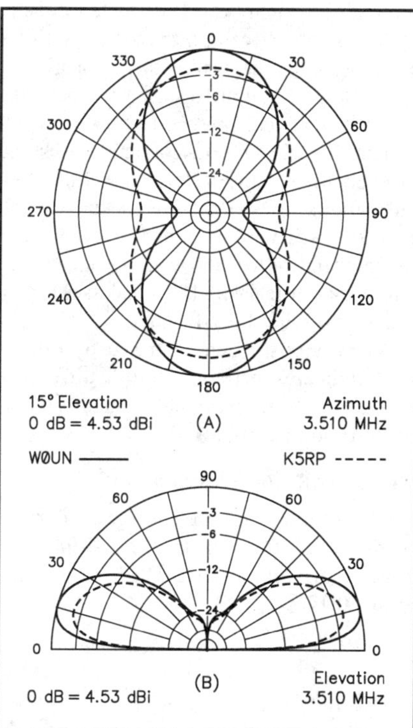

Fig 6—At A, comparison of azimuth patterns at 15° elevation for WØUN two-wire double-magnetic slot (DMS) and K5RP magnetic slot antennas. At B, elevation pattern comparison for same.

tributed (that is, 1:2:1) in the three vertical sections regardless of whether the feedpoint is A, B or C. The impedance is closest to 50 Ω at point C, which from a mechanical standpoint also offers the best location for attaching a coaxial line.

The horizontal span of either the double- or single-wire DMS is approximately 5% less than twice the length of the single K5RP slot. This is perhaps due to the added length of the center vertical section. The main advantage of the single-wire DMS is that the four crossarm supports are not required.

Comparisons

Fig 9 charts the antennas for comparison in the order of performance:

- Double-wire DMS
- Single-wire DMS (Fed at point C in Fig 7)
- Bobtail Curtain
- K5RP Magnetic Slot
- λ/4 ground-mounted vertical with 120 radials
- Elevated λ/4-vertical, with 4 resonant radials 10 feet off ground

At takeoff angles below 15°, the double-wire DMS is clearly the best performer by about 1.0 dB. The Bobtail Curtain and the single-wire DMS perform about the same, but the Bobtail requires 10% wider horizontal space. Additionally, the Bobtail requires an LC matching network with its inherent loss plus a suitable grounding system. The DMS is fed directly with 50-Ω coaxial cable in the corner. Network losses were not included in this comparison.

For those with limited horizontal space, the K5RP Magnetic Slot at 60 feet offers an excellent choice. At low takeoff angles it should offer about the same performance as the Bobtail Curtain. However, it is interesting to compare the K5RP Magnetic Slot even when mounted only 4 feet above ground (see Fig 2) with a λ/4-vertical with 120 radials (see Fig 6). The slot still has the advantage!

As previously demonstrated by others, *NEC* indicates there is very little difference between ground-mounted verticals and elevated-radial verticals.

Design and Construction

The following relationships were derived using *NEC* for constructing either the single- or double-wire DMS at heights above 0.17 λ.

$$L_h = 797/f$$

$$L_v = 78.6/f$$

where

L_h = total length of each horizontal wire, in feet

L_v = height of each vertical wire, in feet

f = center frequency of operation, in MHz

Some pruning of the horizontal spans may be required to obtain resonance. The middle vertical wire(s) is (are) connected to the top and bottom horizontal wires in the center, as shown in Fig 5 and Fig 7.

Anyone constructing one of these designs should be aware that as the maximum current points are in the center and at each end of the vertical sections, at least 0.1 λ separation from all metallic supporting structures should be provided.

I optimized with *NEC* the dimensions for a scaled-down version resonant at 28.4 MHz, mounted with the lower wire 9 feet above ground. This was for our new Novices to use at Field Day. When fed at the corner with 50 Ω coax, the actual antenna measured 1:1 at 28.430 MHz, demonstrating the effectiveness of *NEC*! The 2:1 SWR bandwidth measured approximately ±1%, or 500 kHz. The antenna performed very well, notwithstanding the poor propagation encountered that weekend.

Conclusions

The results of modeling the magnetic slot antenna have been very encouraging. As a low-angle transmitting DX antenna for 80 meters, the double-wire DMS performance is close to that obtained with the W1CF four-square vertical array.[4] No counterpoise or radial system is required for the DMS. No impedance matching is required when designed for either CW or SSB; however, a matching device would be required to cover both CW and SSB portions of 80 meters with a single DMS. Although bidirectional, most serious 80-meter DXers and contesters use separate directional receiving antennas, such as Beverages and loops. The bidirectional characteristics of the DMS should not be considered a major disadvantage.

Notes

[1]Information about the Numerical Electromagnetics Code (*NEC*) may be obtained from ACES Secretary R. W. Adler, Code 62AB, Naval Post Graduate School, Monterey, CA 93943

[2]With the exception of the λ/4 vertical, the Sommerfeld/Norton ground approximation

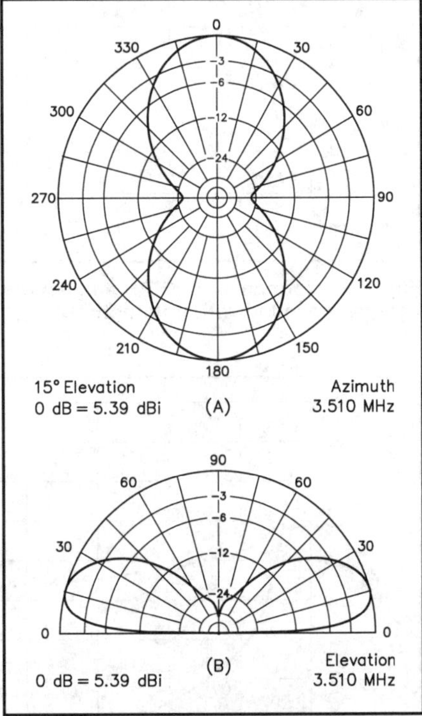

Fig 8—At A, azimuth pattern at 15° elevation for single-wire DMS whose bottom wire is 60 feet above ground with conductivity of 13 mS/m and dielectric constant of 13. At B, elevation pattern for same.

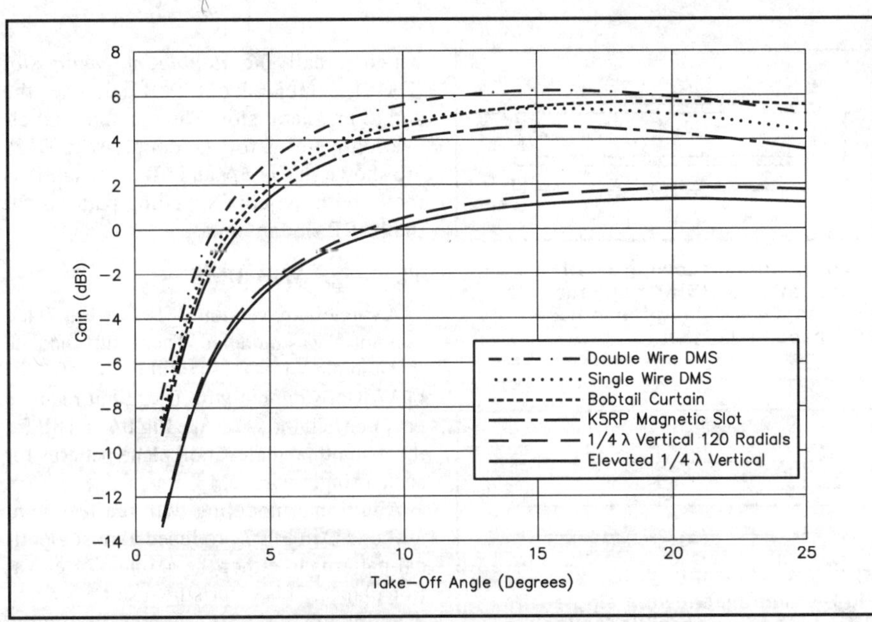

Fig 9—Gain comparison for bottom wire 60 feet off ground for various 80-meter antennas. The two-wire DMS is best of the antennas shown, and about equivalent to a four-square vertical array with no requirement for extensive ground system.

was used at 3.510 MHz. The λ/4 vertical was modeled using an image modified by the Fresnel vertical plane-wave reflection coefficient. Vertical polarization was used in all cases for pattern comparison. In all instances the ground was assumed to have 13 mS/m conductivity and dielectric constant of 10. All conductors were #12 Copperweld wire.

[3]I scaled the 40-meter dimensions in K5RP's article to 80 meters for this model. Fig 4 indicates that the impedance of the DMS is approximately 115 $+j\,0\ \Omega$ at 9 to 10 feet height. These are results similar to what K5RP obtained in actual construction. As indicated in his article, the antenna should be pruned to resonance.

[4]D. W. Atchley, Jr., "Updating Phased Array Technology," *QST*, Aug 1978, pp 22-25.

The BIG Wire Beam for 75 Meters

By Floyd Koontz, WA2WVL
c/o Harris Corp
1680 University Avenue
Rochester, NY 14610

WA2WVL had previously built some large 75-meter wire beams. This time he pulled out all the stops!

Previously, I have shown how to optimize long-boom wire Yagis.[1] I decided to design and build a 9-element wire beam on a 328 foot (100 meter) boom. To allow reversal of the pattern, the final design incorporates relay switching in control boxes on the ground. A special tuning technique using computer-generated frequency shift information, was developed to optimize the antenna as it was finally erected. The antenna was optimized for the best trade-off between gain and pattern in the SSB DX Window (3790 to 3800 kHz).

Computer Design

Yagi design always requires a compromise between factors such as gain, feed impedance and front-to-back ratio. I also wanted the pattern to be reversible by applying a dc control voltage through the feed coax. To achieve maximum gain in either direction, it is necessary to use the full boom length by moving the feedpoint to the opposite end when switching the pattern. It is also necessary to retune the end elements (the reflector and driven element in one direction) to operate as directors when the array is beamed in the opposite direction.

The directors in the center of the antenna must be symmetrically spaced and tuned, since they are not switched. Equal length and equally spaced directors can give near-optimum performance of the antenna as noted by Lawson.[2] For long-boom Yagis there is little advantage to using element tapers or uneven spacings, since gain is dependent mainly on boom length. A good pattern will result from having many directors with the correct average length. [This is true for a narrow-band Yagi design, such as WA2WVL's antenna for the 75-meter DX Window. Achieving wider bandwidths involves uneven element spacings, usually at the expense of some forward gain too.—*Ed.*]

Inverted-V elements were to be used, so that only two support points are needed. I intended to use two 100-foot high towers (Rohn 25), placed 330 feet apart in the direction of the Far East from Rochester, with a catenary rope between them as the boom. The effective height of inverted-V elements is somewhat lower than the top of the V, resulting in a slightly higher takeoff angle compared to a flat-top configuration.

At 3.800 MHz. the computer showed that when the elements are sloped at 45° from 100 feet, the effective height is about 80 feet. Knowing this, a horizontal Yagi was designed at a height of 80 feet using Brian Beezley's *Yagi Optimizer* (*YO*) software.[3] The results from the *YO* analysis were then transferred to Beezley's *MN* software and converted to inverted-V elements with 100-foot apexes.

Since V elements at this height must be longer to achieve the same resonant frequency as horizontal elements, each element length was scaled by an equal percentage to force best performance to occur at exactly 3.800 MHz. After a little tweaking, a final design resulted. I note that Beezley now has a software program called *Antenna Optimizer* (*AO*). This could have been used directly to optimize this inverted-V wire array.

The original design concept was to fit all the elements with small relays up on the catenary boom for pattern reversal. After thinking some about the severe weather here in western New York, this idea was abandoned in favor of using ladder line to allow switching and tuning of the elements from the ground. The simplest way to do this is to make the ladder line exactly ½ wavelength long so that the element impedance would be repeated at the ground. Measurements on a 450-Ω ladder line, hung in the air, confirmed that the loss of this line at a 9:1 SWR was less than 0.3 dB, but when the line was wet the velocity factor decreased from 0.93 to 0.89. The effect of this was to detune the elements when there was rain or ice.

This idea was abandoned in favor of the design shown in **Fig 1**. This scheme uses diamond-shaped quad elements for the two driven elements and two reflectors. All tuning and switching can be done in weather-

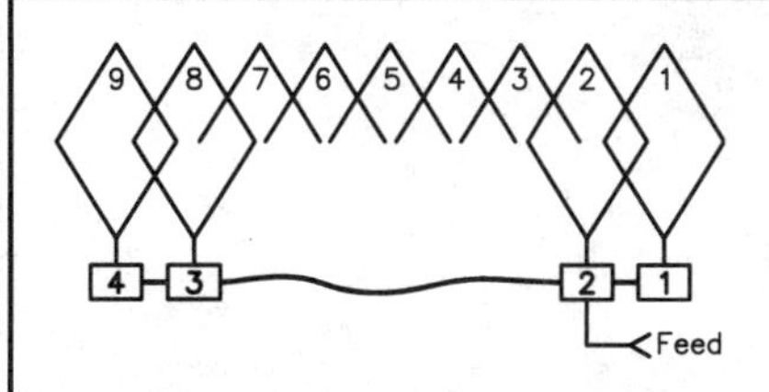

Fig 1—BIG wire beam using diamond-shaped quad elements for reflectors and driven elements. The relay boxes under elements 1, 2, 8 and 9 use coaxial cables for dc feed for switching pattern direction. Director elements 3 through 7 are not switched.

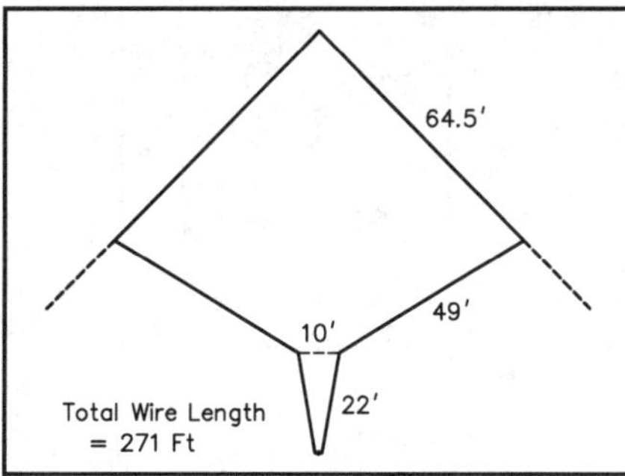

Fig 2—Layout of one of the quad-type elements in a modified delta configuration.

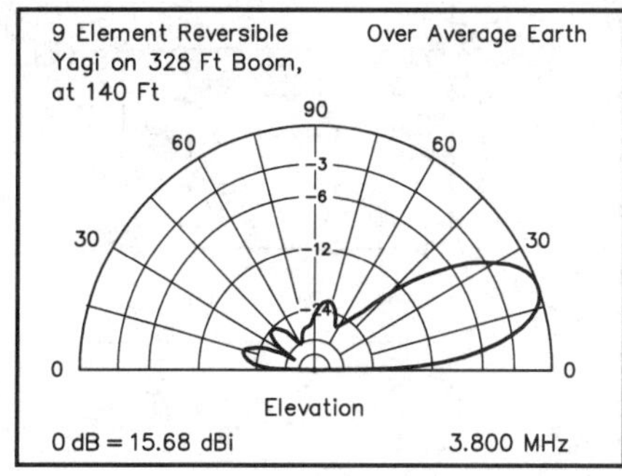

Fig 4—Polar plot of gain for 9-element BIG wire beam, with top at effective height of 140 feet over average earth.

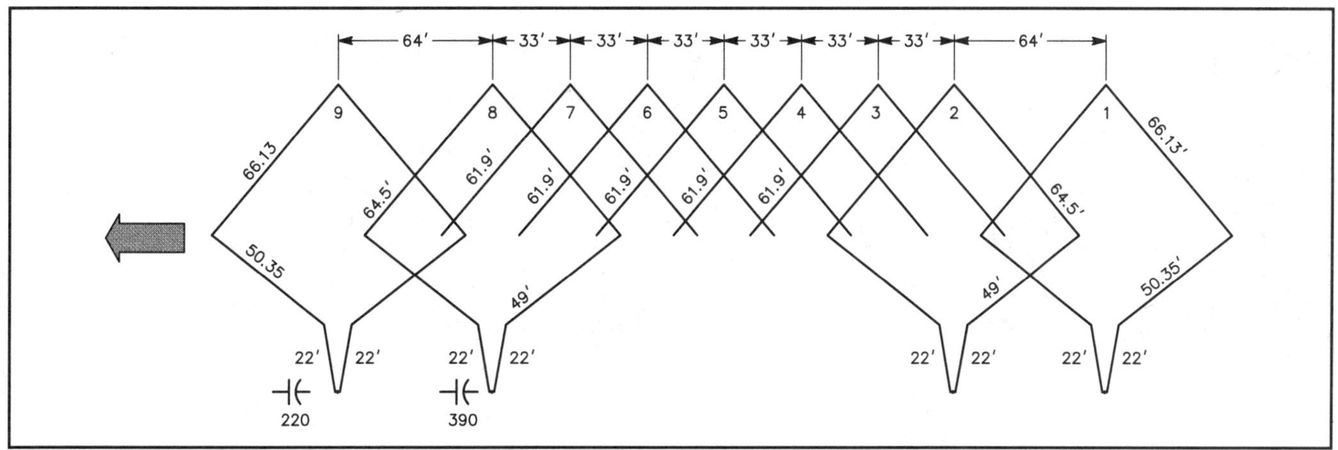

Fig 3—Dimensions for finished 9-element wire beam. Top is suspended from catenary at height of 100 feet.

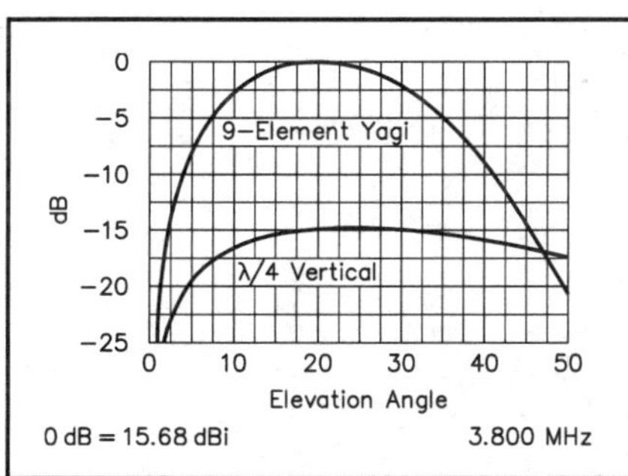

Fig 5—Rectangular plot of gain for 9-element BIG wire beam, compared with response for quarter-wave vertical antenna over same ground.

tight diecast boxes mounted on wood poles six feet off the ground. No open-wire or ladder line would be needed and no extra weight due to relays, water-tight housings or feed lines would have to be supported by the long catenary boom. I used *MN* to calculate how the array would perform if all the elements were quad types, but it was not worth the extra wire and weight needed.

Quad Loops Near Ground

A single driven element was erected and two problems were immediately observed. First, the lower half of the loop drooped near ground and could not be pulled up since the tension in the guy rope was at right angles to the desired wire position. Second, the close proximity to ground of the lower wires gave high ground losses and badly detuned the element. The solution to these problems is shown in **Fig 2**.

The wire is configured in a shape somewhat like a Delta Loop, fed from the bottom with a wide-spaced open-wire line. Since most of the useful radiation at 25° and below comes from the upper half of the loop, the remainder of the 1-λ loop serves to bring the feedpoint near the ground while not introducing significant ground losses. The final dimensions of the driven elements are shown in Fig 2. The guy ropes and the 10 foot rope pulling the lower wires together were made of 1/8-inch braided nylon. The reflectors were constructed similarly, but were a few feet longer.

Computed Performance

Fig 3 shows the computed lengths and the spacings used for all the elements. In order to raise the feed-point impedance to 50 Ω, elements no. 1 and no. 9 (reflectors) were spaced more widely than the other elements. The top half of the quad loops were raised above 45° in slope to add some effective height while still keeping the bottom of the diamond 6 feet off the ground.

The low-angle gain of a horizontally polarized antenna is heavily dependent upon height above ground and the contour of the ground 500 to 5000 feet in front of the antenna.[4] Over flat earth "ground gain" can be as much as 6 dB over free-space gain, when the reflected and direct waves are in phase. When the land slopes away in the desired direction the gain can be enhanced, although it can also be degraded on occasion.[5] My two 100-foot towers are located on the top of a small hill 80 to 100 feet above the surrounding terrain. Assuming thus that the effective height of the BIG wire beam is 140 feet, I computed the gain to be 15.68 dBi at a takeoff angle of 20°. **Fig 4** shows the polar elevation pattern over 180° of elevation. **Fig 5** shows a rectangular plot of the gain of the 9-element BIG wire beam, up to 45° elevation, compared to that of a 1/4-wave vertical over the same kind of ground.

It is very difficult to compute accurately the gain of a horizontal antenna over uneven terrain and "on-the-air" comparisons against other known antennas are needed to fully evaluate performance. In the months since the BIG wire beam was erected, signals from

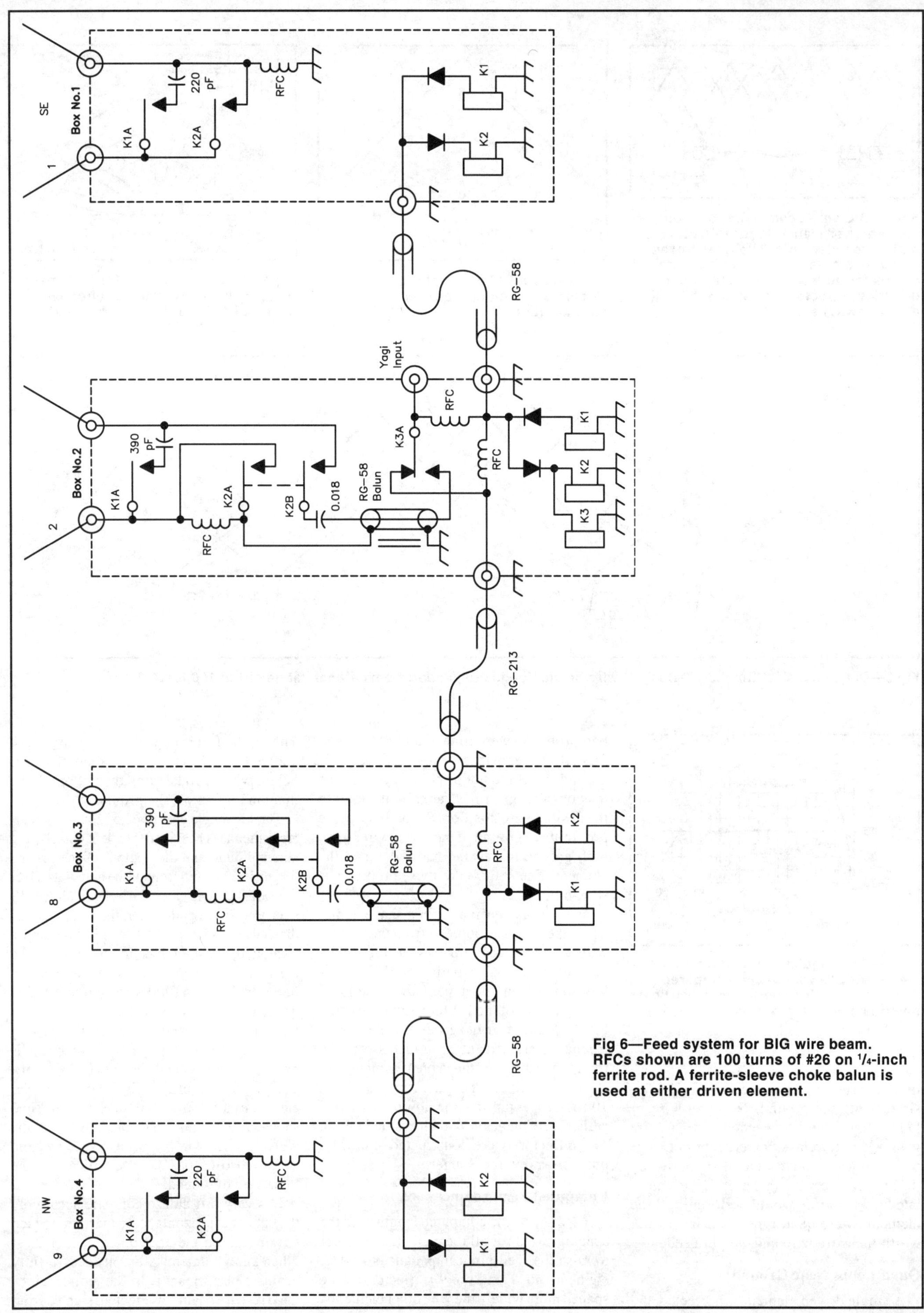

Fig 6—Feed system for BIG wire beam. RFCs shown are 100 turns of #26 on 1/4-inch ferrite rod. A ferrite-sleeve choke balun is used at either driven element.

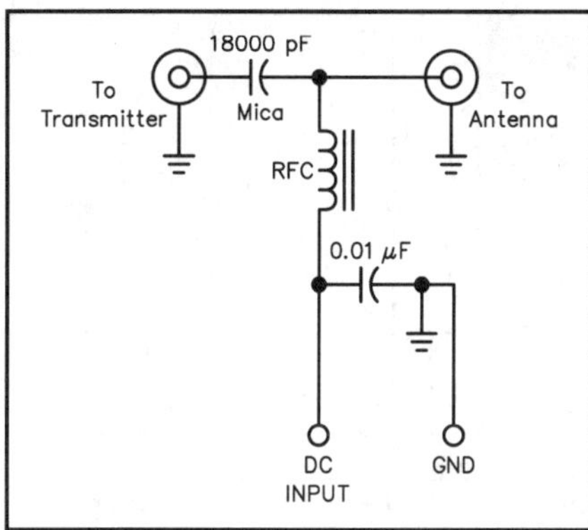

Fig 7—Feed-line adapter at transmitting end to apply dc voltage to the feed coax.

the Far East have consistently been 20 dB louder on it compared to my 6-element vertical phased array. This is true on both long and short paths.

Azimuth beamwidth is 43° at the half-power points. Front-to-back ratio is between 20 to 30 dB below a takeoff angle of 30°, and decreases to 15 dB at 65° elevation. This high-angle lobe is not shown in Fig 4 and 5 due to the limitations inherent in *MININEC*-derived programs where the antenna is close to ground. The front-to-side ratio is about 30 dB.

Construction

A 2½-inch diameter, ¼-inch closed pulley was installed at the top of each 100-foot tower. A 550-foot long piece of 2100 pound test Phillystran cable was fed through the pulleys to act as the catenary boom. Each end of the catenary was extended with polypropylene rope so that the cable between the towers could be lowered to the ground. The polypropylene ropes were 100 feet long so that the rope/Phillystran junction can be allowed to reach the pulley but not pass through. When the boom is pulled up to its final height, the Phillystran is anchored to the tower at one end and tensioned at the other end to about 200 pounds pull using a boat winch. This reduces the sag in the middle of the span to about 10 feet, with a negligible effect on antenna performance.

Hard-drawn stranded #12 copper wire is used for the elements. Small (1 × 2 × ¼-inch) element clamps were made from polyethylene sheet by drilling a small hole at each end for the wire, and two 5/16-inch holes in the middle for a ¼-inch wire-rope clamp. The clamp fits over the catenary with the two nuts on the bottom securing the polyethylene block. This material is excellent for this application, since it does not absorb moisture, become brittle in freezing weather or degrade due to ultraviolet rays from the sun.

The wire element tips are secured with 130 pound test Dacron fishing line. Plastic rope could have been used but the fishing line has much lower wind resistance and weight and is almost invisible.

Feed System

Fig 6 shows the feed system for the BIG wire beam. The switching and tuning circuits are distributed in four diecast boxes. Boxes 1 and 4 are mounted 6 feet up each metal tower, while boxes 2 and 3 are mounted 6 feet high on 4 × 4 wooden poles. The main feed coax goes from the shack to box 2, and then from box 2 to box 3. RG-58 is run from box 2 to box 1 and from box 3 to box 4 for dc feed, as shown in Fig 6.

Applying +24 V to the main coax energizes K2 in box 1, tuning this element as a reflector. It also closes K2 and K3 in box 2, connecting the coax to element 2 as the driven element. In box 3 and 4, capacitors are switched in series with elements 8 and 9 to tune them as directors. Removing voltage from the coax causes elements 1, 2, 8 and 9 to open. This was done to reduce possible interaction with a second Yagi planned for the same tower at a later date.

Applying –24 V to the coax causes elements 1 and 2 to be tuned as directors, with element 8 fed as the driven element and element 9 becoming the reflector. The gain and pattern are the same in either direction, and the center five directors remain unchanged. The four quad-type elements are grounded for dc static buildup and safety against lightning strikes using RF chokes.

Fig 7 shows the feed-line adapter used at the transmitting end to apply dc voltage to the feed coax.

Tuning Procedure for the BIG Wire Beam

One of the goals in putting up this antenna was to verify the accuracy of the computed wire lengths against the actual lengths used. Tuning of the elements is influenced by many things that are difficult to take into account when modeling by computer. The required lengths could differ from the calculated lengths because of ground that isn't flat, unknown or changing ground dielectric constant and conductivity, sag in the catenary and proximity to guy wires and other antennas. The following procedure was used to reduce the effects of these many factors and to arrive at a finished antenna custom-tuned as it was designed.

After the final design was completed using *MN*, all of the parasitic elements were removed in the computer model and the self-resonant frequency of the driven element was found. The real driven element was installed on the catenary and pulled up to full height. The resonant frequency of the driven element by itself had been calculated to be 3818 kHz. By using an impedance bridge I found the resonant point to be 3860 kHz. The wire was lengthened to move the resonance down to 3818 kHz. The feed resistance was calculated to be 105 Ω and measured to be 100 Ω.

Next, the reflector was added to the catenary and pulled up to full height. *MN* predicted that the resonant frequency of the driven element would be 3762 kHz, but it turned out to be 3760 kHz. After adjusting the reflector slightly, the resonant frequency was brought to exactly 3762 kHz. The feed resistance was calculated to be 69.4 Ω and measured 66 Ω. Now I was getting somewhere!

The middle five directors were added to the catenary and pulled to full height. The resonant frequency was expected to be 3804 kHz, and this was what was measured with no length corrections needed. The feed resistance was supposed to be 41.2 Ω and measured 45 Ω. These feed resistances are closer to predicted than one can normally expect but the most important measurements are the frequency shifts as the parasitic elements are added. Later measurements showed that this procedure could be used with all the elements installed on the catenary, so long as the inactive elements were dropped straight down from the catenary.

All that remained was to add the last two directors, which were the same corrected lengths as the first driven element and reflector, and pull the antenna up. The tuning of the last two elements as directors is accomplished using 390 and 220 pF capacitors installed in boxes 3 and 4 (and in 1 and 2 also). Verification of the F/B was made on the air with many stations at different distances and, as predicted by *MN*, averaged 20 to 30 dB. Front-to-back ratio is quite sensitive to element tuning, especially the front two directors, so the measured results confirm that the antenna tuning is very close to the design values.

Other Reversible Designs

Several types of reversible Yagi designs have been studied using the computer and each has advantages and disadvantages. It is generally better to use reflector-length elements at the front and back of a reversible Yagi and retune it as a director higher in frequency using a center capacitor. This technique works better than trying to use coils or stubs to lower the frequency of a director to function as a reflector. A porcelain capacitor, such as one made by American Technical Ceramics Corp (Type ATC-100E), is about the size of a US dime (approximately 1 cm diameter). It has adequate voltage and RF current rating to do the job and adds little weight to the wire element.

Three Elements with Center Feed

This beam has two reflector-length ele-

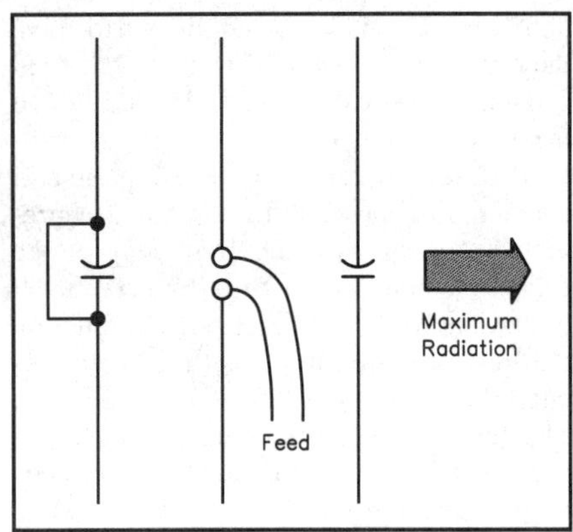

Fig 8—Schematic of 3-element beam with center feed and reversible directionality, depending on which capacitor is shorted out using relays.

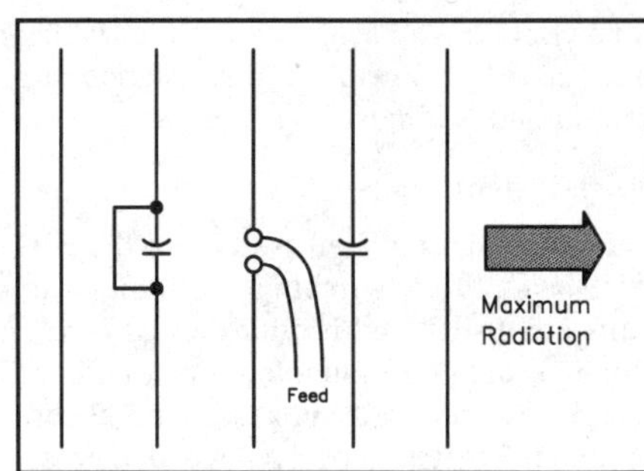

Fig 9—Schematic of wire beam with either 5, 7 or 9 elements with center feed and reversible directionality. Additional directors would be employed at both ends of boom to extend beam for more elements than 5 elements shown here. Only part of boom length is used in each direction.

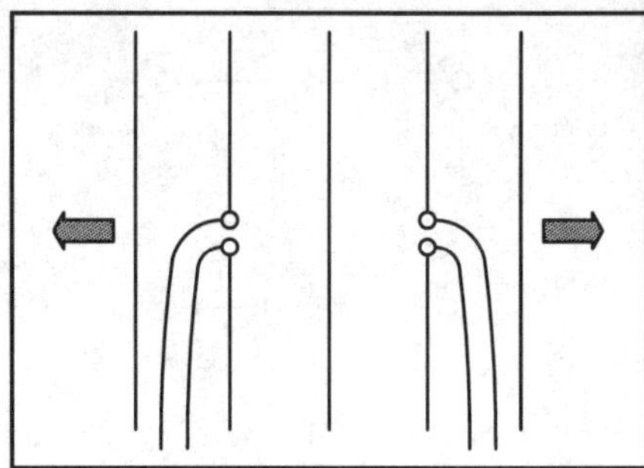

Fig 10—Schematic of wire beam with either 5, 7 or 9 elements with common reflector and separate driven elements. Additional directors would be used at both ends of boom to extend beam for more elements than 5 elements shown here. Only half of boom length is used in each direction in this design.

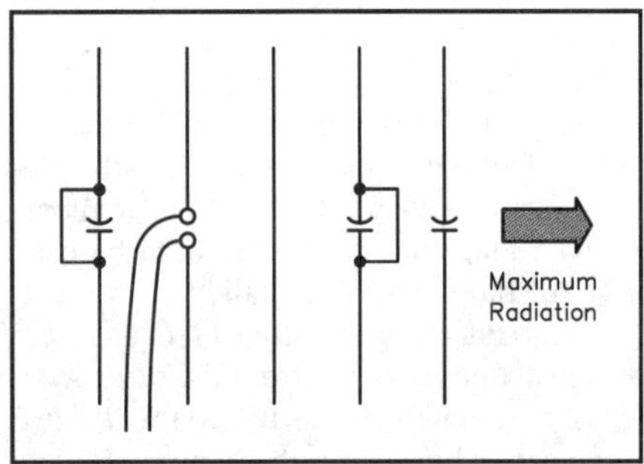

Fig 11—Maximum gain design using 5 or more elements. This is basic design for WA2WVL BIG wire beam.

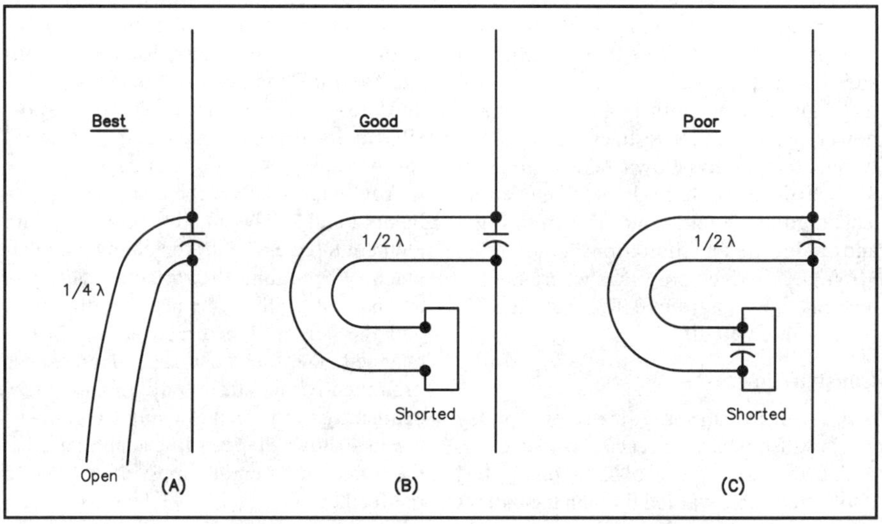

Fig 12—Methods for shorting out capacitors using transmission lines. Method at A has least losses and shortest line length.

ments with equal-value capacitors at the center. One or the other is shorted to tune that element as a reflector. Full boom length is available in either direction and only two elements must be switched to reverse direction. See **Fig 8**.

5, 7 or 9 Elements with Center Feed

This beam is an extension of the 3-element design, in which the center element is driven and the adjacent elements are reflector lengths. All other elements are directors and do not require switching. **Fig 9** shows a 5-element design. Only 2 elements require switching but the effective boom length is less than the total span of the Yagi. For 5 elements, the boom length is 75% of the total; 7 elements cover 67%, and 9 elements cover 62.5% of the total length. Designs for 6 or 8 elements could be used, resulting in a small shift in resonant frequency when changing direction.

5, 7 or 9 Elements with Common Reflector

Another possibility is to use a common reflector in the center of the antenna and switch the feed to either driven element. See **Fig 10**. The disadvantage to this design is that the effective boom length is only 50% of the total span.

5 or More Elements (Maximum Gain)

This design is the one used in the BIG wire beam. Four elements must be switched to reverse the pattern but the full boom is used in either direction. When switching, the feed must be moved to the other end of the antenna and the previous driven element must be tuned as a director, along with the previous reflector. See **Fig 11**.

Switching Methods

Shorting of capacitors placed at the element can be done in several ways, but you must be careful not to introduce significant loss. Three methods are shown in **Fig 12**. These use ladder line or open-wire line and have feed line losses ranging from less than 0.2 dB to more than 1.5 dB. As mentioned previously, small relays could be used up at the element center if the situation is suitable.

Notes and References

[1]F. Koontz, WA2WVL, "The Design and Construction of Wire Yagis." Presented at the Antenna Forum, 1993 Dayton HamVention.

[2]J. Lawson, W2PV, *Yagi Antenna Design* (Newington: ARRL, 1986), pp 2-7 to 2-32.

[3]B. Beezley, K6STI, 3532 Linda Vista Dr, San Marcos, CA 92069, Tel 619-599-4962. Beezley writes antenna-modeling software for the IBM PC or compatibles: *MN* (*MININEC*), *YO* (*Yagi Optimizer*), *AO* (*Antenna Optimizer*) and *NEC for Wires*.

[4]*Handbook on High-Frequency Directional Antennae* (Geneva: International Telecommunication Union, 1966), Fig 78, p 90.

[5]L. A. Moxon, *HF Antennas for All Locations* (London: RSGB), pp 138-141, sold by ARRL in US.

The Tuned Guy Wire—Gain for (Almost) Free

By John Stanley, K4ERO
8495 Hwy 157
Rising Fawn, GA 30738

K4ERO describes a practical low-frequency vertical antenna with gain and good pattern. You may be able to modify an antenna you already have in your backyard.

With the bottom of the sunspot cycle fast approaching, many hams are turning their attention to the "low bands"—40, 80 and 160 meters. As low band DXers know, the low frequencies do not go dead when we have low sunspots—they actually work somewhat better. Now is the time to be thinking about improving your antennas for the lower frequencies.

I'm going to describe an antenna that doesn't use a new technique, but a much underused one: the *tuned guy wire*. Most vertical antennas need guy wires for mechanical reasons, and these guys can be used to your advantage. In the past, attempts to avoid resonances in guy wires may have been overdone. On the HF bands it seems that solid guy wires usually cause few real problems.[1] The idea of deliberately making guy wires resonant to improve antenna patterns deserves more attention. With modern antenna analysis tools, this has become quite easy to do.[2]

Let us assume that you presently have a pretty decent vertical antenna for one of the lower bands, with a decent buried-radial ground system. Even if your vertical is not so decent, keep reading. You can get a front-to-back ratio of 20 dB and more than 3 dB of forward gain from your vertical, with an investment of a weekend's work and a few dollars! Think what that front-to-back ratio would do to your ability to work the weak ones in the presence of strong stateside QRM.

Changing one of your present non-resonant guys to a tuned reflector can accomplish this magic. The following is for 3.8 MHz, but you can scale all dimensions for either 160 or 40 meters. See **Table 1**. Measure out from the base of your vertical about 50 feet. Put down some kind of anchor in the ground. Prepare a wire 60.2 feet long and hang it between the ground anchor and the top of your tower, insulating the wire at both ends. It won't reach, you say. Good. That means that your tower is tall enough, say 50 to 70 feet, for this plan to work. (I assumed a 63-foot tower for my computer model.) Add enough insulated material to reach the top of the tower. Phillystran is ideal and the 20 feet or so you will need for this project should cost under $15.[3] Use any handy plastic rope if you wish, but plan to replace it every 2 to 3 years.

Table 1
Length of Slanted Reflector Wire for Low-Frequency Amateur Bands

MHz Freq	*Reflector Length*
1.8	127 ft 1 inch
1.9	120 ft 4 inches
2.0	114 ft 5 inches
3.6	63 ft 6 inches
3.8	60 ft 2 inches
7.1	32 ft 2 inches
7.3	31 ft 4 inches

Add insulated rope or Phillystran to wire to reach to a point near the top of your vertical antenna. Use a mechanically sound insulator between anchor point and bottom of the tuned wire. Connect the bottom end via a flexible lead through the loading inductor and then to the ground radials.

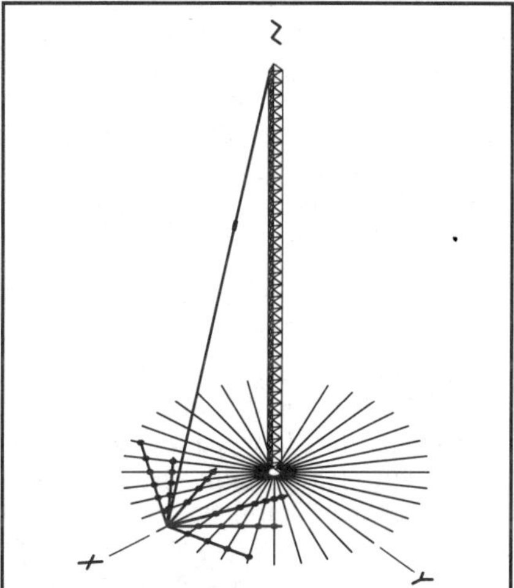

Fig 1—Bird's eye view of K4ERO tuned guy wire technique. Note that ground radials used by the vertical driven element are "shared" by slant-wire reflector element.

Table 2
Coil Specifications for Low-Frequency Amateur Bands

Freq MHz	*Turns*	*Wire size AWG*	*Dia inches*	*Length inches*
1.9	12	14	3	3
3.6	8	14	3	2
7.1	5	14	3	1.5

All dimensions in inches. Other combinations of turns and sizes are also usable, but the coil should not be physically too small if power capability and Q are to be maintained. X_L of coil to be maximum of 100 Ω; only a portion will be used in final antenna, based on actual results of tuning.

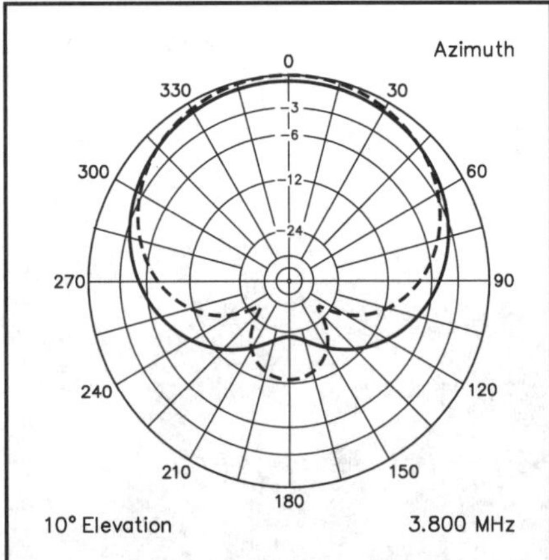

Fig 2—Azimuth patterns for 80-meter K4ERO array, with null adjusted at 180° azimuth by reflector tuning (solid line). Null can be placed in a different direction by retuning reflector, as shown by dashed line.

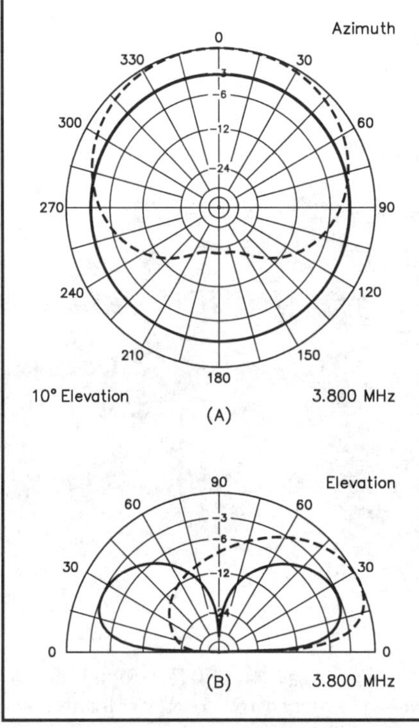

Fig 3—At A, comparison of azimuth patterns for antenna with and without tuned-guy reflector. The radial system is assumed to be excellent in this computation. At B, elevation patterns are shown.

Cut slits in the ground from the anchor point to uncover your radials in 6 different directions. See **Fig 1**. At each point where you cross an existing radial, make a solder connection to a #12 bare wire you have laid in the slit. Tie all the new radials together at the anchor point. Between the radials and the bottom of the slanted wire, connect a coil. This should be adjustable up about +100 Ω of reactance. See **Table 2** for design of the coil. Apply a low power signal to the vertical and measure the field strength in the direction of the sloping wire. Adjust the coil for a minimum signal. A null of between 10 and 20 dB should be attainable, depending mainly on your radial system. An even better method is to use a small signal source placed in the direction you wish to null and tune the reflector for minimum received signal. You may wish to tune for a null at an angle other than directly off the back. For example, you may have an interference source (line noise, etc) at an angle between 180° (off the back) and 90°. This can be dropped into a pattern null by tuning for the dashed-line pattern shown in **Fig 2**. I have done this on an actual antenna with results very close to those calculated. (*NEC* modeling really does work.)

When the desired null is obtained, go to the matching network for the vertical. You should find that the SWR has changed somewhat from what it was before you installed the tuned reflector, perhaps by as much as two to one. This is good, indicating that the sloping wire is indeed affecting the vertical, as it should. You should find that the input impedance has gone more inductive because the tuned radial has made the vertical look longer. This means that you will have to compensate by removing some inductance, adding some series capacity, or in some other way retuning your matching network at the base of the vertical. Of course, if you have a tuner in the rig or shack, you may just want to reset it a bit, as a 2:1 SWR will introduce negligible loss in your system, unless you already have far too much loss in your feed line.

At this point, you may want to check the forward gain. It should be 1.5 to 3.5 dB, again depending on your radial system. See **Fig 3** for a comparison of antenna with and without slanted reflector. Do not check the gain without retuning and readjusting the power of your transmitter, as its output may have fallen off while you were tuning for F/B ratio.

If you have an easy way to disconnect the tuned guy wire from its loading coil, you can quickly go back to the omnidirectional pattern. You may also wish to install a few other tuned guys for other directions and select the one you wish by connecting it to its loading coil, while leaving the others open. The solid 60.2-foot wire will not affect the pattern, provided it is left unconnected at the bottom end.

Very importantly, let me mention that unless you have a lot of experience in tower work and know how to use temporary guys when modifying existing ones and know how to tell when a guy is properly tensioned, or if you have an experienced friend who will work with you to do this project, I recommend that you use separate wires for the reflectors, and DO NOT use the steel guy wires that HOLD UP your tower. If your tower is not self supporting, do not disconnect any of the guys, even temporarily, lest the tower fall. If your tower is normally unguyed, the addition of 3 or 4 tuned "guys" will give it a bit of extra strength for high winds, etc.

Do not compromise the structural soundness by any attempts to get gain. It isn't worth it. Use strong wire, such as the "silky" #12 wire sold by the "Wireman," to help make for a safe installation.[4] Soft copper wires will stretch, and you will have to retune your reflectors from time to time.

Table 3 will give you an idea of what to expect as you tune your guys. Reflector tuning is highly recommended, since it is easy to tune for F/B, which occurs close to the tuning point needed for good gain. Tuning as a director is not recommended, since it tends to lower the radiation resistance of the driven tower. Unless you have an exceptionally good ground system, this can do more harm than good. The ground loss appears as a resistance in series with the radiation resistance, and this will cause the efficiency to drop quickly. If you have only a modest ground system, do not despair. Reflector tuning can actually raise the radiation resistance somewhat, and the effect on pattern is not nullified, just reduced.

Fig 4 illustrates the azimuth and elevation patterns to be expected for a ground system having a series loss equal to 10 Ω, represent-

Table 3
Effect of Coil Tuning on Performance

X_L Ω	R_r Ω	X_s Ω	*Gain dBi*	*F/B dB*	
0	26.2	17.7	1.78	3.0	
5	25.2	18.5	1.91	3.2	
10	24.1	19.5	2.04	3.3	
15	22.8	20.8	2.19	3.5	Director tuning
20	21.8	22.4	2.35	3.5	of guy wire
25	20.0	24.7	2.48	3.4	
30	18.5	27.7	2.56	3.1	
35	17.2	31.8	2.45	2.1	
40	16.7	37.4	1.89	0.4	Bidirectional (Fig 3A)
45	18.1	44.7	2.63	2.3	
50	23.0	52.8	3.39	6.3	
55	32.6	59.2	3.55	12.4	
60	45.3	60.0	3.32	25.2	
65	56.3	54.0	2.95	17.7	Reflector tuning
70	62.3	46.3	2.59	12.0	of guy wire
75	64.3	38.3	2.27	9.2	
80	63.9	32.1	2.00	7.4	

Above values are for a quarter-wave vertical with a good radial system. Other sizes and radial systems will give different results, but in general the results will be similar.
R_r = radiation resistance at base of fed vertical.
X_s = series inductive reactance as seen at base of vertical.
F/B = front to back ratio at 0° and 180°.
(Note that main lobe changes direction as X_L used in tuning passes through 40 Ω.)

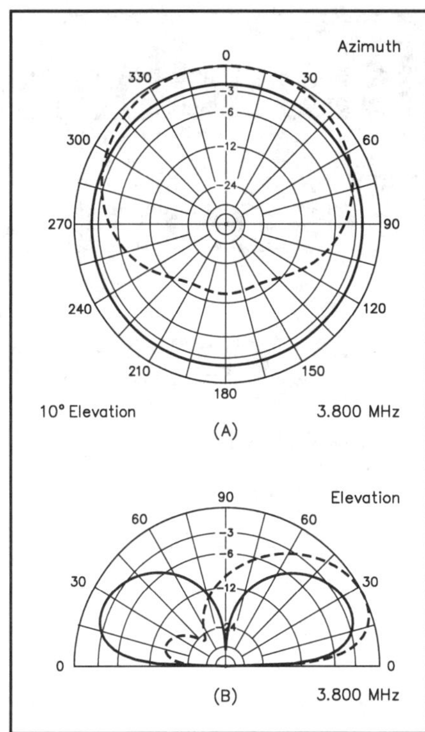

Fig 4—At A, comparison of azimuth patterns for antenna with and without tuned-guy, but with a limited radial system. At B, comparison of elevation patterns.

ing a radial system of 24 radials about 1/8 wave long. Compare these results with the patterns to be expected with a very good radial system as seen in Fig 3. Some may question how using only 6 radials for the reflector can avoid excessive loss. It should be observed that the function of radials is to shield the E-fields produced by the wires from causing currents to flow in lossy ground. This shielding is made possible by a wire spacing that is small in terms of a wavelength. From any given point on the ground's surface, the induced currents can "find" a nearby wire to flow in, rather than flowing in the resistive earth. If the wire it finds is going at right angles to its desired direction of flow, it will follow that wire until it finds a wire going in the right direction, that is, toward the base of the wire or tower from which the original E-field came. Thus, the six widely spread wires are "borrowing" the original extensive radial system to provide return paths.

Radials on the "back" side of the reflector are not very important for two reasons. First, the slope of the reflector makes the E-fields much stronger on the front side of the reflector, and second, the lack of radials in the back direction may actually contribute to the F/B ratio. This is another reason why the reflector tuning is preferred to the director tuning. In the director case, it would seem to be desirable to add additional radials in the direction of the beam; that is, beyond the tuned guy and away from the driven tower.

When using a sloping reflector, you will find that the usual null of a standard vertical at 90° elevation is lost. This could be an advantage or a disadvantage, depending on your needs. If you have a lot of potential QRM from within a few hundred miles of your QTH, you may find that the signals from short skip, which your present vertical rejects, are now a problem. On the other hand, if you use the lower bands for local net operation or rag-chewing as well as DX, you may be able to take down your low band dipole! Only very close-in signals, say a hundred miles from your QTH, will be increased by this change in the response at higher angles. For stations more than a few hundred miles away, the reduction in QRM off the back should be very noticeable.

The changes in patterns and SWR with frequency are shown in **Fig 5**. Since most low-band DXers use a relatively small portion of the band, retuning may not be necessary, although changing from 80-meter CW to 75-meter SSB will require you to readjust the base coils on your reflectors for best results.

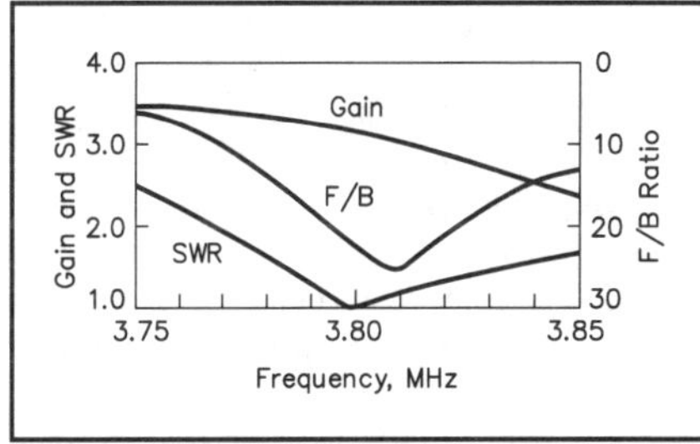

Fig 5—Variation of SWR, gain and F/B vs frequency across 75-meter DX band, excellent radial system assumed.

Notes

[1]Lew Gordon, K4VX, "The Effect of Continuous, Conductive Guy Wires on Antenna Performance," *QST*, Aug 1993, p 22.

[2]The graphics and tables included in this article were computed using *MN*, version 4.5, by K6STI. Other implementations of *NEC* and *MININEC* show similar results.

[3]See *The ARRL Antenna Book*, 15th edition, p 3-13 or *QST*, Dec 1976, p 13.

[4]The Wireman can be reached at 800-433-9473. Phillystran is sold by Texas Towers, 800-272-3467. Both sources are frequent *QST* advertisers.

Broadband Matching with the Transmission Line Resonator

By Frank Witt, AI1H
20 Chatham Road
Andover, MA 01810-5772

Signal transport and broadband matching are provided simultaneously by a uniquely useful device called the Transmission Line Resonator.

Introduction

The primary objective of the feed line in an antenna system is to transport signals between the antenna and the transmitter. The feed line may also be used to increase the match bandwidth observed at the transmitter end of the antenna system. Match bandwidth is defined as the width of the frequency band over which the SWR is less than some prescribed value. One mechanism for increasing the match bandwidth is the *Transmission Line Resonator*, which I call the TLR.

I have already described several specific applications of TLRs in wideband 80-meter antennas.[1,2] The generalized theory behind the TLR is presented here, so that the idea may be applied to other applications.

The TLR is a segment of feed line that is resonant at midband. In its simplest form, it is a multiple of an electrical half wavelength. However, it will be shown that TLRs that are odd multiples of a λ/4 can add useful features to the antenna system. Of course the TLR must be connected to the antenna and, in its simplest form, must have the right characteristic impedance to be effective. An optimum impedance transformation must be made at the transmitter end, since most transmitters are designed to work into a nominal 50-Ω load. In the latter part of this article, I show how to get around these limitations.

After I outline the basic TLR concept, I will review the underlying theory of broadband matching using lossy networks as a foundation for a more detailed explanation of the TLR. An example will be used to illustrate the design technique.

About SWR, SWR Meters and Generator Resistance

SWR stands for *standing wave ratio*. To measure SWR on a feed line, we use an SWR meter with a reference resistance equal to the characteristic impedance of the feed line. For example, to measure the SWR on an RG-213 50-Ω feed line, an SWR meter is connected between the transmitter and the feed line. To get an accurate reading, the SWR meter must be designed to work in a 50-Ω environment. That's what I mean when I say that the meter has a "50-Ω reference resistance."

However, the SWR meter has another use. A 50-Ω SWR meter connected between the load and the transmitter will tell how close the load is to 50 Ω. If the load impedance is 50 Ω resistive, the SWR meter will read 1:1. Transmitters and power amplifiers are designed to provide full power output when the SWR meter reading is less than some specified value, often 2:1, although some solid state transceivers begin to reduce the output power when the SWR exceeds 1.5:1. In a sense, the SWR meter provides a measure of the *suitability* of the load for the transmitter.

It is important to realize the following:

- The SWR meter reading should be *independent* of the *power output* of the transmitter. (Meters are somewhat power-dependent, so my statement applies only over the range where the meter gives an accurate reading.) Be suspicious if your SWR reading changes drastically with change in output power. Either the meter is at fault or the antenna system has a nonlinear element.
- The meter reading is *independent* of the *output impedance* of the transmitter. The SWR meter reading depends only on the impedance seen looking in the direction of the load.
- The load need not be a terminated transmission line with a characteristic imped-

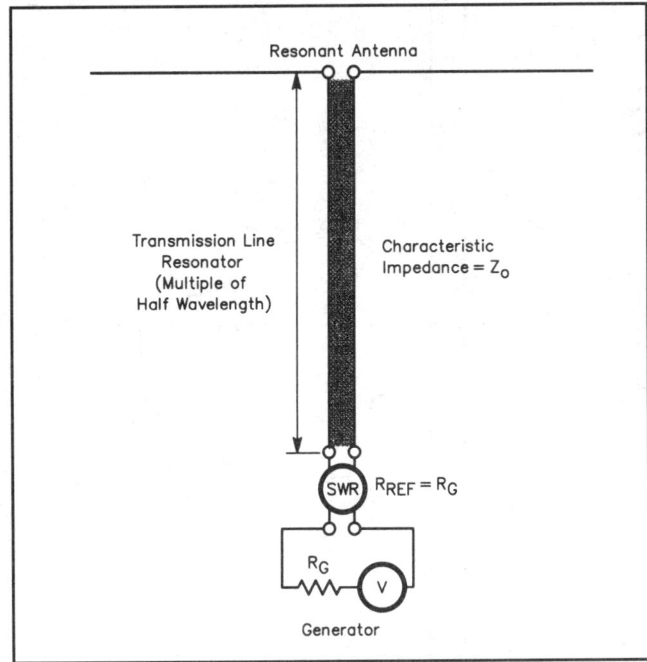

Fig 1—A broadband antenna system that uses the simplest form of TLR. The reference resistance for the SWR measurement is R_G.

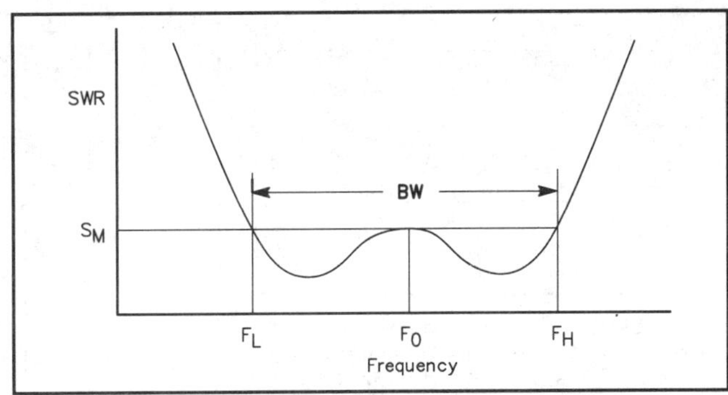

Fig 2—W-shaped SWR versus frequency characteristic that results when a parallel-LC matching network is used.

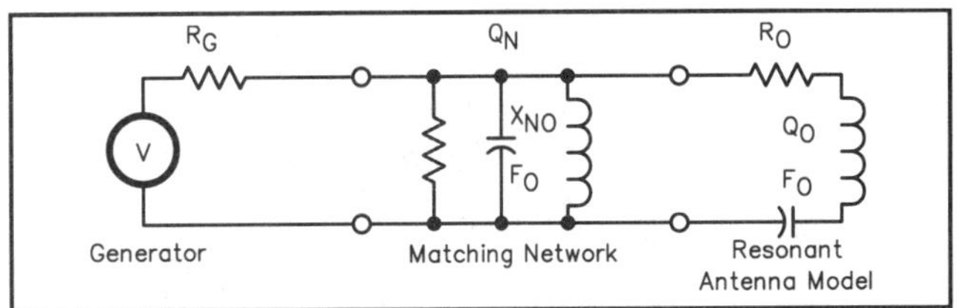

Fig 3—Assumed equivalent circuit for broadband antenna system. The antenna and matching network have the same resonant frequency.

ance equal to the reference resistance of the meter.

- Assume that a low-loss impedance-transforming device, such as a λ/4 Q-section or a transformer, has an impedance transformation ratio, N. The SWR measured at either port of the device will be the same if the reference resistance of the SWR meter at the high-impedance end of the device is N times the reference resistance of another SWR meter at the low-impedance end.

These facts should be kept in mind as you read on. I find it useful to visualize the antenna system with a generator. The diagrams show generators with output impedances equal to R_G. An SWR meter next to that generator will be assumed to have a reference resistance equal to R_G. To reiterate, the reading on that meter will not change even if the generator output impedance is changed from the value, R_G.

The Basic Idea

Before going into details, I'll review the TLR broadband matching technique. Shown in **Fig 1** is a rudimentary broadband antenna system. It consists of a resonant antenna with a feed-point resistance at resonance, R_0. It is fed by a low-loss feed line of characteristic impedance Z_0, whose electrical length is a multiple of λ/2. The generator resistance, R_G, is deliberately made larger than R_0. For low-loss lines at resonance (frequency = F_0), the SWR is approximately R_G/R_0, since a feed line of this length provides a replica of the load impedance at resonance. Off resonance, the reactance of the TLR compensates for the reactance of the antenna. The appropriate value of Z_0 must be selected to provide this compensation. In what follows, I discuss the determination of the proper Z_0 and some ideas for achieving the right value of R_G.

Broadband Matching with Lossy Networks

To understand the TLR, it is useful to review some earlier results, which were first presented in the reference of Note 3. In that paper, it was shown that broadband matching of a resonant antenna, such as a half-wave dipole, may be accomplished with the addition of a lossy parallel-tuned LC circuit as a matching network. With the aid of the matching network, the 2:1 SWR bandwidth can be increased by a factor of about 2.5. In practical cases, especially when transmission lines are used in the matching network to approximate the behavior of a parallel-tuned LC circuit, it is important to consider the losses in the matching network.

Instead of the familiar bowl-shaped SWR versus frequency characteristic, the addition of the matching network yields the W-shaped characteristic of **Fig 2**. The maximum SWR over the band is S_M. The band edges are F_L and F_H. Thus the match bandwidth, BW, and center frequency, F_0, are

$$BW = F_H - F_L \qquad \text{(Eq 1)}$$

$$F_0 = \sqrt{F_H \times F_L} \qquad \text{(Eq 2)}$$

It is useful to define a normalized match bandwidth, B_N.

$$B_N = \frac{BW}{F_0} Q_0 \qquad \text{(Eq 3)}$$

where Q_0 = antenna Q.

Fig 3 shows the pertinent equivalent circuit for many broadband antenna systems. The antenna model assumed is a series RLC circuit. The model will be refined later, but this is a useful starting point. Note that only three parameters, resonant frequency, resistance at resonance and antenna Q, are needed to define the antenna. See the references of Notes 4 and 5 for a description of a method for finding the resonant frequency, resistance and Q of a resonant antenna. Refer to Fig 3 for a definition of the terms used in the following equations.

The goal is to have the SWR over the band be as low as possible. There is a best or minimum value of S_M that may be achieved over the entire band. It depends on the desired match bandwidth, and the Qs of the antenna and matching network, and is given by the following formula:

$$S_{M\,min} = \frac{\sqrt{B_N^2 + 1} + \sqrt{B_N^2 + 1 + \frac{2Q_0}{Q_N}\left(1 + \frac{Q_0}{2Q_N}\right)}}{2\left(1 + \frac{Q_0}{2Q_N}\right)} \qquad \text{(Eq 4)}$$

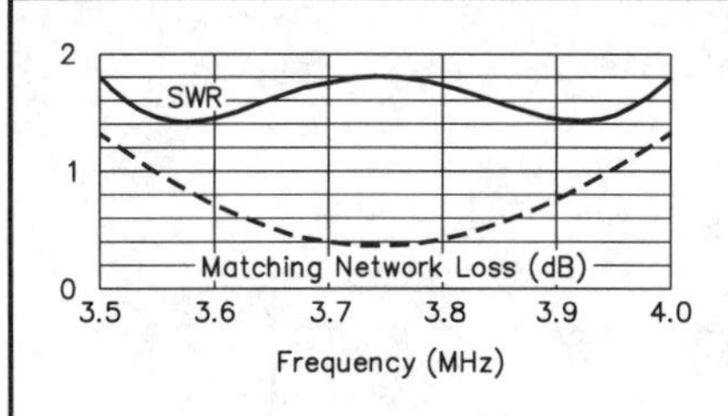

Fig 4—SWR and matching network loss for parallel-LC matching network and series-RLC resonant antenna model. Matching network Q = 40.45 and antenna Q = 13. Note that S_M is exactly 1.8, as calculated from Eq 4. Also, the band-edge loss is 1.32 dB, as predicted by Eq 7.

where Q_N = matching network Q.

Now the reactance at resonance of the matching network inductor or capacitor, X_{N0}, in ohms, may be calculated.

$$X_{N0} = \frac{R_0}{Q_0}\left[\left(1+\frac{Q_0}{2Q_N}\right)S_{M\,min}^2 - \frac{Q_0}{2Q_N}\right] \qquad \text{(Eq 5)}$$

where R_0 = antenna resistance in ohms.

The generator resistance, R_G, is calculated using

$$R_G = \frac{S_{M\,min} R_0 Q_N X_{N0}}{R_0 + Q_N X_{N0}} \qquad \text{(Eq 6)}$$

The loss in the matching network is always greatest at the edges of the band. This loss, L_{MNE}, in dB, is given by

$$L_{MNE} = 10\log\left[1+\frac{R_0}{Q_N X_{N0}}\left(B_N^2+1\right)\right] \qquad \text{(Eq 7)}$$

Now all the information needed to design an optimum matching network is available. An example will illustrate the procedure. The example is an 80-meter dipole. The following parameters are assumed:

F_L - 3.5 MHz
F_H = 4.0 MHz
R_0 = 57.2 Ω
Q_0 = 13
Q_N = 40.65

These are near the values that were used in the reference of Note 1. Q_N = 40.65 is the calculated value of the Q of a resonator made from RG-213 or similar coaxial cable in the 80-meter band. The use of this value will make it easy to compare calculations made in the various sections of this article. The reason for using R_0 = 57.2 , which is typical for low-height 80-meter dipoles, will become clear later. The results are

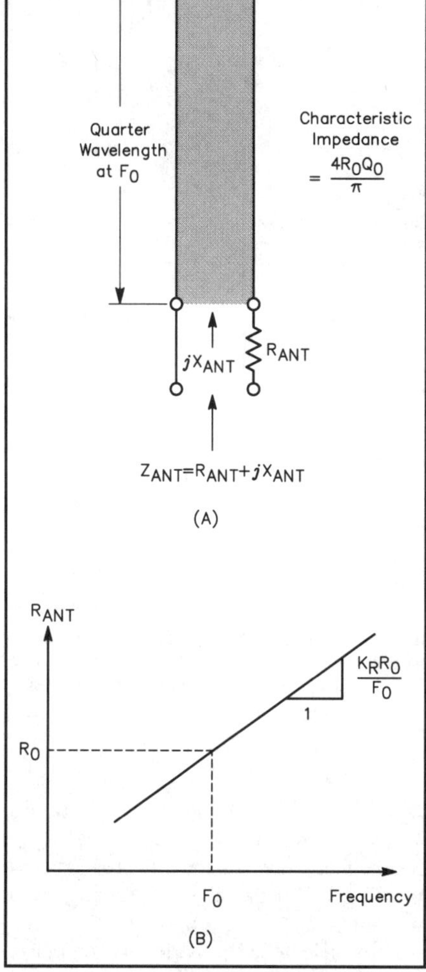

Fig 5—Resonant antenna equivalent circuit. This model is more accurate than the series-RLC equivalent circuit.

BW = 0.5 MHz	(from Eq 1)
F_0 = 3.742 MHz	(from Eq 2)
B_N = 1.737	(from Eq 3)
S_{Mmin} = 1.8	(from Eq 4)
X_{N0} = 15.9 Ω	(from Eq 5)
R_G = 94.8 Ω	(from Eq 6)
L_{MNE} = 1.32 dB	(from Eq 7)

Shown in **Fig 4** is the plot of SWR and matching network loss versus frequency for this example. The reference resistance for the SWR calculation is R_G or 94.8 Ω. If the desired load resistance for the transmitter is 50 Ω, then some means is needed to transform from 94.8 to 50 Ω. The inductor and capacitor in the matching network would each have a reactance of 15.9 Ω at 3.742 MHz.

Note that in a real application the loss of the transmission line would be added to the loss shown in Fig 4. In the TLR case shown later, some of the signal transport function is imbedded in the matching resonator. When comparisons are made, this fact should be kept in mind.

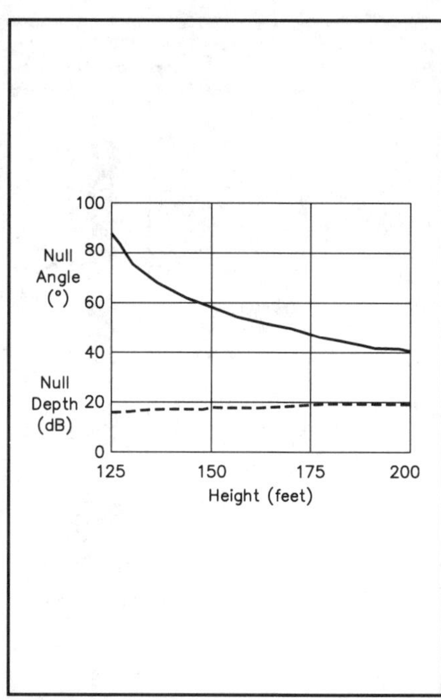

Fig 6—A resonant circuit realized by lines whose lengths are multiples of λ/4. Z_N is the input impedance of the line.

An Improved Antenna Model

There are two problems with the resonant antenna model shown in Fig 3. The actual antenna resistance varies with frequency. Also, the antenna may be viewed as an "opened-up" transmission line, so a transmission-line model for antenna reactance is usually more representative than a series-LC model. The improved resonant antenna model is shown in **Fig 5**.[6] It consists of a frequency-dependent resistor in series with an open-circuited lossless λ/4 transmission line.

The resistance is assumed to vary linearly with frequency. Over the relatively narrow frequency range, this assumption gives useful results, although there is some curvature in the actual resistance versus frequency characteristic. Antenna impedance is calculated as follows:

$$R_{ANT} = R_0\left[1+K_R\left(\frac{f}{F_0}-1\right)\right] \qquad \text{(Eq 8)}$$

$$X_{ANT} = -\frac{4R_0Q_0}{\pi}\cot\frac{90f}{F_0} \qquad \text{(Eq 9)}$$

where f = frequency in same units as F_0.

In addition to F_0, R_0 and Q_0 defining the antenna, K_R specifies the frequency dependence of resistance.

$$K_R = \frac{\text{Proportional change in resistance}}{\text{Proportional change in frequency}} \qquad \text{(Eq 10)}$$

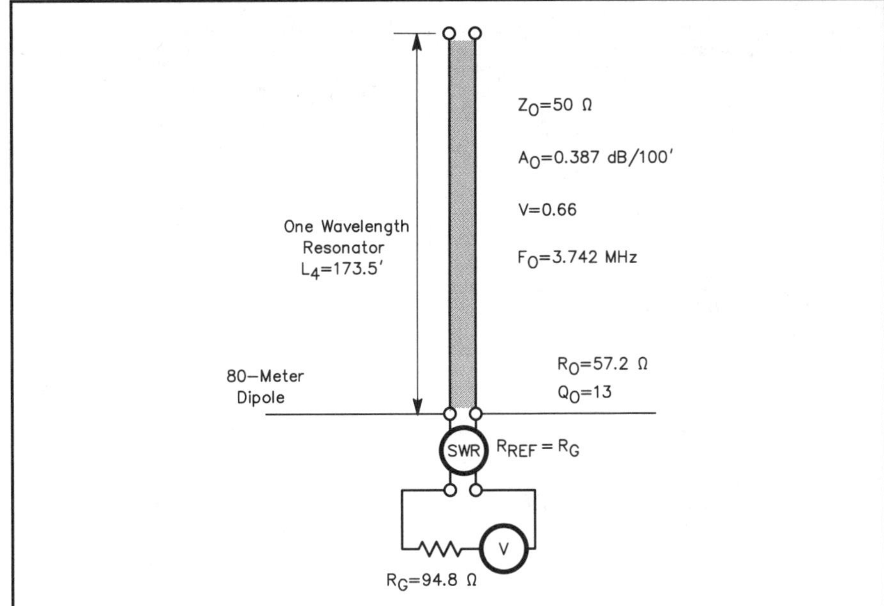

Fig 7—80-meter dipole example with a 1-λ resonator as the matching network.

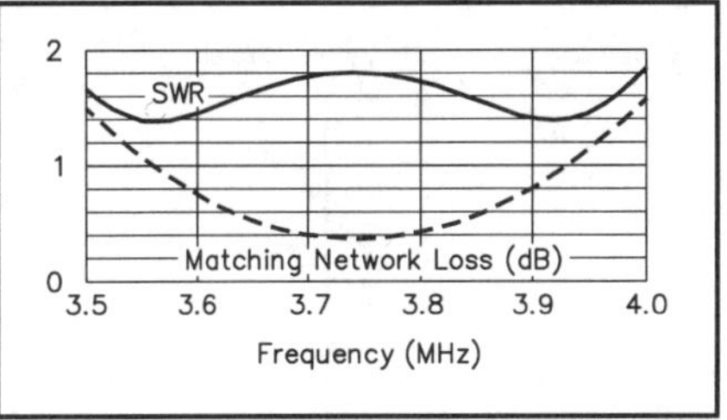

Fig 8—SWR and matching network loss for matching network made from 1 λ of 50-Ω transmission line. The antenna model is the one shown in Fig 5 with K_R = 0. Matching network Q = 40.65 and antenna Q = 13.

Table 1
Resonant Lines Providing Essentially Identical Reactance Compensation

n	*End*	Z_0 (Ω)	L_n *(feet)*
1	Short	12.5	43.4
2	Open	25	86.7
3	Short	37.5	130.1
4	Open	50	173.5
5	Short	62.5	216.9
6	Open	75	260.2

K_R can take on positive or negative values, and equals zero when the resistance is constant. A typical value for a half-wave dipole is K_R = +3.5.

The Transmission Line as a Resonator

A convenient way to realize the parallel-tuned circuit required for the matching network is to use transmission lines that are multiples of a quarter wavelength.[7,8] The odd-multiple λ/4 lines must be shorted at the far end and the even-multiple λ/4 lines must be open circuited.

Shown in **Fig 6** is the equivalence between short- and open-circuited transmission lines and a parallel-LC circuit. The length of each line is n electrical quarter wavelengths. The following equations provide the mathematical relationships:

$$X_{N0} = \frac{4Z_0}{n\pi} \quad \text{or} \quad Z_0 = \frac{n\pi X_{N0}}{4} \qquad \text{(Eq 11)}$$

$$Q_N = \frac{2.774F_0}{A_0 V} \qquad \text{(Eq 12)}$$

$$L_n = \frac{245.9nV}{F_0} \qquad \text{(Eq 13)}$$

where

Z_0 = characteristic impedance in Ω
A_0 = matched line loss in dB per 100 feet
V = velocity factor, and
L_n = line length in feet .

Now we can find the parameters of the transmission line required in our example. The line is assumed to have the same matched loss per unit length and velocity factor as RG-213 (which has a Z_0 = 50 Ω).

V = 0.66
A_0 = 0.4 dB/100 feet at 4 MHz.

We need the matched loss at 3.742 MHz. Since the line loss is proportional to $\sqrt{\text{frequency}}$,

$$A_{0F2} = A_{0F1}\sqrt{\frac{F_2}{F_1}} \qquad \text{(Eq 14)}$$

where A_{0F1} and A_{0F2} are the matched line losses at frequencies F_1 and F_2, respectively.

Hence,

A_0 = 0.387 dB/100 feet at 3.742 MHz

Q_N = 40.65 (from Eq 12)

Q_N of this value was used earlier. Note that the Q of the matching network is independent of n. Earlier it was calculated that

X_{N0} = 15.9 Ω (from Eq 5)

and therefore,

$$Z_0 = \frac{n\pi \times 15.9}{4} = 12.5n\,\Omega \qquad \text{(from Eq 11)}$$

Table 1 lists some resonant lines that provide essentially identical reactance compensation. For example, a broadband antenna system using 1 λ of 50-Ω cable (n = 4) is shown in **Fig 7**. (It is now clear why R_0 was chosen to be 57.2 Ω. This choice makes Z_0 = 50 Ω in this example. The accommodation of other values of R_0 will be covered later.) Note that a feed line has not yet been added to the system. **Fig 8** shows the SWR and matching network loss for this case. The dipole model is the transmission line model of Fig 5. The antenna resistance is assumed to be constant over the band (K_R = 0). The effect of a frequency-dependent antenna resistance will also be covered later.

All the cases given in Table 1 have very similar SWR and loss characteristics. Notice that longer lines have higher values of Z_0. Further, examination of Fig 8 reveals that the SWR and matching network loss are essentially the same as for the antenna system that uses the parallel-LC matching network of Figs 3 and 4.

Incidentally, one way of realizing a characteristic impedance of 25 Ω is to parallel two equal-length 50-Ω lines. Thus, a half-wavelength 25-Ω TLR made in this fashion will have the same Q_N and loss as a full-wave 50-Ω TLR.

The Transmission Line Resonator—Using the Matching Network as the Feed Line

The broadband antenna system of Fig 7 has a desirable SWR characteristic, but the feed line is not yet present. Can the line used for the matching network also be used as a part of the feed line? It can if the line length is a multiple of λ/2 (n even). Even if n is odd, it may be used, but this case will be covered later. A property of a feed line whose length is a multiple of λ/2 is that its input impedance is a near replica of the impedance at the far end. This is exactly true for lossless lines at the frequencies where the λ/2 condition is

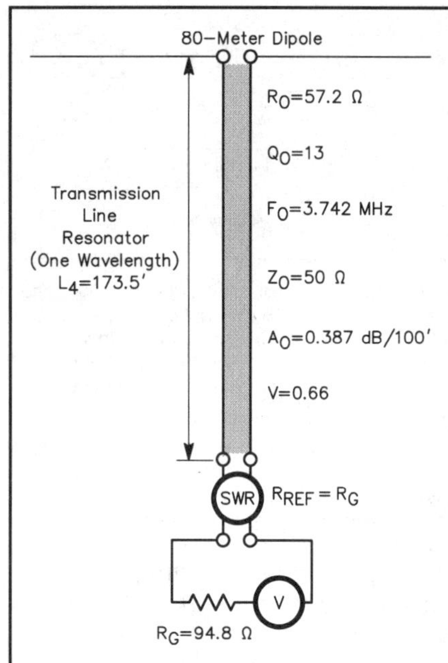

Fig 9—The antenna is at the far end of the 1-λ line. The line does double duty—broadband matching and signal transport.

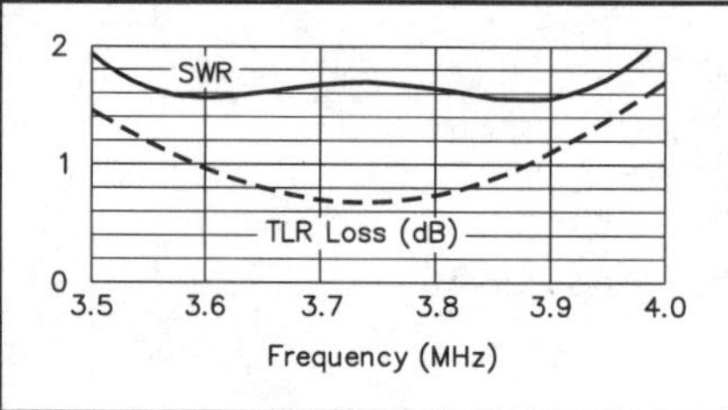

Fig 10—SWR and feed-line loss for a 1-λ 50-Ω feed line. The antenna model is the one shown in Fig 5 with K_R = 0. Matching network Q = 40.65 and antenna Q = 13.

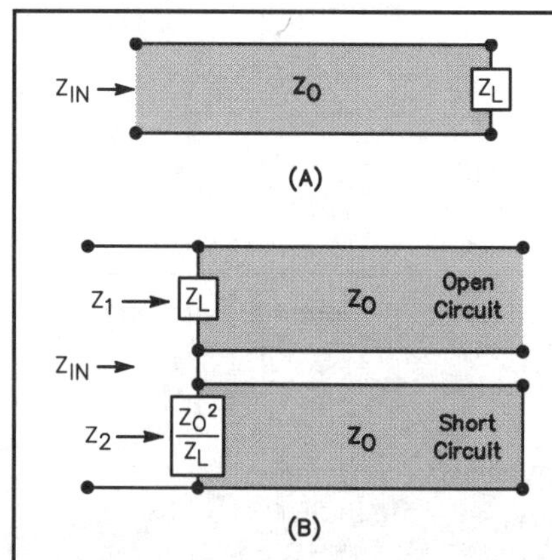

Fig 11—A terminated transmission line and an equivalent circuit that has the same input impedance, Z_{IN}. All of these transmission line segments have the same length and are identical in other respects.

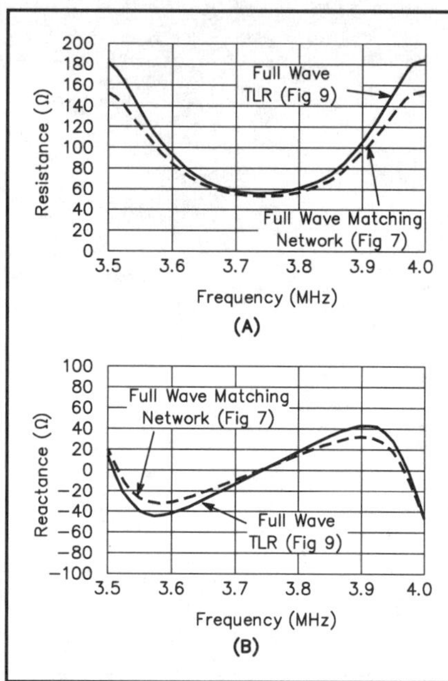

Fig 12—At A, the real and at B, the imaginary parts of Z_1 (dashed line) and Z_{IN} (solid line). They are similar, but not the same.

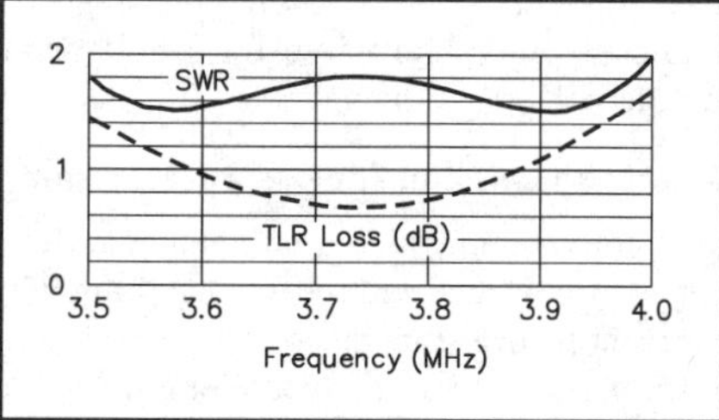

Fig 13—The SWR and loss characteristic for the configuration of Fig 9 when R_G is changed to 102 Ω. Note that this change in generator resistance leads to nearly the same SWR as that shown in Fig 8. The feed-line loss is unchanged from that of Fig 10.

true. For lines with loss and for other frequencies, the situation is not so simple. Fortunately, practical lines have low enough loss so that the property mentioned above is nearly true at resonance. Off resonance is another story. However, it is just this fact that makes the TLR work.

Let's see what happens when the antenna is moved to the far end of the full-wave stub of Fig 7. This configuration is shown in **Fig 9**. The SWR and loss for this case are given in **Fig 10**, which should be compared with Fig 8. Now we have picked up both a feed line 1 λ long and *broadband* matching. Note that the feed line loss is quite acceptable, and the band edge loss has hardly suffered.

One should not expect the configurations of Figs 7 and 9 to have identical SWR characteristics. **Fig 11** shows why. Shown in Fig 11A is a transmission line terminated in a load impedance, Z_L. Through some mathematical manipulations, it is possible to show that the circuit of Fig 11B is equivalent—equivalent in the sense that the input impedance, Z_{IN}, is the same in both cases. This equivalence is true for any length line, but we are specifically interested here in the case where the line length is a multiple of λ/2. Notice that the upper part (Z_1) of Fig 11B is identical to the matching network of Fig 7. The lower part (Z_2) has a much lower impedance than the upper part.

This is graphically illustrated in **Fig 12**, where the frequency dependencies of impedances Z_{IN} and Z_1 are plotted for the example we have been examining. This result makes sense, since the short circuit at the end of the line translates to a low impedance for Z_2 around resonance. Away from resonance, the impedance, Z_0^2/Z_L, cooperates by becoming smaller.

The original SWR characteristic of Fig 8 can be approached by simply increasing the value of the generator resistance, R_G, from 94.8 to 102 Ω. This results in the SWR and loss characteristic of **Fig 13**. From a practical point of view, this degree of design refinement is not very useful, but it is instructive to know how the SWR characteristic may be controlled.

Departure from the Optimum Design

Up to this point, the example has been limited so that we could have an optimum broadband match with one wavelength of RG-213 50-Ω cable. This meant that the required antenna resistance at resonance was 57.2 Ω. Further, we ended up with an optimum generator resistance of 102 Ω. Also, the antenna resistance has been assumed to be constant over the band. What about a real-world situation, where the antenna resistance is some different value and varies with frequency? What does one do to make the antenna system approximate a 50-Ω load, since most modern transmitters want to see such a load?

Fortunately, it is not necessary that the broadband match be *optimum* to realize a useful design. If both the antenna resistance at resonance and the generator resistance are increased a bit, a 50-Ω cable 1-λ long will still provide a good match over the entire 80-meter band. Such an arrangement is shown in **Fig 14**. The effective generator resistance is made to be 112.5 Ω by using a 75-λ Q-section ($75^2/50$ = 112.5). The antenna parameters are as follows:

R_0 = 65 Ω (instead of 57.2 Ω)
K_R = +3.5 (instead of 0)
Q_0 = 13
F_0 = 3.742 MHz

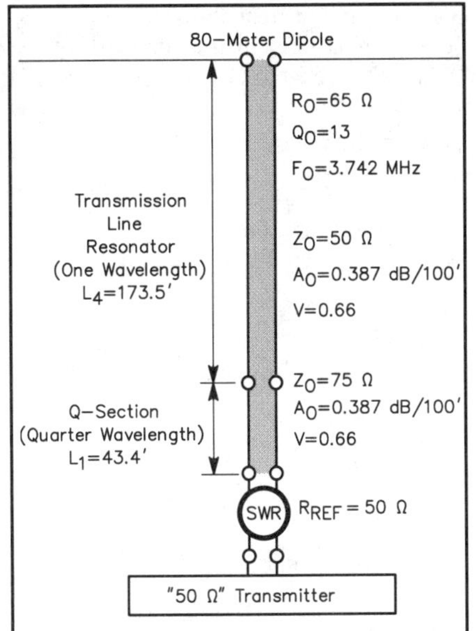

Fig 14—A practical slightly non-optimum design. The Q-section provides an equivalent 112.5-Ω generator resistance.

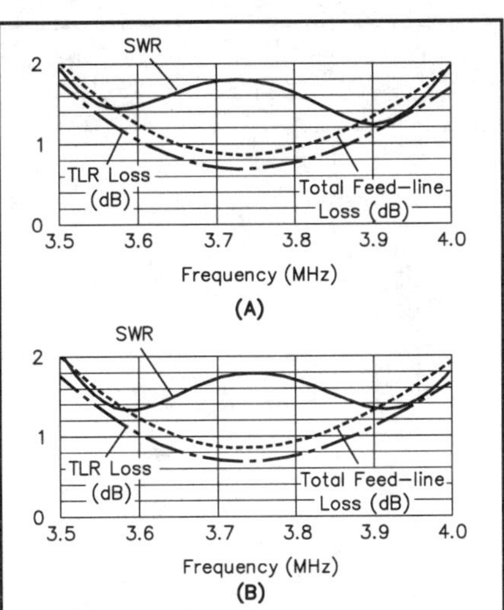

Fig 15—SWR and loss for the configuration of Fig 14. The loss contributed by the TLR is the dash-dot line and the loss of the TLR plus that of the Q-section is the dotted line. By decreasing the length of the TLR by 1.5 feet , the tilt of the SWR in A was removed, as can be seen in B.

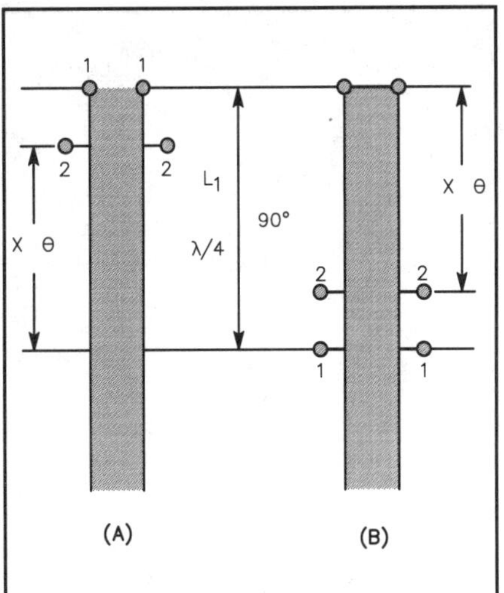

Fig 17—Transformer action in the TLR. The definition of transformer terminals depends on whether the TLR end is open- or short-circuited. θ is the distance between the minimum of the voltage standing wave (at resonance) and the connection point, expressed as an electrical angle. The distance x is that same distance expressed in feet.

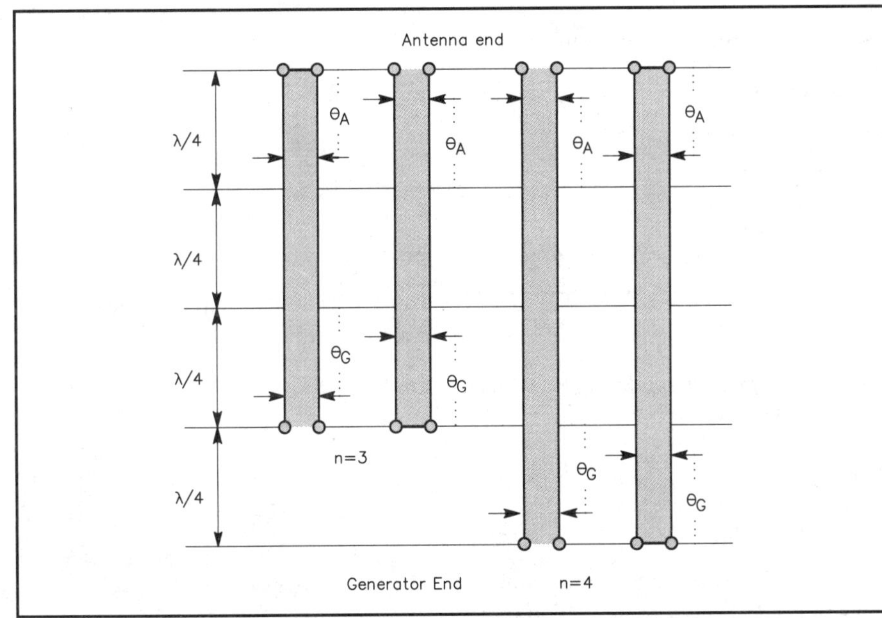

Fig 16—Some TLR possibilities. When n is odd, there must be a short at one end. When n is even, both ends are either open or shorted. The arrowheads designate the connection points to the lines. $θ_A$ and $θ_G$ are defined in the text.

The SWR and loss for the configuration of Fig 14 are given in **Fig 15A**. Note that the SWR exhibits some tilt over the band. This effect is primarily due to the variation of the antenna resistance over the band (52 Ω at 3.5 MHz to 82 Ω at 4.0 MHz). As seen in Fig 15B, this tilt may be removed by decreasing the length of the TLR from 173.5 feet to 172 feet. The tilt could also have been removed by decreasing the dipole's resonant frequency from 3.742 MHz to 3.715 MHz. Notice that the antenna system is not quite centered in the 80-meter band, but this can be easily corrected (if desired) by translating the dipole and TLR resonant frequencies downward by the same amount. In this case, the tilt and centering adjustments would have no practical value, but the techniques for making these corrections are worth noting.

The example given in this section is the same as the one shown in Fig 3B of Note 1. It is based on the dipole data presented by Walt Maxwell, W2DU, in his book, *Reflections,*[9] representing many practical 80-meter dipole installations.

The TLR Transformer

In what I have described thus far, the application of the TLR has been limited by the suitability of available coaxial cable types. This limitation leads to the relatively long (1 λ) TLR of the previous example. In some applications, the length may be needed anyway, just to get from the shack to the antenna, but often a feed line of 100 feet or so will suffice. For low power applications, when RG-58 and RG-59 cables are used, a λ/2 TLR yields a very reasonable broadband solution.[10]

The key cable parameter for realizing a resonator is its characteristic impedance. Fortunately, one property of the TLR that can remove this "available characteristic impedance" constraint is its ability to act as a transformer. This has not been exploited thus far in this article.

Shown in **Fig 16** are a variety of TLR possibilities. In earlier sections, all connections to the TLR have been at the ends of the line. This meant that n was an even number, and the TLR was a multiple of λ/2. As seen in the figure, odd values of n may also be used, but a short must appear at one end of the line. Further, n-even cases may have a short at both ends of the TLR.

By making the antenna and/or generator connection to the TLR at intermediate points along the line, transformer action may be achieved.[11] This feature, which is used in the coaxial resonator match,[12,13] opens up the possibility for enhanced op-

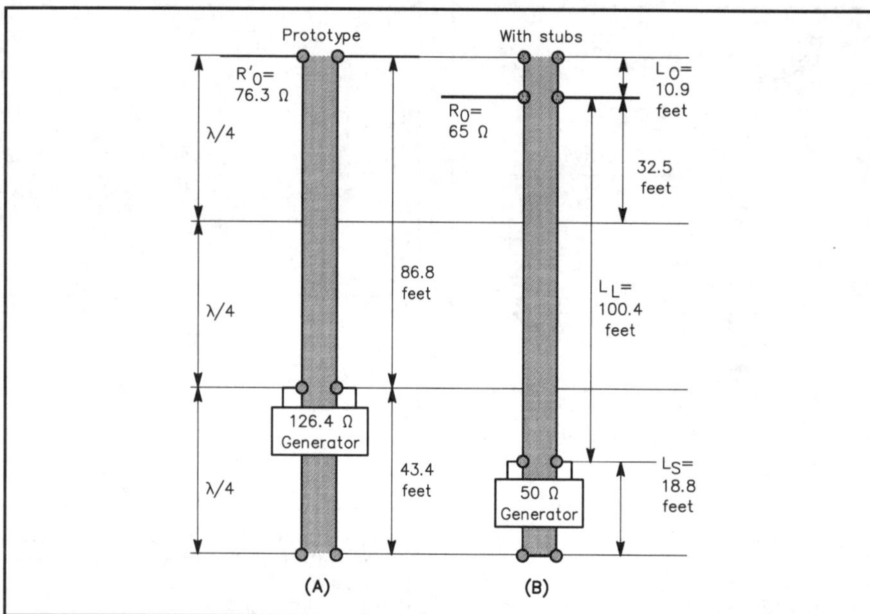

Fig 18—An optimized antenna system with a 3/4-λ TLR made from RG-213 coaxial cable. The prototype system at A is a convenient intermediate step in the design process. At B is the configuration with TLR transformers. It has an open stub, L_O, a link, L_L (which serves as the feed line), and a shorted stub, L_S.

Table 2
Calculated and Optimized Parameters Using TLRs

	Calculated	*Optimized*
Dipole resonant frequency (F_0)	3.742 MHz	3.710 MHz
Open stub (L_O)	10.9 feet	13.1 feet
Shorted stub (L_S)	18.8 feet	20.1 feet
Link (L_L)	100.4 feet	101.0 feet

timization of the antenna system. The nature of the transformer action is seen in **Fig 17**, where the ends of the transformer are designated 1-1 and 2-2. The impedance ratio of the transformer, N_Z, is approximated by

$$N_Z = \sin^2\theta \qquad \text{(Eq 15)}$$

where θ = distance between the minimum of the voltage standing wave and the connection point (expressed as an electrical angle).

At TLR ends, the minimum of a voltage standing wave appears at a short or at a quarter wavelength from an open. The low-impedance end of the "transformer" is always at the connection point, shown as end 2-2 in Fig 17. The same meaning is ascribed to θ_A and θ_G in Fig 16, where the subscripts designate the antenna and generator ends, respectively.

An example will make the power of the TLR transformer clear. Suppose we want to broadband match the antenna of the previous example to a 50-Ω transmitter. We want to feed the antenna with RG-213 50-Ω cable, and the total distance between the antenna and transmitter is less than 100 feet. The antenna parameters are

$R_0 = 65\ \Omega$
$K_R = +3.5$
$Q_0 = 13$
$F_0 = 3.742$ MHz

Since we know the characteristic impedance of the TLR is 50 Ω, we will design a "prototype system" with a 3/4-λ TLR (n = 3). Since n is odd, the TLR must be shorted at one end and we will place the short at the transmitter end of the line. The intermediate step of a prototype system will enable us to design a system with the final TLR, but without transformer action. Then the proper connection points will be calculated. **Fig 18** shows the prototype system and the system with built-in transformers.

Note that the prototype antenna is connected to the open end of the TLR. The prototype generator is connected at a multiple of λ/2 from the antenna. Here, at resonance, an approximate replica of the antenna will appear. We will use primed variables to indicate prototype antenna system elements. The prototype antenna differs from the actual antenna only in impedance level, so only R_0 will change. K_R, Q_0 and F_0 remain the same.

Starting with the 3/4-λ TLR, we have

$X_{N0} = 21.2\ \Omega$ (from Eq 11)

The prototype resistance at resonance, R_0', is calculated by rearranging Eq 5.

$$R_0' = \frac{X_{N0}Q_0}{\left(1+\dfrac{Q_0}{2Q_N}\right)S^2_{M\,min} - \dfrac{Q_0}{2Q_N}} \qquad \text{(Eq 16)}$$

Earlier results are repeated here.

$Q_N = 40.65$ (from Eq 12)
$S_{M\,min} = 1.8$ (from Eq 4).

Now we calculate

$R_0' = 76.3\ \Omega$ (from Eq 16)
$R_G' = 126.4\ \Omega$ (from Eq 6).

The electrical angles (in degrees) and connection locations (in feet) may be calculated from

$$\theta_A = \sin^{-1}\sqrt{\frac{R_0}{R_0'}} \qquad \text{(Eq 17)}$$

$$\theta_G = \sin^{-1}\sqrt{\frac{R_G}{R_G'}} \qquad \text{(Eq 18)}$$

$$x_A = \frac{\theta_A}{90}L_1 \qquad \text{(Eq 19)}$$

$$x_G = \frac{\theta_G}{90}L_1 \qquad \text{(Eq 20)}$$

The lengths of the stubs and link of Fig 18 are given by

$$L_O = L_1 - x_A \qquad \text{(Eq 21)}$$

$$L_S = x_G \qquad \text{(Eq 22)}$$

$$L_L = L_n - L_O - L_S \qquad \text{(Eq 23)}$$

Since L_1 = 43.4 feet and $L_n = L_3$ = 130.1 feet, we have

$\theta_A = 67.4°$ (from Eq 17)
$\theta_G = 39.0°$ (from Eq 18)
$L_O = 10.9$ feet (from Eq 21)
$L_S = 18.8$ feet (from Eq 22)
$L_L = 100.4$ feet (from Eq 23)

These dimensions are summarized in Fig 18. The final result for the design with the TLR transformers is a feed line with a length of 100.4 feet. An open stub of 10.9 feet is connected at the antenna terminals, and a shorted stub of 18.8 feet is connected at the generator terminals. If required, additional 50-Ω cable may be added between the 50-Ω generator and the point of connection to the shorted stub and link.

The calculated SWR and loss of the antenna system with TLR transformers are shown in

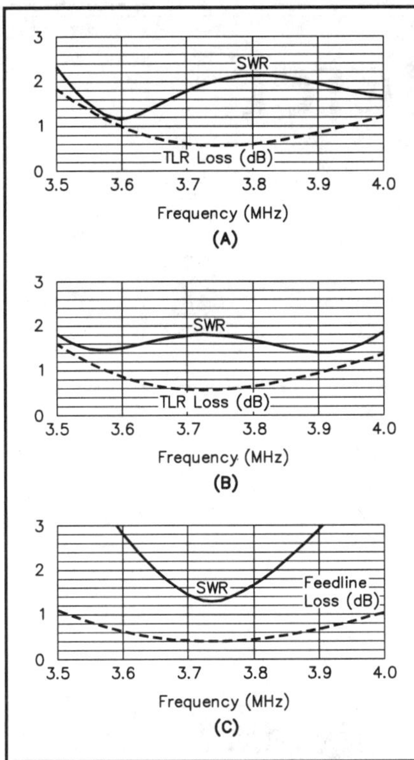

Fig 19—The SWR and loss of the TLR transformer antenna system. At A, the results are shown with the calculated line lengths. At B, the improvement after the lengths were optimized is shown. At C, for comparison, results are shown with the same setup as at A, but with the stubs removed. In all figures, K_R = 3.5.

Fig 19A. It was mentioned earlier that the stub calculations were an approximation. Fig 19A bears this out, but the changes necessary to achieve the optimized results are relatively small. By changing the lengths and dipole resonant frequency slightly, the target optimized SWR characteristic of Fig 19B results. See **Table 2**. Note the great similarity between the SWR plot of Fig 19B and the SWR characteristics of Figs 4 and 8. Those earlier cases were without a feed line. The losses are comparable, even though a feed line (the link) is a part of the TLR transformer antenna system.

A very desirable feature of the design shown in Fig 18 is that grounding the shield of the RG-213 coaxial cable for lightning protection automatically grounds both legs of the dipole. The presence of the shorted stub produces this result.

To demonstrate the significant improvement in match bandwidth that the TLR provides, the SWR and loss of same dipole fed with 100.4 feet of RG-213 cable is shown in Fig 19C. The loss added by the two stubs used to obtain match bandwidth enhancement is negligible (0.2 to 0.5 dB).

Other Applications of the TLR

The TLR has been applied to the 80-meter dipole. It is ideally suited to that application since this amateur band has a large percent bandwidth (13.4%). The quest for an antenna that does not require an antenna tuner to work both Phone and CW on 80 meters has interested many experimenters over the years. The 160-meter band also has a large percent bandwidth. Resonant antennas for these lower-frequency bands (such as dipoles, monopoles, inverted Ls and loops) are good candidates for broadband matching using the TLR.

Shortened antennas, which are tuned to resonance with reactive or linear loading, have reduced match bandwidths. The same principles described here may be used to achieve match bandwidth enhancement.

The TLR might be applied to broadband-match a beam antenna, where the designer must settle for a compromise among gain, front-to-back ratio and SWR bandwidth. With the TLR, the match bandwidth might be made wide enough that the gain and front-to-back ratio could be optimized better.

A TLR transformer may be used by itself to achieve a low-loss narrow-band impedance transformation, functioning like a $\lambda/4$ Q-section. It may be integrated into the antenna system design and is a powerful tool when a narrow-band impedance transformation is needed.

Summary

Transmission lines are most commonly used for signal transport between the antenna and the transmitter and/or receiver. For some applications, significant benefits can be obtained by taking advantage of the properties of resonant lines. The TLR is a tuned transmission line that can provide substantial match bandwidth enhancement, as well as signal transport.

When any broadband matching scheme is described, the losses in the antenna system should be revealed. Thus, all examples presented show not only the SWR properties, but also the resultant antenna system loss. The broadbanding provided by the TLR is primarily the result of reactance cancellation and not from losses. Many other broadbanding schemes cannot make this claim.

In its simplest form the TLR is a multiple of a half wavelength. Although this form is useful in many applications, match bandwidth improvement is limited by available cable types, and optimum broadbanding may not be possible. However, the compromise non-optimum result may still prove to be very satisfactory.

The opportunity to achieve the optimum broadbanding result is present when TLR transformers are used at one or both ends of the TLR. Although additional optimization may be required because of the approximations used for the transformers, the end result is an elegant solution to the challenge of obtaining a low-loss broadband match to a resonant antenna.

I want to thank Warren Bruene, W5OLY, and Jim Evans, NV1W, for taking the time to clarify for me the issues of output impedance of a transmitter and mismatch loss. As a study of the material in this paper reveals, the actual transmitter output impedance does not enter into the design of a broadband matching network. What does matter is the intended load impedance for the transmitter. The reading on the SWR meter should be viewed as a measure of the *suitability* of the load impedance as a termination for the transmitter.

Notes

[1]F. Witt, "A Simple Broadband Dipole for 80 Meters," *QST*, Sep 1993, pp 27-30, 76.

[2]F. Witt, "Optimizing the 80-Meter Dipole," elsewhere in this book.

[3]F. Witt, "Optimum Lossy Broadband Matching Networks for Resonant Antennas," *RF Design*, Apr 1990, pp 44-51 and Jul 1990, p 10. The term, Z_N, used in the original paper for the matching network impedance level, has been replaced in this paper by the term, X_{N0}. The change to matching network *reactance* at resonance is more descriptive.

[4]F. Witt, "The Coaxial Resonator Match," *The ARRL Antenna Compendium, Vol 2* (Newington: ARRL, 1989), pp 116-117.

[5]F. Witt, "How to Design Off-Center-Fed Multiband Wire Antennas Using that Invisible Transformer in the Sky," *The ARRL Antenna Compendium, Vol 3* (Newington: ARRL, 1992), p 74.

[6]F. Witt, "Broadband Dipoles—Some New Insights," *QST*, Oct 1986, p 34. Also see Note 2, section entitled "Dipole Parameters."

[7]G. L. Matthaei, L. Young, E. M. T. Jones, *Microwave Filters, Impedance Matching Networks, and Coupling Structures* (New York: McGrawHill, 1964), pp 214-217.

[8]R. D. Straw, editor, *The ARRL Antenna Book* (Newington: ARRL, 1994), 17th ed, pp 24-12 to 24-13.

[9]M. W. Maxwell, *Reflections—Transmission Lines and Antennas* (Newington: ARRL, 1990), p 15-19.

[10]See Note 1, Fig 5B.

[11]H. P. Westman, editor, *Reference Data for Radio Engineers* (Howard W. Sams & Co., Inc., 1968), 5th ed, pp 22-13 to 22-15.

[12]F. Witt, "The Coaxial Resonator Match and the Broadband Dipole," *QST*, Apr 1989, pp 22-27.

[13]See Note 4, pp 110-118.

Optimizing the 80-Meter Dipole

By Frank Witt, AI1H
20 Chatham Road
Andover, MA 01810-5772

A simultaneous look at signal propagation, radiation characteristics, antenna parameters and matching options can yield an optimum design for you.

Many articles and parts of books have been written about the horizontally polarized 80-meter half-wave dipole.[1-5] Much of the attention stems from the strong dependence on the presence of ground and on the relatively wide percentage bandwidth of the 80-meter band. However, no other treatment has simultaneously looked at signal propagation, antenna parameters, radiation characteristics and feed-system design. Now this is possible because of the powerful computing capability available to the serious antenna experimenter.

Two computer programs have recently come into the reach of anyone with a current personal computer and the willingness to spend a few bucks. The antenna simulation program *NEC/Wires*[6] accurately calculates the feed-point impedance over real ground, as well as radiation properties such as patterns and gain. It runs under Microsoft-DOS and requires a 386-based PC with an 80387 math coprocessor, a 486DX or Pentium PC.

The other program is *Mathcad*,[7] which can be used to analyze the simulation results from *NEC/Wires*. Results of the analysis may be presented as graphs. *Mathcad* also requires a 386- or 486/Pentium-based PC. At least 4 Mbytes of memory and Microsoft *Windows* are required. The role of *Mathcad* may be played by a spreadsheet such as Microsoft *Excel*, *Lotus 1-2-3* or *Quattro Pro*, but I have found *Mathcad* to be a better match to the application, partly because of its natural handling of complex numbers and its convenient graphing capability.

Use of these programs is not required by the reader to optimize an 80-meter dipole, since I summarize the salient results here. I describe various properties of the half-wave dipole and considerations for its use. Two often overlooked properties, antenna Q and the frequency dependence of resistance are also covered. These properties are important when maximum match bandwidth is wanted. With the information provided here, 80-meter enthusiasts can design an antenna that best meets their needs.

This paper concentrates on the horizontal half-wave dipole. For an excellent overview of antenna alternatives for 80 meters, I recommend John Devoldere's book, *Low-Band DXing*.[8]

What is Optimum?

What is an "optimum" 80-meter dipole? The optimization is influenced by your operating habits and objectives and by physical and financial constraints. Here are most of the factors to be considered in the optimization:

- Signal arrival and takeoff angles. For near-in contacts, higher-angle radiation is appropriate. For DX contacts the signals arrive at low angles. Antenna height determines the radiation pattern, with higher heights more appropriate for longer distances.
- Achievable height. Local restrictions, available space and cost will affect the height used. It is important to know the effects of increasing or decreasing the height of the antenna. The signal level can be increased if the ground slopes downward in the direction of the other station.
- Signal discrimination. Some broadside-to-endfire directivity is available at low elevation angles, but high angles are nearly omnidirectional.
- Available components. The designer must consider the system components. Transmitters and power amplifiers are designed to work best into a nominal 50-Ω unbalanced load. SWR bridges and directional power meters usually provide valid data only when the reference resistance is 50 Ω. For coax-cable feed lines,

50-Ω and 75-Ω characteristic impedances are the most common. Antenna tuners are useful, but they can conceal losses in the feed system and make tuneup take longer. Also, antenna tuners have losses of their own. Ladder-line feed systems have the promise of low feed-system loss, but they require an antenna tuner that can handle a balanced load efficiently.

- Match bandwidth at the transmitter. If you operate over only a few kHz of the 80-meter band, then only a small match bandwidth is required. However, to operate SSB and CW, some attention to match bandwidth in the design can eliminate the need for an antenna tuner and reduce feed-line loss.
- Losses. Ground losses are significant only at low antenna heights. Antenna-wire losses are usually negligible. Losses in the feed system must be carefully considered.
- Operation on other bands. Sharing the feed line with other paralleled dipoles is sometimes a desirable feature.

You can see that optimization can be complex, depending on how many of the above items are important to you. Usually, the design will be a compromise and the optimization will result from a considered weighting of all of the factors.

Signal Arrival Angle

Dean Straw, N6BV, has done an extensive study of signal arrival angles for the HF bands using a program called *IONCAP*. A summary of his analyses appears in the 17th edition of *The ARRL Antenna Book*.[9] **Table 1** makes use of his data for the 80-meter band to show signal arrival angles at locations in the US from various DX sites. His data spans the entire 11-year solar cycle.

An important observation is that between any two locations on the globe, the signals arrive over a narrow band of angles. Further, in many cases more than one peak exists, but each distribution tends to be sharply peaked. The angles shown in Table 1 are the centers of 5°-wide bands. The percentages indicate the percent of time the signals arrive within that band. For example, the 20°/82% "first peak" entry for signals arriving in W1 from Europe means that 82% of the time the signals arrive at angles between 18° to 22°. There is no "second peak" for this path.

As another example, take the path between W5-locations and Europe. The table reveals that 55% of the time the signals arrive between and including 11° and 15°. From the "second peak" entry, 28% of the time signals arrive 21° ± 2°. This 10° total range accounts for 83% of the time!

Table 1 clearly shows that the signals involved in DX QSOs on 80 meters arrive at very low angles. When the antennas are horizontally polarized, as is the case with the 80-meter dipoles discussed here, maximum signal strength at these angles requires very high antennas. For example, in order to have maximum gain at an elevation angle of 20°, the antenna height over flat ground must be about 180 feet. Fortunately, the variation of gain with height is gradual at low angles, and even for a height of 80 feet the signal drop-off from the maximum value is about 5 dB. This is a significant amount, but not enough to be a contact-killer most of the time.

For domestic contacts, the arrival angles are higher than those shown in Table 1. However, *IONCAP* results indicate that lower angles than commonly thought occur for these shorter-distance contacts. For example, contacts between Boston and either Chicago (817 miles) or Miami (1257 miles) have average peak angles of 20° and 26°, respectively. For shorter distances, say Boston to Washington, DC (339 miles), the average peak angle is 50°.

So what is the conclusion? Clearly, an antenna that favors low elevation angle propagation is preferred if long-distance contacts are the primary objective. For horizontal antennas, this means that the antenna should be high in the sky. A horizontal antenna optimized for DX by high placement will usually also produce strong signals for shorter-distance contacts, because the shorter paths are very efficient and produce S9+ signals most of the time. Very high antennas do produce nulls for higher-angle propagation, but for 80 meters such antennas must be over 170 feet before a null occurs at an elevation angle of 50°.

Height and Sloping Ground Considerations

For level ground, maximum gain at the

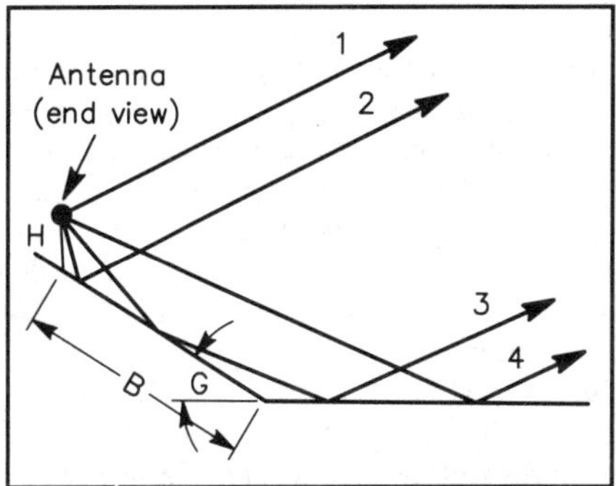

Fig 1—Up to four rays can contribute to radiation at the same elevation angle. These are the direct ray from the antenna (1), the ray reflected from the sloping surface (2), the ray reflected from the level surface (3) and the ray reflected from both surfaces (4).

Table 1
Statistical Elevation Angle/Percentage of Occurrence for Various 80-Meter Propagation Paths

	W1	*W2*	*W3*	*W4*	*W5*	*W6*	*W7*	*W8*	*W9*	*W0*
					First Peak					
Europe	20°/82%	19°/79%	18°/82%	15°/77%	13°/55%	10°/78%	12°/86%	16°/78%	17°/74%	15°/59%
South America	15°/90%	14°/87%	15°/90%	15°/86%	13°/81%	11°/100%	14°/52%	14°/97%	9°/29%	13°/89%
Far East	12°/71%	13°/67%	13°/67%	14°/70%	14°/63%	15°/73%	16°/63%	14°/68%	14°/71%	14°/56%
South Africa	12°/56%	11°/100%	11°/100%	12°/100%	6°/30%	*	*	10°/100%	10°/100%	6°/25%
South Asia	10°/60%	*	*	*	*	10°/100%	14°/100%	*	*	*
South Pacific	7°/100%	7°/33%	*	*	7°/75%	12°/100%	10°/100%	6°/40%	5°/43%	6°/53%
					Second Peak					
Europe	None	None	None	23°/13%	21°/28%	16°/22%	None	None	25°/13%	24°/26%
South America	None	None	None	None	None	None	18°/57%	None	14°/71%	None
Far East	18°/23%	18°/27%	18°/26%	18°/30%	18°/36%	25°/13%	25°/12%	18°/29%	18°/32%	23°/23%
South Africa	18°/43%	None	None	None	*	*	*	None	None	*
South Asia	18°/40%	*	*	*	*	None	None	*	*	*
South Pacific	None	*	*	*	*	None	None	*	*	*

* Signal levels below analysis threshhold.

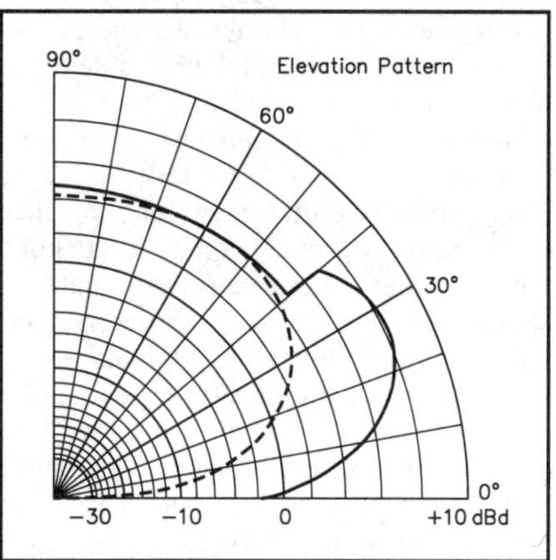

Fig 2—Elevation patterns over sloped ground with a level extension (solid line) and over level ground (dashed line). The step in gain at 40° occurs because of the abrupt change in ground slope.

low elevation angles shown in Table 1 requires high antenna height, which can be expensive, inconvenient, impractical, illegal or even impossible. What can be done?

Some of us live on a hill or on sloping ground. If the ground slopes downward from the antenna site in the direction of the distant station, what improvement in low-angle gain is available? See **Fig 1**, which shows a simplified picture of a sloping ground situation. Here the slope angle is G with respect to level ground. No slope continues forever, so we assume that the slope continues from a point directly under the antenna for a distance B before the ground levels off. The height of the antenna is H. We are concerned with the antenna pattern to the right of the antenna. Using a simple ray analysis, the elevation pattern in a direction perpendicular to the dipole may be calculated. It is necessary to consider all four of the rays shown in Fig 1.

Using *Mathcad*, I studied the ground profile model in Fig 1 and discovered that for many practical values of H, B and G a significant improvement in gain at low elevation angles is possible for the 80-meter band. This results from a fortuitous reinforcement of some of the signal components from rays shown in Fig 1.

For 80 meters, an "optimum" ground profile exists if the ground break (from sloped to level ground) occurs so that the double-reflected ray (shown as Ray 4 in Fig 1) just barely *does not* occur. I call this $B_{optimum}$.

$$B_{optimum} = \frac{H\cos 2G}{\sin G} \qquad \text{(Eq 1)}$$

To see how much improvement is possible, let's look at a specific example where the height H = 80 feet and the downslope

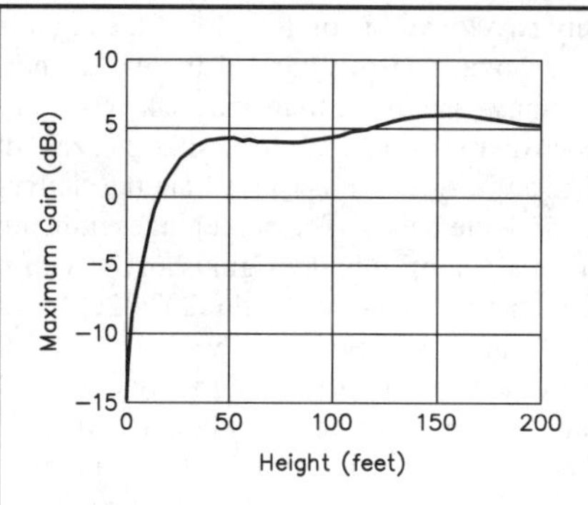

Fig 3—Maximum gain in dBd for 80-meter horizontal dipole over flat ground.

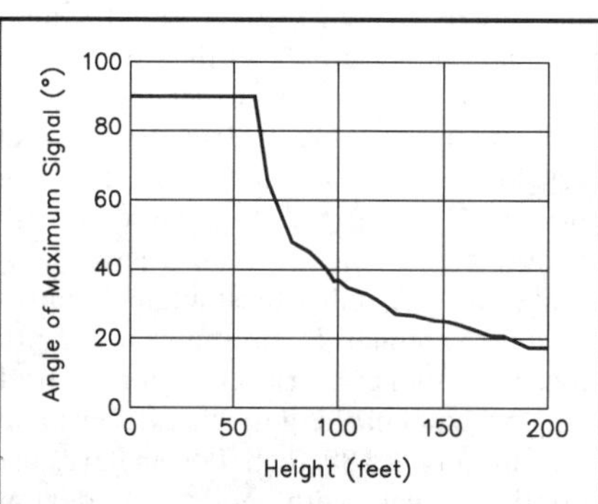

Fig 4—Elevation angle in degrees for maximum gain for 80-meter horizontal dipole over flat ground.

G = 20°. Then $B_{optimum}$ = 179 feet. For a frequency of 3.75 MHz, **Fig 2** shows the elevation pattern broadside to the dipole for a horizontal half-wave dipole 80 feet above level ground versus the "optimum" case described above. Perfectly conducting ground is assumed, but the improvement would be about the same over average ground, because at glancing signal angles the earth is an efficient reflector of horizontally polarized signals. Note the substantial gain advantage for the sloped-ground case for elevation angles between 0° and 40°. For example, the gain advantage for an elevation angle of 15° is about 8 dB. For other values of ground slope, the improvement will be for elevation angles from 0° to 2G.

Of course, we usually can't reshape our antenna site to fully take advantage of this "optimum" property, but I introduce the idea to provide some food for thought. Departures from the "optimum" condition can yield worthwhile improvement, but each case must be studied individually. In Eq 1, $B_{optimum}$ and H could just as well have been expressed in units of wavelength. Thus the elevation patterns for a 160-meter dipole at 160 feet and for a 40-meter dipole at 40 feet would be like the ones shown in Fig 2. For 160 meters, $B_{optimum}$ = 358 feet and for 40

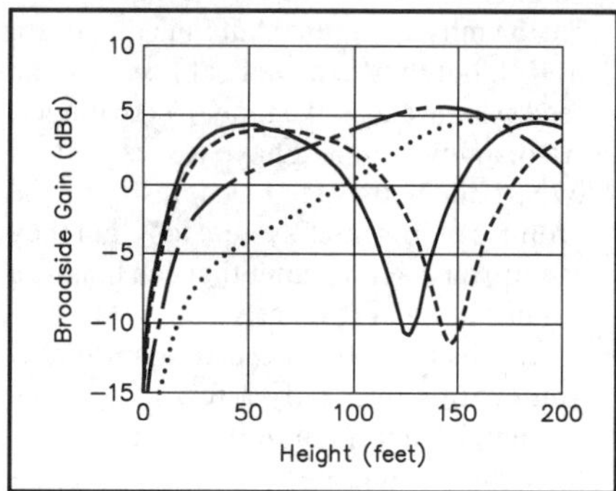

Fig 5—Broadside gain for elevation angles of 90° (solid), 60° (dashed), 30° (long dash-dotted), and 15° (dotted) for 80-meter horizontal dipole over flat ground.

meters, $B_{optimum}$ = 90 feet. Unfortunately, the same effect that gives the desirable property shown in Fig 2 for 80 meters can lead to an undesirable elevation pattern (nulls where you don't want them) on the higher frequency bands for a given height.

This analysis is based on ray analysis and assumes that the ground is smooth and free of obstructions. For an accurate result, the full extent of the wave front must be considered. However, a good approximation of the effect of sloping ground may be obtained from this simple ray analysis. The closer your site is to the model, the more indicative will be your results.

Radiation Pattern and Gain

In what follows, antenna system properties will be shown for a half-wave dipole made from #12 wire, located over average level ground (dielectric constant = 13 and conductivity = 5 mS/m). The antenna is 127.7 feet long, resonant at 3.75 MHz in free space. I use the same length throughout the analysis. The performance variation with height will be highlighted, since changes in height provide the greatest impact. However, the effects of different wire sizes, ground characteristics and drooping wires will be described as well. Where it makes sense, the properties of a half-wave dipole in free space are also shown for comparison.

Because of wave reinforcement (caused by ground reflection), the maximum gain is about 4 to 6 dBd (referenced to a dipole in free space) for any height above 30 feet. See **Figs 3** and **4**. This gain occurs broadside to the dipole at an elevation angle of 90° (straight up) up to a height of 60 feet. As seen in Fig 4, for greater heights the elevation angle for maximum gain drops off sharply.

Other views of antenna performance are shown in **Figs 5**, **6** and **7**, where the gain, nulls and directivity are presented. For design optimization, I recommend that you

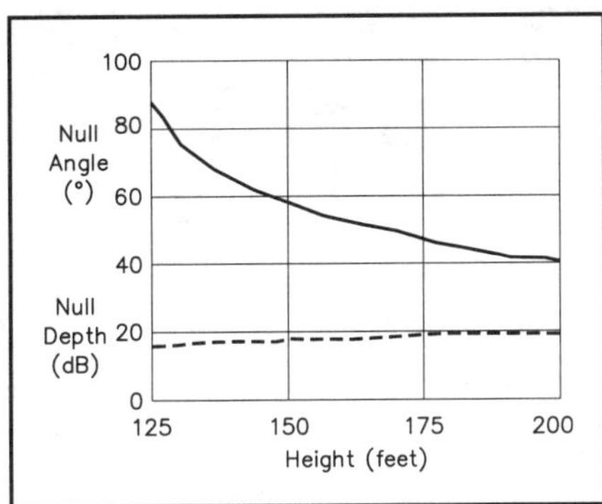

Fig 6—Null angle (solid line) in degrees and depth (dashed line). Null depth is expressed in dB down from the maximum gain for the same antenna height. This is for 80-meter horizontal dipole over flat ground.

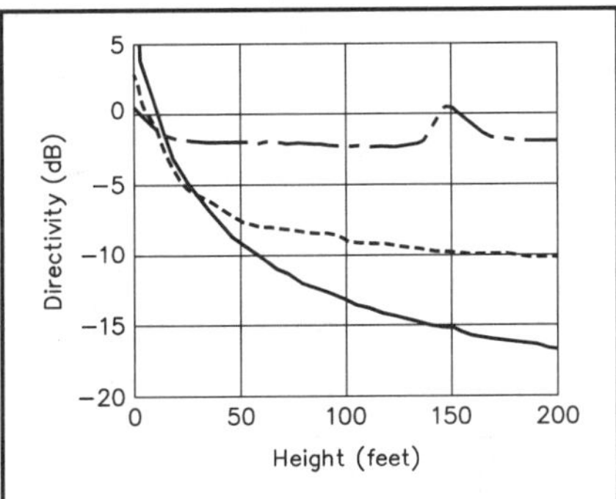

Fig 7—Endfire directivity expressed in dB for elevation angles of 15° (solid), 30° (dashed) and 60° (dash-dot). This is the signal level off the ends of the dipole relative to the broadside signal level. This is for 80-meter horizontal dipole over flat ground.

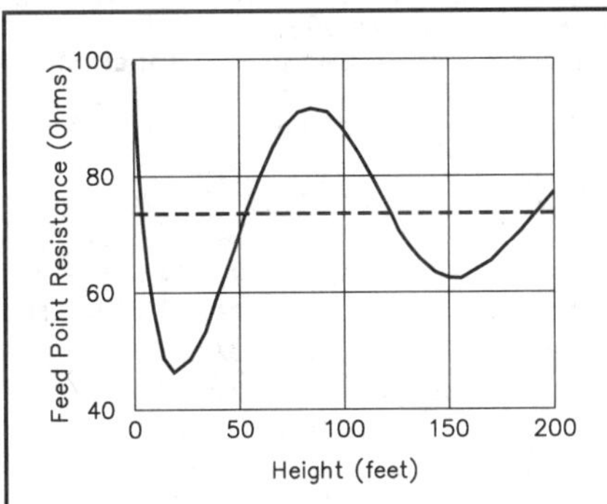

Fig 8—Feed-point resistance (solid line) for 80-meter horizontal dipole at resonance versus height over flat ground in feet. The free-space value is shown as a dashed line.

carefully examine Fig 5. Broadside gain is shown for elevation angles of 90°, 60°, 30° and 15°. Significant signal loss occurs for heights under 15 feet. High-angle radiation peaks at heights between 30 to 75 feet. Good all-round short- and medium-skip perfor-

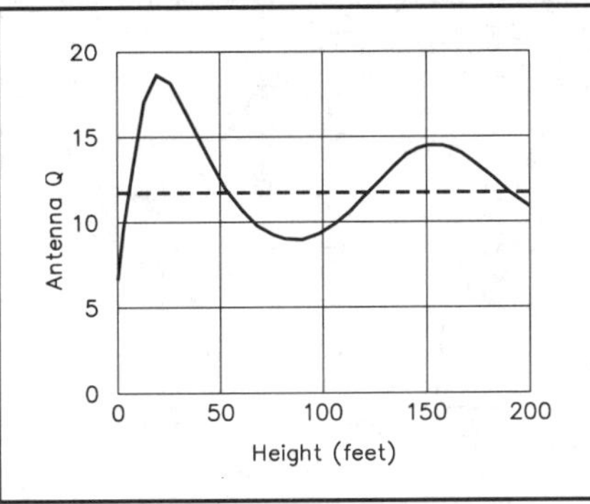

Fig 9—Antenna Q for 80-meter horizontal dipole (solid line) versus height over flat ground. The free-space value is shown as a dashed line.

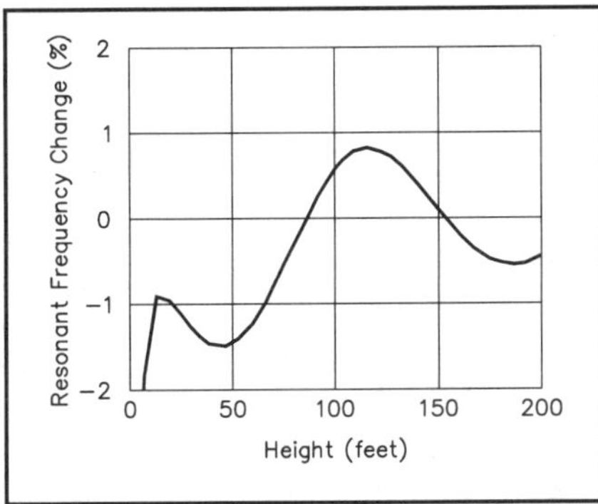

Fig 10—Effect of the presence of ground on resonant frequency expressed as percent offset from the resonant frequency of a dipole in free-space. Again, this is for 80-meter horizontal dipole over flat ground.

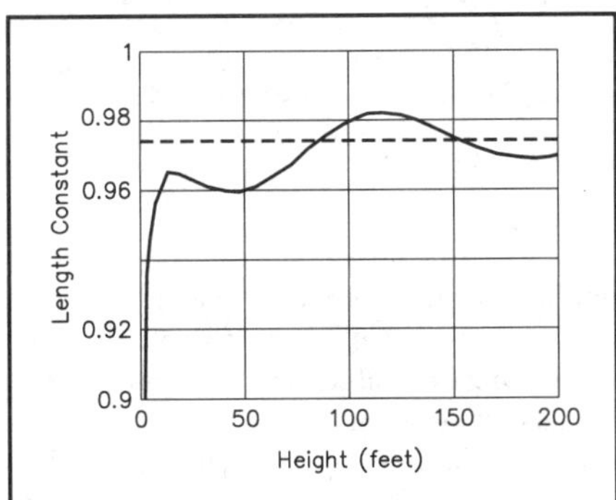

Fig 11—The length constant, K (solid line) for 80-meter horizontal dipole. Multiply the free-space half-wavelength by this factor to find the dipole length. The free-space value of K is shown as a dashed line.

mance exists for heights between 40 and 130 feet. Above 120 feet there are deep nulls in the high-angle radiation. The position of these nulls and their depth is shown in Fig 6.

As Fig 7 shows, high-angle radiation is omnidirectional. The broadside-to-endfire directivity for lower angles increases with height. Don't count on much directivity unless the signal arrives at a low angle.

Dipole Parameters

Only four parameters are needed to describe accurately the feed-point impedance of a half-wave dipole over the entire 80-meter band. I calculated these parameters using *NEC/Wires* data. Just how this is done is shown in the Appendix. The four parameters are:

- R_0, the feed-point resistance at resonance. See **Fig 8**. R_0 is not only the familiar radiation resistance, but also includes the effective wire resistance and induced ground resistance.
- Q_0, the antenna Q at resonance. Antenna Q is proportional to the ratio of the energy stored in the antenna and its surroundings to the power radiated and dissipated. The wide gyrations of Q_0 with height, as seen in **Fig 9**, are primarily due to the changes in R_0. Q_0 is inversely proportional to match bandwidth, so a lower value of Q_0 is desirable, unless it is caused by wire or ground losses.
- F_0, the resonant frequency. It is always expressed in units of MHz throughout this article. **Fig 10** shows how the antenna is detuned by nearby ground. F_0 may be varied by changing the length of the dipole. In practice, changes in the other three dipole parameters at resonance will not be greatly affected by small length changes. A related characteristic is seen in **Fig 11**, where the length constant, K, is shown. K is used for cutting a dipole to length relative to a half-wavelength in space. Note that K = 0.95, the most commonly quoted value for K,[10] does not occur for practical antenna heights on 80 meters. Some types of "end effects," such as the shape of the wire looped through the end insulators, have not been included in this analysis. In most practical installations, amateurs trim the lengths of their dipoles to achieve a desired F_0.
- K_R, the resistance coefficient. The antenna resistance varies with frequency and K_R is a measure of how much the resistance changes with frequency. For example, K_R = +3 means that if the frequency increases 1% the resistance will increase 3%. Since the percentage bandwidth of the 80-meter band is 13.3%, this would mean that the resistive part of the antenna impedance varies 40% over the band. In practice, K_R can be ignored unless one is trying to attain maximum match bandwidth. The variation of K_R with antenna height is shown in **Fig 12**. K_R is the primary reason that dipole SWR versus frequency plots are asymmetrical.

A useful equivalent circuit for a resonant antenna is a frequency-dependent resistor in

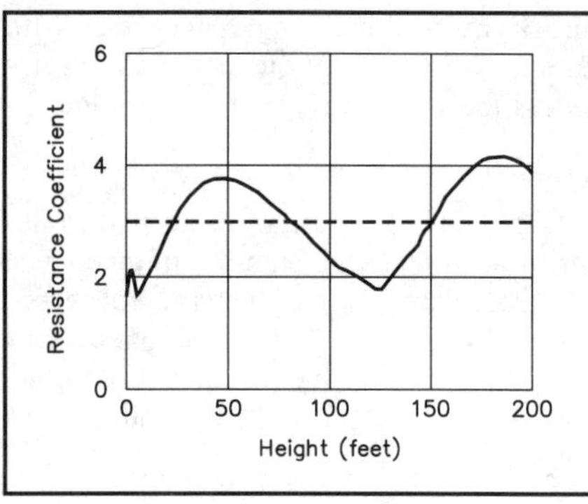

Fig 12—K_R (solid line) versus height for 80-meter horizontal dipole. The free-space value is shown as a dashed line.

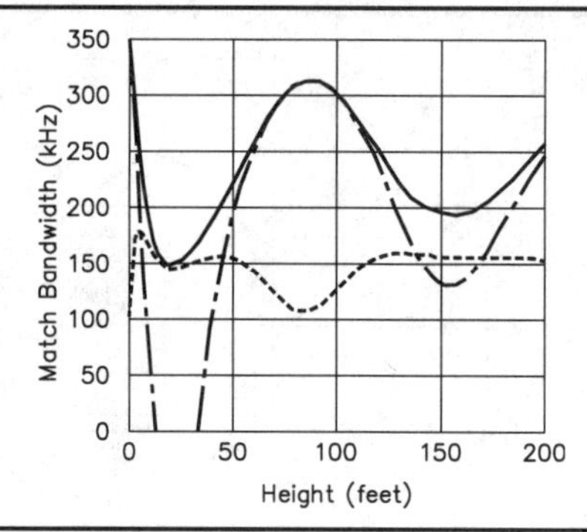

Fig 13—Match bandwidth in kHz for 80-meter horizontal dipole versus antenna height over flat ground. The solid curve shows the maximum match bandwidth possible without a matching network. The dash-dot curve is the match bandwidth referenced to 112.5 Ω and the dashed curve applies to a 50-Ω reference.

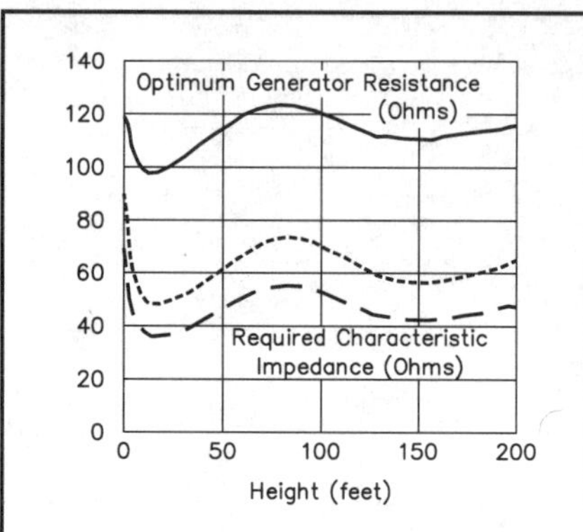

Fig 15—Optimum generator resistance (solid line) and characteristic impedances for TLR matching when the cable has the same loss as RG-213 coax. The characteristic impedances for 1-λ (dashed line) and 3/4-λ TLRs (long-dash line) are plotted also. Curves are for 80-meter horizontal dipoles over flat ground.

series with a lossless open-circuited quarter-wave transmission line.[11] In terms of the above dipole parameters, the frequency-dependent antenna impedance, $Z_A(f)$, is given by:

$$Z_A(f) = R_0\left[1 + K_R\left(\frac{f}{F_0} - 1\right)\right] - j\frac{4R_0Q_0}{\pi}\cot\frac{90f}{F_0} \quad \text{(Eq 2)}$$

This equation gives an accurate mathematical description of the impedance of a half-wave dipole over the entire 80-meter band. The very same equation is useful for many other resonant antennas. See the appendix for a formula for $Z_A(f)$ that eliminates the need for finding the antenna Q and K_R and also takes into account the antenna-wire size.

Match Bandwidth

Any half-wave dipole resonant in the 80-meter band will radiate with essentially the same pattern and gain over the entire band. Our goal should be to transfer energy efficiently from the transmitter to the dipole over the part of the band we choose to operate. Another design constraint is present because modern transmitters and power amplifiers are optimized for 50-Ω unbalanced loads. As seen in Fig 8, antenna resistance at resonance ranges far from 50 Ω at heights providing good radiation patterns. So what can we do?

We want to optimize the match bandwidth at the transmitter. However, it is useful to look at match bandwidth at the antenna for comparison purposes. Even wider match bandwidths are seen at the transmitter because of loss in the feed system. In **Fig 13**, the solid curve shows the maximum 2:1 SWR bandwidth possible at the antenna when no matching network is used. This maximum value is achieved when the reference impedance is 25% greater than R_0, the feed point resistance at resonance.[12]

Also shown in Fig 13 is the 2:1 SWR bandwidth that would be measured at the antenna if the reference resistances for the SWR measurements were 50 and 112.5 Ω. These are the values of common interest because they occur when the antenna is fed directly with 50-Ω coax or through a 75-Ω Q-section ($75^2/50 = 112.5$). Note that direct feed with 50-Ω coax gives the maximum match bandwidth (about 150 kHz) when the antenna height is around 25 feet. However, for the more desirable (because of radiation pattern) heights of 60 to 120 feet, the use of the Q-section provides impressive match bandwidths (250 to 300+ kHz). Over the same range of heights, there is a dip in the match bandwidth if the antenna is directly fed with 50-Ω line.

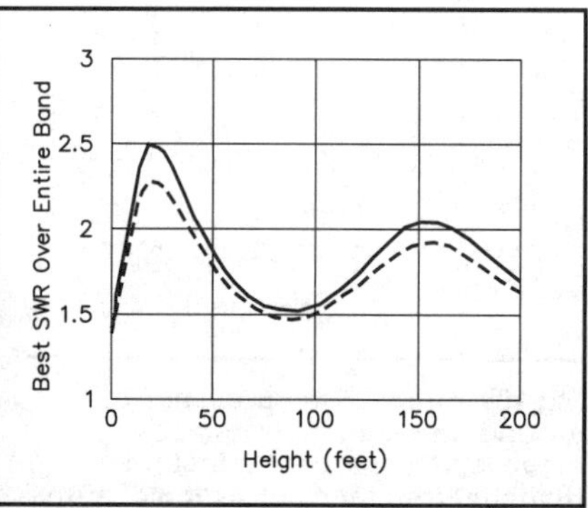

Fig 14—The maximum SWR over the full 500 kHz of the 80-meter band when a TLR matching network is used. The solid line applies to the use of 1/2 inch Hardline and the dashed line results when cable with the loss characteristics of RG-213 is used. These are for 80-meter horizontal dipoles situated over flat ground.

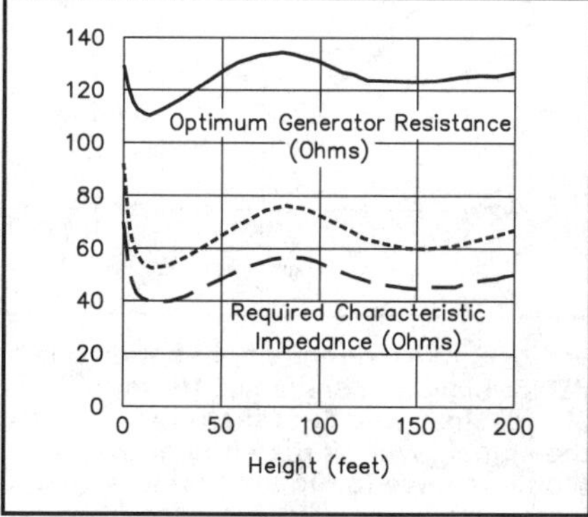

Fig 16—Optimum generator resistance (solid line) and characteristic impedances for TLR matching when the cable has the same loss as 1/2-inch Hardline. The characteristic impedances for 1-λ (dashed) and 3/4-λ (long-dash) TLRs are plotted as well.

Broadband Matching with the Transmission Line Resonator

You can significantly increase the match bandwidth over that shown in Fig 13 with a remarkably simple modification to the feed system. The approach makes use of the transmission line resonator or TLR.[13,14] For all heights above 40 feet, the SWR using the TLR can be made less than 2.1:1 over the entire 80-meter band. The maximum SWR over the band is shown in **Fig 14** for two cases: a TLR made from cable that has the same loss as RG-213 coax and one made with 1/2-inch Hardline. For design purposes, you must know the optimum generator resistance and the characteristic impedance of the TLR. These are plotted in **Figs 15** and **16** for RG-213 cable and 1/2-inch Hardline, respectively. For all heights, it is strictly serendipity that the optimum characteristic impedance value is near either 50 or 75 Ω. The required generator resistance is readily obtained by taking advantage of the imped-

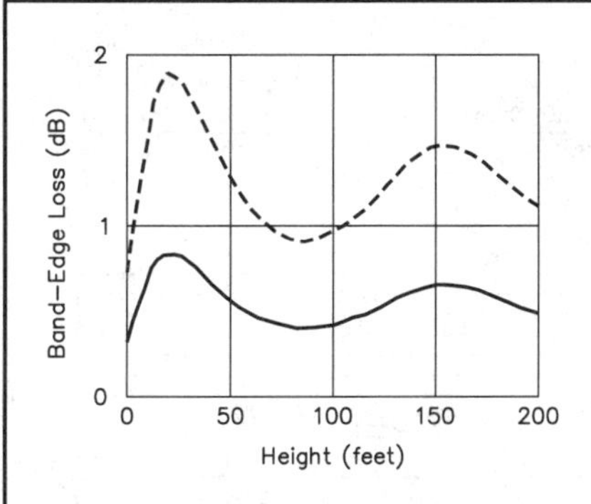

Fig 17—Band-edge loss (in dB) of a TLR when it is used as a resonator to compensate for the reactance of a resonant antenna. The effective loss is less since the TLR is used for signal transport as well. The solid line applies to the use of 1/2-inch Hardline and the dashed line results when cable with the loss characteristics of RG-213 is used.

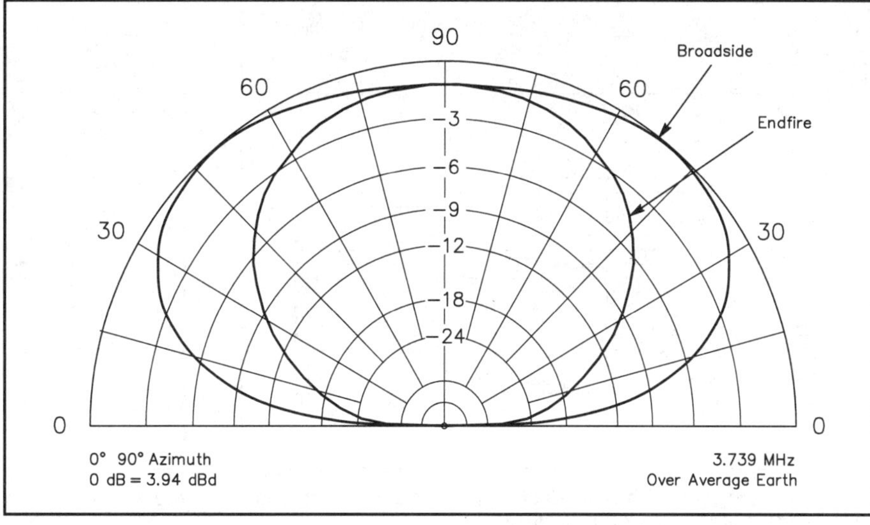

Fig 18—Broadside and endfire radiation patterns for the selected "optimum" 80-meter dipole over flat ground.

ance-transforming property of the TLR, or by incorporating a Q-section in the design. These details will become more clear later when some specific designs are described.

Note that Figs 15 and 16 are very similar, although they apply for cables of significantly different loss. At 4 MHz, cables in the same family as RG-213 (RG-8, RG-11, etc) have a matched-line loss of 0.4 dB/100 feet and a velocity factor of 0.66, while 1/2-inch Hardline has a matched-line loss of about 0.14 dB/100 feet and a velocity factor of 0.81. From the examples, you will see that the generator resistance and characteristic impedance do not have to be "on the money." The cable *lengths* are critical, however.

The probability of having a lightning strike is minimized by having all elements of the antenna system at dc ground potential. The 3/4-λ TLR provides an excellent way to do this, since one end of the TLR must be shorted. By connecting the shield of the feed system to a ground rod outside the shack, both legs of the dipole will always be grounded.

Losses

The five sources of loss in a dipole antenna system are:

- Wire loss. This is negligible for the wire sizes normally used. For #12 copper or copper-clad steel wire, the loss is only about 0.1 dB.
- Ground loss. Ground causes loss from signal absorption. For heights less than 30 feet, the effect is strong enough to induce an additional resistance at the feedpoint. For all heights, the reflected wave is attenuated by the lossy ground. For horizontal polarization, this kind of loss decreases as the elevation angle decreases.
- Feed-system loss. The loss in the transmission line and in any matching network, such as a TLR, must be considered. TLR loss, which is maximum at the band edges, is shown in **Fig 17**. Fig 17 applies where the TLR is used in shunt with the antenna terminals and a separate feed line is used. Since the TLR can also provide signal transport, the effective loss is usually much less than that shown in Fig 17.
- Antenna tuner loss. Some antenna tuners have as much loss as the feed line.
- Mistermination loss. If no antenna tuner is used and/or if the output network of the transmitter is fixed-tuned, departure of the transmitter load impedance from $50 + j0\ \Omega$ can lead to reduced power transfer, amplifier power-down or both.

You should evaluate all these sources of loss as a part of the optimization process. The equations used for generating Figs 13-17 are given in the Appendix.

The Optimum Design

There is no single optimum design. However, we now have enough information to examine various trade-offs. I'll select a particular height as an example and describe the antenna system properties there. I'll use an antenna that is very competitive in stateside contacts and that also performs well for the ham who is going for DXCC endorsements. It is equally well suited for CW, SSB and AM contacts, since it can provide a low SWR over the entire 80-meter band.

The height selected is 80 feet, over flat ground. At this height, the gain overhead and the broadside gain at 30° are equal and the maximum broadside gain of 3.9 dBd occurs at 48° elevation. The broadside and endfire elevation patterns are shown in **Fig 18**; the good all-round radiation pattern is evident. For a half-wave dipole, a presentation like Fig 18 provides lots of useful information.

At the height of 80 feet, the antenna Q is a minimum value (Fig 9), so the match bandwidth is a maximum value (Fig 13). The clear advantage of feeding with a Q-section over using direct 50-Ω feed may be seen by comparing **Figs 19** and **20**. The match bandwidth is actually greater than that predicted in Fig 13, due to the loss of the feed system between the antenna and the point of SWR measurement.

In these examples, the total feed-line length was 113.1 feet so that the TLR feed system (shown in **Fig 21**) may be directly compared. If required, additional 50-Ω feed line may be added to the left of the SWR meter.

All that is required to obtain the excellent match of a 3/4-λ TLR, seen in Fig 21, is the addition of a 21.6 foot shorted stub to the already-present 50-Ω cable in Fig 19. A complete TLR would also have an open stub at the antenna. Even if this open stub is omitted, the TLR achieves an SWR of better than 1.55:1 over the band. A detailed treatise on designing TLR matching networks is provided in the reference of Note 15.

It is important to know the loss of any feed system. Feed-system loss is quite tolerable—the difference between the three cases shown in Figs 19 through 21 is less than 0.6 dB. For those situations where there is a very long feed-line run, many amateurs use surplus CATV 1/2-inch 75-Ω Hardline. For a TLR using 1/2-inch Hardline, Fig 14 shows that an SWR of 1.53:1 can be achieved over the entire 80-meter band. Rather than employing the full power of the TLR, you might want to consider using a one-wavelength TLR and a Q-section.

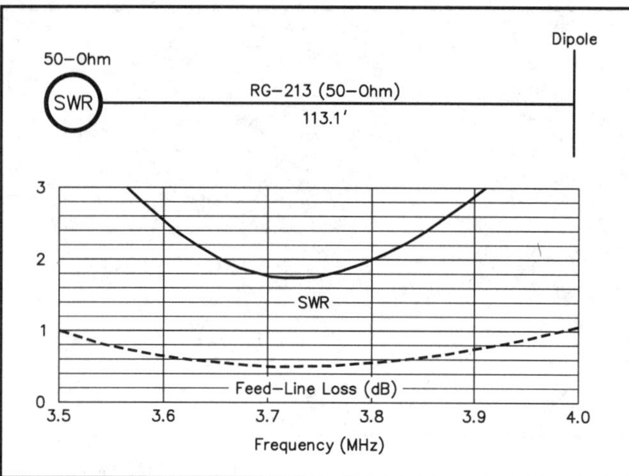

Fig 19—Direct feed with RG-213 50-Ω coax. As usual, this is for 80-meter dipole over flat ground.

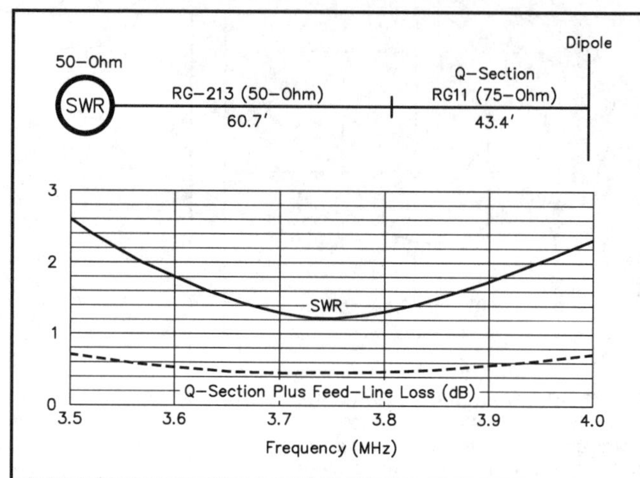

Fig 20—The 75-Ω Q-section provides a reference resistance of 112.5 Ω at the antenna.

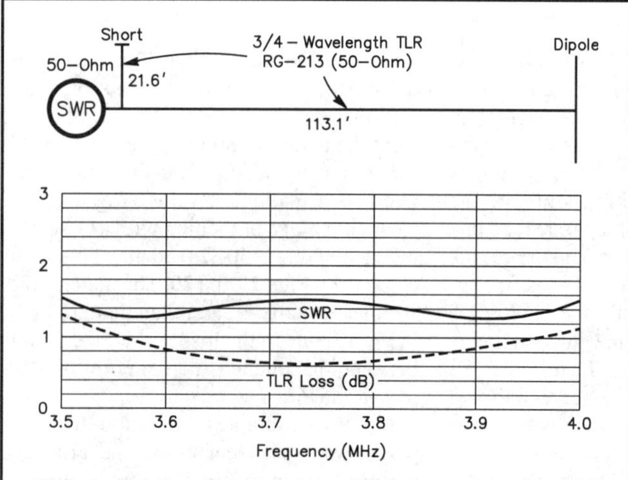

Fig 21—Dipole fed with ¾-λ TLR made from RG-213 50-Ω coax. This is identical to the configuration of Fig 19 with a shorted stub added. Notice that the addition of the stub yields an impressive improvement in match bandwidth.

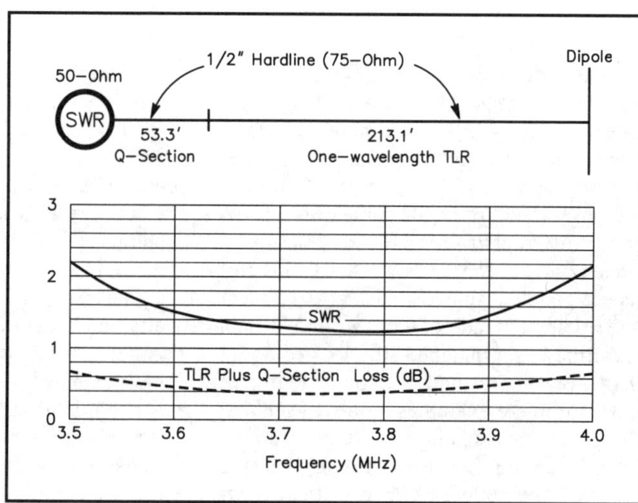

Fig 22—A 1-λ TLR for long runs. The 75-Ω Q-section provides a reference resistance of 112.5 Ω at the transmitter end of the TLR.

Fig 22 shows this simpler form. Note the very low loss possible for a 266.4-foot coaxial feed line!

An Optimum Design for DX

Earlier I mentioned that sloping ground in the direction of the desired signal can significantly increase the gain at low elevation angles. See the section entitled "Height and Sloping Ground Considerations." In that section, which assumed perfectly conducting ground, the example used was an antenna at 80 feet over ground sloping downwards at an angle of 20°. In such a case, in order to obtain the antenna SWR bandwidth properties, you should consider the antenna as though it were mounted over level ground at a slightly lower height. Look at the dipole and matching parameters for this antenna in Figs 8 to 17 at a height of 75 feet (80 cos 20° = 75.2). For example, if the antenna is matched with a 75-Ω Q-section, the 2:1 SWR bandwidth is about 300 kHz (See Fig 13). This is an approximation, because of the break in the ground slope, but the effect is usually negligible.

Variations

If better low-angle performance and more gain are desired, a greater height should be selected. Use the data from Figs 1 through 7 to establish whether the added cost and effort are justified. The same figures will reveal what one gives up if lower heights are used.

What about other variables, such as wire size, ground characteristics and drooping wires? To get a feel for these effects, I have examined the effects on the performance of our 80-foot-high half-wave dipole when these parameters are changed. The exact conclusions may have to be modified for other heights, but the general trends will be the same.

Wire Size Effects

The effect of antenna wire size is small enough to be of no practical consequence. Even for copper wires as thin as #18, the wire loss is only 0.15 dB for an 80-meter dipole. See **Table 2** for the antenna wire loss for various wire sizes. The list includes cases where ladder lines are used as the antenna wires. There is essentially no effect on radiation pattern, and the gain is reduced by only the small wire loss. Because of skin effect, no difference in electrical performance will be seen between copper and copper-clad wire. The thickness of the copper layer should be at least 4 mils. Further, no noticeable difference will result from solid, stranded, insulated or uninsulated wire. The wire gauge and type chosen will depend primarily on the structural properties wanted.

There is an excellent approximation for the wire loss in dB, L_W, which depends on the assumption that the wire is copper and the current distribution over the dipole has a cosine shape (a very good assumption):

$$L_W = -10\log\left[1 - \frac{1}{4.08DR_0\sqrt{F_0}}\right] \approx \frac{1.07}{DR_0\sqrt{F_0}} \quad \text{(Eq 3)}$$

where

D = wire diameter in inches

F_0 = resonant frequency in MHz.

Since the wire diameter and antenna feed-point resistance appear as a product, the same loss will result if you use wire of twice the diameter on an antenna with half the feed-point resistance. The value of R_0 from Fig 8 may be used to find the wire loss. Another point worth mentioning is that for the same wire size and antenna resistance, the wire loss of a resonant half-wave dipole antenna decreases as the frequency increases. This runs counter to intuition, since skin effect causes resistance to increase as frequency increases. The explanation is that change in wire loss with dipole length is proportional to $1/F_0$, while the skin effect loss is proportional to $\sqrt{F_0}$. The net effect is that the loss is proportional to $\frac{1}{\sqrt{F_0}}$.

It is well known that increased match bandwidth can be obtained by using thicker antenna wires or by making the antenna from ladder line. Just how much is the effect? See **Table 2**, which also shows how the 2:1 SWR bandwidth is affected by wire size. For up to #10 wire, there is very little match bandwidth improvement. The table shows that antennas made from 450-Ω (#18 wires spaced 0.8 inch) or 600-Ω (#12 wires spaced 6 inches) ladder lines are equivalent to 1/4-inch or 1-inch diameter copper wires, respectively. The improvement shown by using very thick wire or by using parallel-wire feed lines as antenna wires is hardly worth the effort. If you ever tried to put up (and keep up) an 80-meter folded dipole, you'll know what I mean.

If you really want increased match bandwidth, I recommend that you use a reliable wire antenna and a Q-section or the TLR matching method. The effectiveness of those methods is also shown in Table 2. For any wire size shown, with a TLR made from RG-213 cable, an SWR of 1.53:1 or better is possible over the entire 80-meter band.

Table 2

Wire Loss, Maximum 2:1 SWR Bandwidth and Maximum SWR Over Band Using TLR for Different Wire Sizes and Types

Antenna Wire Size and Type	*Wire Loss (dB)*	*Max Bandwidth (kHz)*	*SWR over Band with TLR*
#18	0.15	289	1.53
#16	0.12	295	1.52
#14	0.1	301	1.5
#12	0.08	308	1.48
#10	0.06	315	1.47
0.25" dia.	0.02	349	1.4
0.5" dia.	0.01	383	1.35
1" dia.	0.01	425	1.29
300-Ω twin lead	0.08	335	1.43
450-Ω twin lead	0.08	352	1.39
600-Ω ladder line	0.04	425	1.29

Ground Characteristics

We are stuck with the ground beneath our antennas, but it is at least educational to know how much the properties vary with ground conditions. **Table 3** shows the change in performance over desert sand (dielectric constant = 7 and conductivity = 1 mS/m), compared to perfectly conducting ground, closely approximated by saltwater.

For our 80-foot-high horizontal antenna, the main changes are in the antenna gain at high and medium elevation angles. There is a difference of 2.2 dB extra overhead gain for perfect ground versus desert. At low elevation angles, 30° and lower, the type of ground makes no practical difference for a horizontal 80-meter dipole. Except at very low heights (less than 25 feet), the change in R_0 and Q_0 with ground conditions is not significant.

The Inverted V

A problem with a horizontal half-wave dipole is the physical support system required. If it is supported only at the ends, the weight of the feed line at the center of the antenna will cause sag and high tension in the wire. A three-point support system, which includes center support would be ideal, but this is not always feasible. This problem has contributed to the high popularity of the inverted-V half-wave antenna,

Table 3

Variation of Broadside and Endfire Gain for 80-Meter Dipoles Over Different Types of Ground. Horizontal and Inverted-V Configurations are Shown

	R_0 (Ω)	*Q_0*	*Broadside Gain (dBd)*				*Endfire Gain (dBd)*			*BW (kHz)*	
			90°	*60°*	*30°*	*15°*	*60°*	*30°*	*15°*	*112.5 Ω*	*50 Ω*
						Horizontal Dipole at 80'					
Desert sand	85.8	9.8	2	2.9	2.5	−1.4	0.5	−6.8	−13.1	285	131
Average ground	91.3	9.1	2.8	3.7	2.8	−1.4	1.6	−5.5	−13.5	308	108
Perfect ground	95.7	8.6	4.2	4.7	2.9	−1.8	2.9	−4.8	−15.5	322	79
						Inverted V's over Average Ground					
Inverted V @ 80'	55	15.8	2.7	2.8	0.7	−4	1.3	−3.3	−7.1	63	158
Inverted V @ 103'	56	15.2	1.1	2.5	2.4	−1.6	0.4	−3.5	−7.4	27	163

requiring only a single high support and two anchor points on the ground.

What, if anything, does one give up by using an inverted-V antenna? In ON4UN's book, *Low Band DXing*, he provides a comparison of the inverted V and the horizontal dipole.[16] He says, "To summarize, it can be said that the inverted V is a fairly good compromise antenna. At the major wave angles, inverted Vs give up about 1 to 3 dB to the horizontal dipole. The high angle nulls are much less pronounced than with the straight dipole." Further, even if the horizontal dipole height is at the average height of the inverted V, "the lower dipole still outperforms the inverted V at all angles in the broadside direction. The difference is minimal however, ranging from only 0.5 dB at 30° to 2.5 dB at 90°. This confirms that the inverted V really is a poor man's dipole, but the trade-off is minimal."

I have found that *NEC/Wires* simulations roughly confirm ON4UN's summary. The endfire gain of the inverted V is better than a horizontal dipole at the lower elevation angles.

What about the effect on match bandwidth when the inverted V is used? **Table 3** shows a half-wave dipole at 80 feet, our "optimum" example, compared with two inverted-V antennas. Both the inverted Vs are mounted over average ground and have included angles of 90°. The first has its apex at 80 feet and the second has its apex at 103 feet, producing an average height of 80 feet.

Let's focus initially on the horizontal dipole at 80 feet and the inverted V whose apex is at 80 feet. Drooping the wires lowers the resistance at resonance (91.3 Ω becomes 55 Ω) and increases the antenna Q (9.1 becomes 15.8). This makes direct feed with a 50-Ω line more appropriate, but the maximum match bandwidth is knocked almost in half. (A horizontal dipole fed with Q-section has a 2:1 SWR bandwidth of 308 kHz, while this drops to only 158 kHz for the inverted V fed with a 50-Ω line.) As may be seen in Table 3, the match bandwidth comparison is virtually the same when the apex of the inverted V is moved to 103 feet.

Using the Antenna on Other Bands

All of the feed systems in Figs 19 to 22 are compatible with sharing the feed line with a parallel-connected 40-meter dipole. The Q-section is transparent on 40 meters, and the particular TLR dimensions shown provide a very good SWR over the 40-meter band. Some interaction between the dipoles will be observed, but can be overcome by a minor readjustment of the wire lengths. The 80-meter dipole should be optimized first and then the 40-meter wires trimmed. This order usually results in satisfactory tuning in a single iteration.

Optimizing 160- and 40-Meter Dipoles

The information in Figs 3 to 12 may be used as a guide for optimizing 160 and 40-meter half-wave dipoles. Ground effects do change over the 1.8 to 7.3 MHz frequency range, but rough estimates of dipole parameters and properties may be obtained by doubling the height numbers for 160 meters and halving them for 40 meters. For example, the data for an 80-meter dipole at 80 feet will be about the same as those for a 160-meter dipole at 160 feet. The same data will approximate the characteristics of a 40-meter dipole at 40 feet. The validity of these statements is illustrated in **Table 4**, where I assumed average ground and #12 antenna wire. This useful comparison is not valid when the antenna height is less than 0.15 λ because of the stronger frequency dependence of ground effects at those heights.

Summary

By looking at several properties of the 80-meter dipole simultaneously, we can arrive at an optimum design for a particular set of circumstances. Although variation of these properties with height is highlighted (because height is the dominant causal factor), the influence of other effects, such as wire size, ground and site properties and drooping wires, is also covered.

Maximization of match bandwidth, while accounting for all losses, is treated. Equations from which the various design curves were derived are presented so that the results may be extended to other applications. Further, some new broadly applicable formulas for wire loss, antenna impedance and Q have been presented.

I am grateful to Dean Straw, N6BV, for making available to me his *IONCAP* results. The narrow range of the signal arrival angles for a given signal path has led to the concise presentation of Table 1, and will help amateurs to better understand the trade-offs involved in the optimization process.

Also, I thank Jim Evans, NV1W, for his thoughtful comments on the manuscript and for proposing the term, "mistermination loss."

APPENDIX

Using NEC/Wires to Find Q_0 and K_R

To determine the antenna Q at resonance, find the antenna impedance above and below resonance and use the following formula.

$$Q_0 = \frac{F_0\left[X_A\left(F_0 + \frac{\Delta f}{2}\right) - X_A\left(F_0 - \frac{\Delta f}{2}\right)\right]}{2R_0\Delta f} \qquad \text{(Eq 4)}$$

where

$X_A(f)$ = antenna reactance at frequency f

Δf = frequency change in same units as F_0.

Similarly, compute the resistance coefficient, K_R, from:

$$K_R = \frac{F_0\left[R_A\left(F_0 + \frac{\Delta f}{2}\right) - R_A\left(F_0 - \frac{\Delta f}{2}\right)\right]}{R_0\Delta f} \qquad \text{(Eq 5)}$$

where $R_A(f)$ = antenna resistance at frequency f.

A good value for Δf on 80 meters is 0.2 MHz.

Antenna Impedance Formula

Eq 2 in the text provides a formula for the antenna impedance, $Z_A(f)$, which is accurate over the entire band. To use Eq 2, R_0, F_0, Q_0 and K_R must be known. Sometimes only R_0 and F_0 are known, say through SWR measurements and a knowledge of the matched-line loss. The following equation gives a very good approximation for $Z_A(f)$.

$$Z_A(f) = R_0\left[1 + 3\left(\frac{f}{F_0} - 1\right)\right] - j119.5\left[\ln\frac{8110}{DF_0} - 1\right]\cot\frac{90f}{F_0} \qquad \text{(Eq 6)}$$

Table 4

Characteristics of Dipoles Over Flat Ground

	R_0 (Ω)	Q_0	Max Gain (dBd)	Max Gain Angle	Broadside Gain (dBd) 90°	60°	30°	15°	Endfire Gain (dBd) 60°	30°	15°
40-m dipole, 40'	87.7	8.9	3.7	52°	2.9	3.6	2.5	−1.8	1.3	−6.4	−13.7
80-m dipole, 80'	91.3	9.1	3.9	48°	2.8	3.7	2.8	−1.4	1.6	−5.5	−13.5
160-m dipole, 160'	94.1	9.5	4.1	49°	3	3.9	2.9	−1.4	1.9	−5.2	−13.5

where

D = diameter of wire, in inches

F_0 = resonant frequency, in MHz.

Note that the value 3 has replaced K_R in the real part of $Z_A(f)$, since that is a good average value to use. The imaginary part of Z_A in Eq 6 was inspired by a formula found in some antenna texts and handbooks for wire antennas in free space.[17] I have found that for all heights above 0.05 λ (about 13 feet on 80 meters) and for wires up to 1 inch in diameter, the imaginary part of Z_A is within 5% of the value given by Eq 2. Note that the imaginary part of Z_A in Eq 6 is independent of antenna height.

The significance of Eq 6 is that with knowledge of only R_0, F_0 and D, a formula for antenna impedance near resonance is available to the antenna experimenter. It is valid at HF and perhaps beyond.

The match bandwidth equations in the next section require knowledge of Q_0. From Eqs 2 and 6, an estimate of Q_0 is:

$$Q_0 \approx \frac{93.9\left[\ln\frac{8110}{DF_0} - 1\right]}{R_0} \qquad \text{(Eq 7)}$$

Match Bandwidth Equations

In the references of Notes 18 and 19, the design technique for matching resonant antennas with transmission lines was described. The equations in these references and data from *NEC/Wires* were used to generate the curves of Figs 13 through 17.

Fig 13

A very simple relationship between match bandwidth and antenna Q follows.

$$BW_{MAX} = \frac{750F_0}{Q_0} \qquad \text{(Eq 8)}$$

where

BW_{MAX} = Maximum 2:1 SWR match bandwidth (in kHz) at the antenna attainable without a matching network

To find the actual 2:1 SWR match bandwidth, BW_2 (in kHz) with a reference resistance, R_{REF}, it is necessary to consider two cases:

Case 1: $R_{REF} > R_0$

$$BW_2 = \frac{1000F_0}{Q_0}\sqrt{\frac{2.5R_{REF}}{R_0} - \left(\frac{R_{REF}}{R_0}\right)^2 - 1} \qquad \text{(Eq 9)}$$

Case 2: $R_0 > R_{REF}$

$$BW_2 = \frac{1000F_0R_{REF}}{Q_0R_0}\sqrt{\frac{2.5R_0}{R_{REF}} - \left(\frac{R_0}{R_{REF}}\right)^2 - 1} \qquad \text{(Eq 10)}$$

Fig 14

The simplest network for broadband matching a resonant antenna is a parallel LC circuit tuned to F_0 and connected across the antenna terminals. An equivalent form of the matching network is a resonant transmission line of appropriate length, which has been called the *transmission line resonator* or TLR.

Before finding out how much benefit may be derived from a TLR matching network, it is useful to define the normalized bandwidth, B_N:

$$B_N = \frac{BW \times Q_0}{1000F_0} \qquad \text{(Eq 11)}$$

where BW = the width of the band (in kHz) over which a match is desired (500 kHz on 80 meters)

Although it is not plotted in Fig 14, the best match attainable over the band with a lossless matching network, S_{M0}, is of interest.

$$S_{M0} = \sqrt{B_N^2 + 1} \qquad \text{(Eq 12)}$$

When the matching network is made from a transmission line, the line loss must be considered. Since commonly used transmission lines have a loss that increases as $\sqrt{f}$, we need only the loss at one frequency, say F_{REF}. Then the loss at F_0 is given by:

$$A_0 = A_{REF}\sqrt{\frac{F_0}{F_{REF}}} \qquad \text{(Eq 13)}$$

where A_0 and A_{REF} are the losses in dB/100 feet at F_0 and F_{REF}, respectively.

The Q of the resonator is

$$Q_N = \frac{2.774F_0}{A_0V} \qquad \text{(Eq 14)}$$

where V = velocity factor of the cable.

Now the best match attainable with a lossy transmission-line matching network is calculated from:

$$S_{M\,min} = \frac{\sqrt{B_N^2+1} + \sqrt{B_N^2 + 1 + \frac{2Q_0}{Q_N}\left(1+\frac{Q_0}{2Q_N}\right)}}{2\left(1+\frac{Q_0}{2Q_N}\right)} \qquad \text{(Eq 15)}$$

Figs 15 and 16

To design a TLR as a broadband matching network, it is necessary to know the reactance at resonance of each element of the equivalent LC matching network.

$$X_{N0} = \frac{R_0}{Q_0}\left[\left(1+\frac{Q_0}{2Q_N}\right)S_{M\,min}^2 - \frac{Q_0}{2Q_N}\right] \qquad \text{(Eq 16)}$$

Three parameters define the TLR: the length, the characteristic impedance of the cable and the optimum generator resistance. The required characteristic impedance depends on the electrical length, which is n quarter wavelengths.

$$Z_{0TLRn} = \frac{n\pi X_{N0}}{4} \qquad \text{(Eq 17)}$$

When n is odd, the TLR is shorted at one end and when n is even the TLR is either shorted or open at both ends. The optimum generator resistance is given by:

$$R_{GOPT} = \frac{S_{M\,min}R_0Q_NX_{N0}}{R_0 + Q_NX_{N0}} \qquad \text{(Eq 18)}$$

Fig 17

The loss at the band edges if the TLR is used as a resonator and not for signal transport is:

$$L_{TLRE} = 10\log\left[1 + \frac{R_0}{Q_NX_{N0}}\left(B_N^2+1\right)\right] \qquad \text{(Eq 19)}$$

The effective loss will be less than this value when the TLR is also used for signal transport.

Notes

[1]J. Belrose, G. Royer and L. Petrie, "HF Wire Antennas over Real Ground: Computer Simulation and Measurement," AGARD Lecture Series No. 165, Modern Antenna Design Using Computers and Measurement: Application to Antenna Problems of Military Interest, Sep 1989.

[2]G. Hall, "The Earth Detunes My Antenna," *QST*, Oct 1993, pp 41-44.

[3]M. W. Maxwell, *Reflections—Transmission Lines and Antennas* (Newington: ARRL, 1990), Chapters 15 and 18.

[4]F. Witt, "A Simple Broadband Dipole for 80 Meters," *QST*, Sep 1993, pp 27-30, 76.

[5]J. Devoldere, *Low-Band DXing* (Newington: ARRL, 1987), 1st ed, pp II-2 to II-7, and J. Devoldere, *Antennas and Techniques for Low-Band DXing* (Newington: ARRL, 1994) 2nd ed, Chapters 5 and 8.

[6]*NEC/Wires 1.5* antenna simulation software, Brian Beezley, 3532 Linda Vista Dr, San Marcos, CA 92069.

[7]*Mathcad* 5.0, MathSoft, Inc., 201 Broadway, Cambridge, MA 02139.

[8]See Note 5, Chapter 2 of 1st ed and Chapters 7 to 13 of 2nd ed.

[9]R. D. Straw, Ed., *The ARRL Antenna Book* (Newington: ARRL, 1994), 17th ed, pp 23-19 to 23-28.

[10]See Note 9, p 2-4.

[11]F. Witt, "Broadband Dipoles—Some New Insights," *QST*, Oct 1986, p 34.

[12]F. Witt, "Match Bandwidth of Resonant Antenna Systems," *QST*, Oct 1991, pp 21-

25 and A. Griffith and F. Witt, "Match Bandwidth Revisited," *QST*, Jun 1992, Technical Correspondence, pp 71-72.

[13]See Note 4.

[14]F. Witt, "Broadband Matching with the Transmission Line Resonator," elsewhere in this volume.

[15]See Note 14.

[16]See Note 5, p II-41 to II-44 of 1st ed.

[17]For example, see R. C. Johnson, *Antenna Engineering Handbook*, 3rd ed (New York: McGraw-Hill, 1993), p 4-4.

[18]F. Witt, "Optimum Lossy Broadband Matching Networks for Resonant Antennas," *RF Design*, Apr 1990, pp 44-51 and Jul 1990, p 10.

[19]See Note 14.

Antenna Modeling

An Adventure in Antenna Modeling

By Brian Beezley, K6STI
3532 Linda Vista Dr
San Marcos, CA 92069

Other articles in this compendium describe antenna designs or technical ideas that work. This one is about an idea that failed. K6STI tells us why he bothered to write about it.

The Problem

In general, long-distance HF communication depends on signals launched and received at low elevation angles. Increasing low-angle radiation can make a dramatic difference in DX signal levels. If you've never operated a station with very high antennas or one located on a hilltop, you may not appreciate just how much difference low-angle radiation can make on long-distance paths. Signal enhancement can be very great.

Low-angle radiation is hard to come by at ordinary locations. For typical antenna heights over average ground at flat, residential locations, the ground-reflected wave nearly cancels the direct wave for both horizontal and vertical polarization at low angles. This results in little useful signal.

Radiation at low angles increases as you raise antenna height because path length for the ground-reflected wave increases. Cancellation doesn't begin to occur until lower angles. But even on 10 meters, heights above 100 feet are needed to bring the first radiation lobe well below 10° elevation. The situation is much worse at lower frequencies.

(Things are different for vertical antennas radiating over saltwater. In this case the phase of the reflected wave doesn't shift enough to cancel much direct radiation until about 1° elevation. Verticals over saltwater have tremendous amounts of radiation at very low angles.)

No practical way has ever been found to increase the low-angle radiation of an antenna without increasing its height. It's an unsolved problem. If a solution were found, it could have great benefit. A solution might make possible reliable, long-distance communication at high signal-to-noise ratio with low power and low antennas.

Fooling Mother Nature

The nature of ground reflection is somewhat different for horizontal and vertical antennas at small angles. For verticals, the magnitude and phase of the ground-reflection coefficient vary over a wide range with angle and ground quality. For horizontals, the magnitude is close to 1.0 and phase is close to 180° for all but the very worst ground.

One day it occurred to me that if somehow you could reverse the phase of radiation from a horizontal antenna at negative elevation angles, the ground-reflected wave would then be in phase with the direct wave. Cancellation wouldn't occur; in fact, you'd actually reinforce low-angle radiation. This would be a great and profound trick to play on ground.

Such an antenna system would need up/down asymmetry; otherwise, it would radiate the same wave above and below the horizon. I pictured a simple horizontal dipole. It seemed that placing another dipole below the first with the same current magnitude but opposite phase would do the trick. Above the horizon the wavefront from the upper dipole arrives first. Below the horizon the lower dipole wins. Paradoxically, signals cancel at the horizon, just where we want to maximize response. In fact, the antenna is just a W8JK beam pointed straight up in the air and straight down toward ground. This seemed liked a ridiculous way to achieve low-angle radiation, but I was intrigued enough by the underlying principle to turn on my computer and model the antenna.

I decided to start with a 40-meter dipole 50 feet above ground. I thought this height could be achieved by most amateurs. I used a *MININEC*-based program to model the dipole over average-quality earth (dielectric constant = 13, conductivity = 5 mS/m). **Fig 1** shows the elevation pattern. Note the poor response at low angles, as expected.

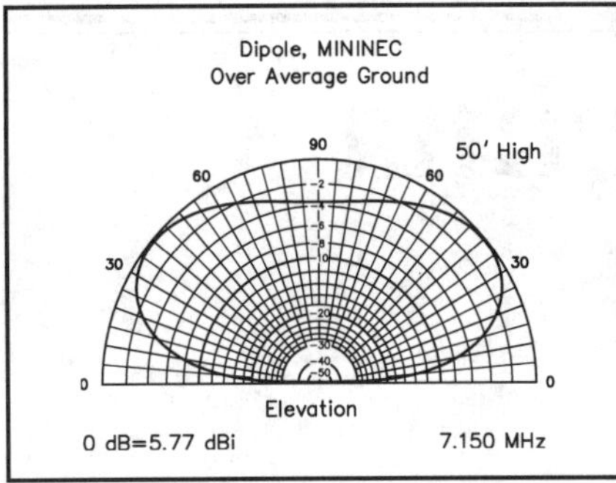

Fig 1—Elevation pattern for half-wave 40-meter dipole, 50 feet over average ground, with conductivity of 5 mS/m, dielectric constant of 13. This was computed using *MININEC* method of moments program. At this height, the same pattern results when *NEC2* code is used.

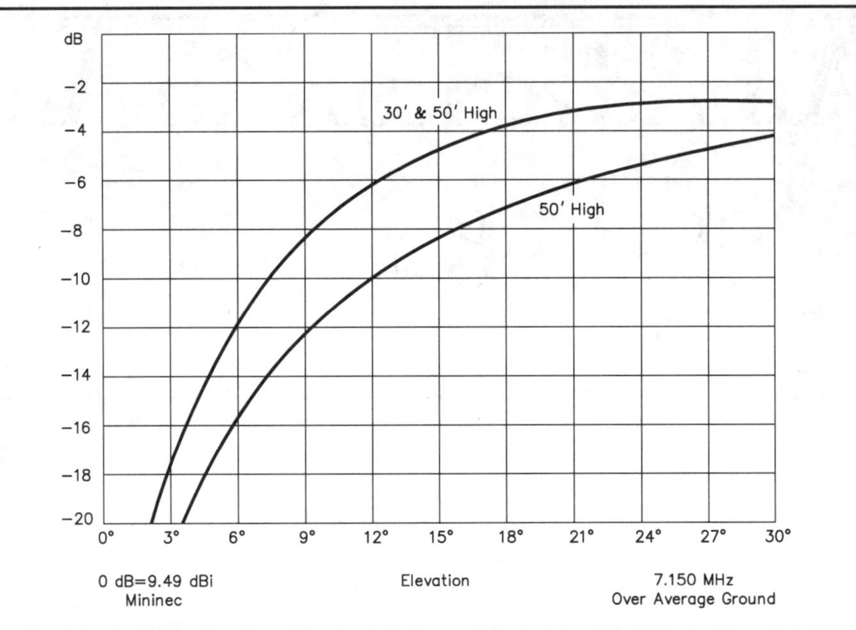

Fig 3—Same comparison as in Fig 2, except using graph rectangular coordinates for greater resolution of difference for elevation angles between 2° to 30°.

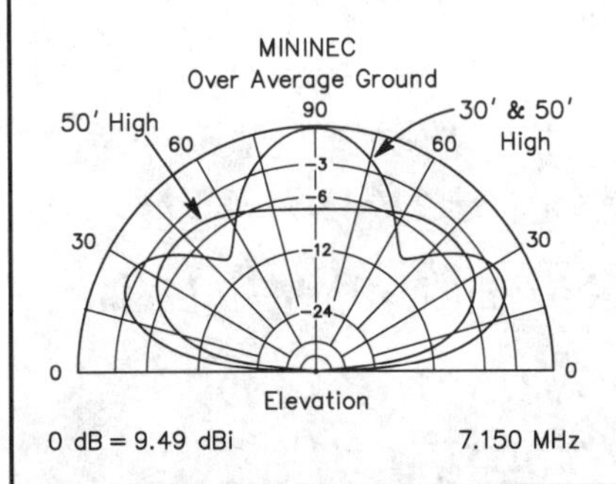

Fig 2—Elevation pattern for phased dipoles located at 50 and 30 feet with equal amplitude feed currents, but with lower dipole's feed current 180° out of phase with top dipole. For comparison, the elevation response of Fig 1 is overlaid at same scale. The pattern for phased dipole was computed using *MININEC*.

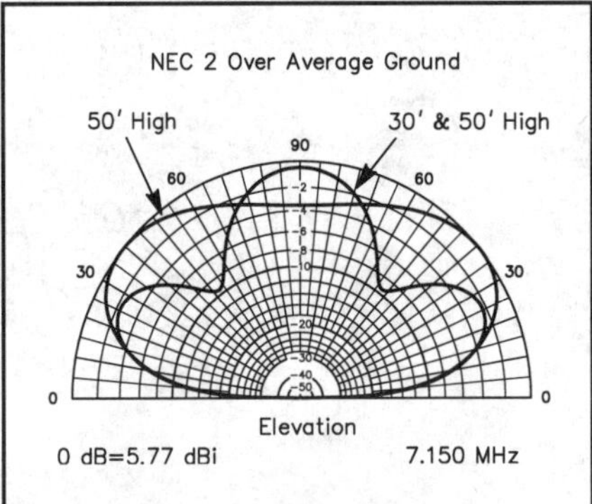

Fig 4—Elevation pattern for phased dipoles at 50 and 30 feet with same feed arrangement as in Fig 1, computed using *NEC2* with Sommerfeld-Norton ground model. For comparison, the elevation response of dipole in Fig 2 is overlaid at same scale. Reality settles in—the trick antenna has no advantage over the dipole.

I then added a second dipole at 30 feet, directly below the first. I fed both dipoles using current sources and flipped current phase in the lower antenna. I used current sources so that differences in antenna impedance wouldn't alter the desired wire currents.

Fig 2 shows the result. Amazingly, the crazy thing worked! It's easy to see the improved response below 35° elevation. **Fig 3** shows an expanded view in rectangular coordinates. The trick antenna has several dB gain over the dipole all the way down to the horizon. It's the first antenna I've ever seen that maximizes gain by dropping a null in the desired direction and pointing its main beam elsewhere!

Reality Check

The sequence of events described above took just a few minutes. As I sat staring at the patterns, I wondered how it was possible that an amateur playing around on his home computer could so easily discover a new antenna principle that had escaped the attention of professionals for the last century. This thought quickly brought me back to earth and motivated me to probe my idea and the model for defects.

First I looked at input impedances. They were low (the impedance of the lower dipole even was negative due to mutual coupling) but they weren't unreasonable. I had no idea how to make a feed system to supply the required currents, but I was sure that one could be devised.

Next I looked at wire losses. I had used lossless conductors for the first model. With the low impedances, I thought it was time to model real conductors. I specified #12 copper wire and reran the models. The single dipole showed only 0.05 dB loss, but the trick antenna lost 1.76 dB. This was disappointing. Now the trick antenna would require thick conductors as well as a complex feed system. Nevertheless, even with some conductor loss the antenna still had substantial gain over the single dipole.

I started to think about limitations of the modeling algorithm. *MININEC* can provide good results for many situations, but it has several limitations (some not so obvious) that can lead you astray if you're not careful. One of these limitations involves modeling antennas over ground.

MININEC uses earth dielectric constant and conductivity when calculating radiation patterns, but it assumes a perfectly conducting groundplane when computing wire currents. Essentially this means that *MININEC* ignores losses due to currents induced in the earth by the antenna. Ordinarily this limitation doesn't affect antennas much unless their conductors are below about 0.2 λ. For wires lower than this height, *MININEC* may give results which are grossly in error. 0.2 λ is about 28 feet on 40 meters. The lower dipole was pretty close at 30 feet and *MININEC's* ground-loss error builds up gradually in this region, not abruptly at 0.2 λ.

Moreover, this particular antenna fired right into ground. I thought this might increase ground losses, and *MININEC*

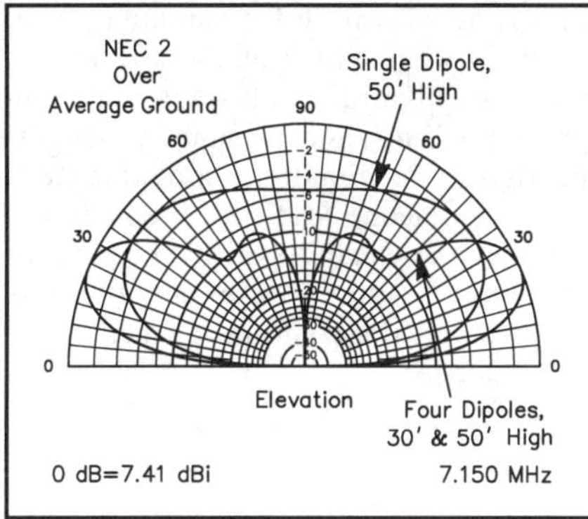

Fig 5—Elevation pattern for phased four dipoles at 50 and 30 feet. For comparison, the elevation response of dipole in Fig 1 is overlaid at same scale. Now, there is some low-angle gain. Note that the feed system necessary to actually achieve the equal feed currents would be complex because of the mutual impedance between the closely spaced dipole elements and the closeness to ground.

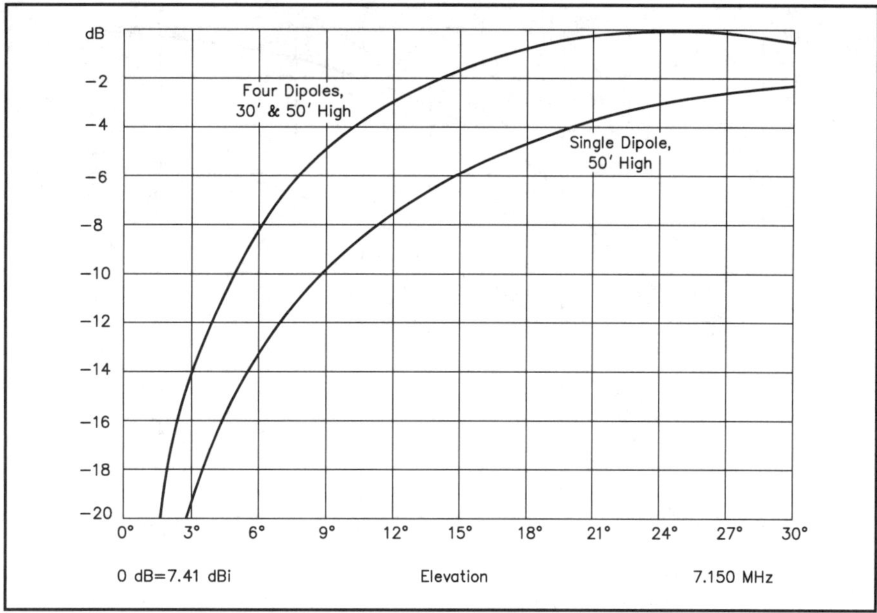

Fig 6—Same comparison as in Fig 5, but in rectangular graph coordinates for more resolution at low elevation angles.

wouldn't be able to include these losses in its results. However, I did have another program which could. This was *NEC*, the *Numerical Electromagnetics Code*, which had inspired the simpler *MININEC* algorithm. *NEC* can provide accurate models for antennas with conductors down to within several wire radii of ground.

When I reran the models using *NEC's* highly accurate Sommerfeld-Norton ground option, reality reasserted itself as shown in **Fig 4**. Even though modeled with lossless conductors, the trick antenna showed no advantage whatsoever over the dipole.

Second Attempt

I let the problem go at this point and turned my attention back to more mundane things, like making a living. But every once in a while, in idle moments, my thoughts would drift back to the problem. I kept turning it over in my mind, wondering if there wasn't some clever way around the difficulties.

Occasionally I'd sneak a try at something on the computer. I tried increasing the height of the lower dipole to 40 feet. This brought a genuine half-ohm input impedance (short-boom W8JK beams are notorious). I tried varying the phase of the lower dipole away from 180°. I tried moving the lower dipole in front of and behind the upper antenna. Nothing helped.

I started mentioning the problem to others, hoping that someone might see something I'd missed. But many of those I spoke with had big antenna farms filled with giant Yagis on stratospheric towers with loads of low-angle radiation. I had to reformulate the problem in terms of "low" 200-foot-high dipoles on 160 meters just to get anyone to pay attention.

Finally, I sat down one day and thought directly about the problem. If ground losses were the root of the difficulty, how could I reduce them? I wasn't prepared to use ground screens. I wanted a simple solution. Well, instead of radiating directly into the ground and incurring losses, why not throw a pattern null in that direction?

I took my dipole pair and replicated it 30 feet forward. I fed the second pair just as I had the first, except that I swapped current phases. The upper dipole of the new pair had the same phase as the lower dipole of the first. This should throw a null at ground while preserving reverse phase below the horizon. I didn't even bother modeling with *MININEC*. I went straight to *NEC*.

Fig 5 and **Fig 6** show the results. This time *NEC* said "yes!" The doubly tricky antenna provides about 4 dB gain at low angles over a dipole without increasing antenna height. *NEC* has limitations of its own, but none that I knew about applied to this model.

Taking Stock

So what do we have? We've increased low-angle radiation substantially without raising antenna height. But this costs a much bigger antenna, much larger and sturdier supports, four times the wire, a nasty, complex feed network, and yes, some very low impedances.

But is it all worth it? Not when a simple wire Yagi at 50 feet easily outperforms the trick antenna! See **Fig 7** and **Fig 8**.

In my love of the devious and the clever, I had thrown common sense completely to the wind. An ordinary 2-element Yagi, which does not rely on faking

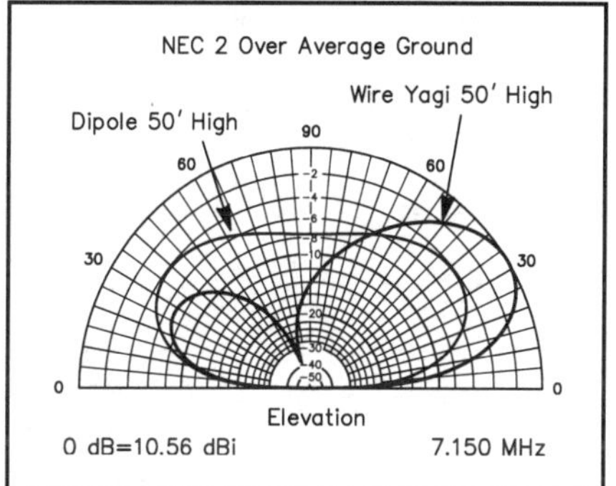

Fig 7—Elevation pattern of 2-element parasitic wire Yagi at 50 feet height, compared with dipole in Fig 2 at same height. Now, the Yagi has more gain at all elevation angles in its forward lobe. At 30°, the difference in favor of the Yagi is just over 5 dB; at 10° the difference is 6 dB.

out ground, provides about 6 dB gain over a low dipole at low angles simply and economically. The Yagi isn't specially designed to enhance low angles like the trick antenna, but it provides gain there nonetheless. And the Yagi gain is much easier to come by.

It's simply not true that there's no way to increase low-angle radiation without increasing antenna height. Any modification that increases forward gain will do the trick.

In the end I decided to write up this fiasco for several reasons. First, I wanted to demonstrate how foolish it's possible to become when you get carried away with computer modeling. Powerful software is no substitute for common sense. Second, I wanted to point out how easy it is to draw invalid conclusions when you ignore the limitations of antenna-modeling algorithms. Finally, I hope this story inspires someone to discover something genuinely new and useful with computer modeling. Despite failure, pursuing this idea was a real adventure for me.

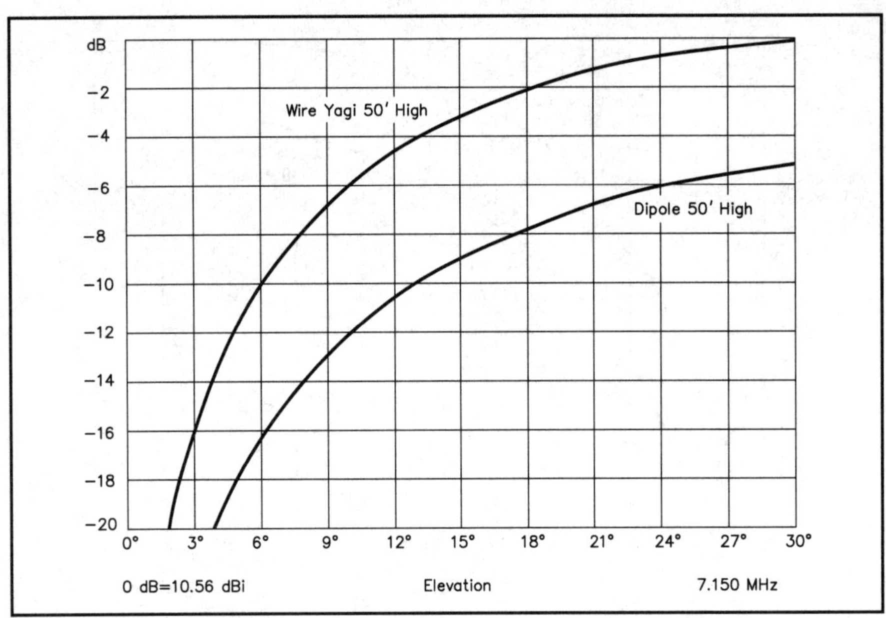

Fig 8—Same comparison as in Fig 7, but in rectangular graph coordinates for more resolution at low elevation angles.

Antenna Matching with L-C Resonant Circuits

By Wilfred N. Caron
2945 Camp Joy Road
Grants Pass, OR 97526

Wilfred Caron explores resonant-circuit techniques to achieve broadband antenna matching.

Introduction

In order to excite an antenna most efficiently, either its impedance should be approximately the same as the characteristic impedance of the transmission line feeding it, or else matching networks must be provided to accomplish any needed impedance transformation. If the antenna is to be broadband, that is, useful over a wide range of frequencies, the impedance of the antenna must be made to match the transmission-line characteristic impedance, or be nearly so.

Many books and periodicals have been published describing impedance matching using series and shunt lumped or distributed-constant matching elements. Sometimes, due to the nature of the antenna impedance, it is impossible to match broadband antennas within satisfactory limits with simple matching systems alone. It may be necessary to resort to resonant circuits, along with appropriate shunt or series matching elements. Unfortunately, there is not much literature available describing the use of resonant circuits.

This paper describes the use of series and parallel L-C resonant circuits in broadband antenna matching. The first part demonstrates matching of a VHF antenna designed to operate between 110 to 140 MHz. The impedance curve of this antenna requires a series-resonant L-C circuit. The second part describes a short, fat vertical antenna designed to operate between 4.5 and 12.0 MHz. Its impedance curve requires a parallel L-C resonant circuit. The objective in both cases is to match the antenna so that it meets a specified SWR limit over its operating frequency range.

Basic Principles of the Series-Resonant Circuit

From the standpoint of practical antenna matching procedures, the essential properties of a perfect series-resonant circuit with no losses are:

1. There is only one frequency for a given series LC combination where resonance occurs.

$$f_r = \frac{1}{2\pi\sqrt{LC}}$$

where f_r = resonant frequency.

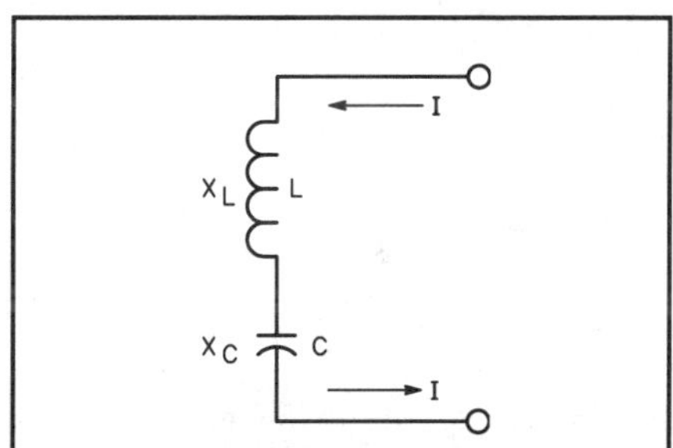

Fig 1—Lossless series-resonant circuit. X_L is inductive reactance and X_C is capacitive reactance.

2. See **Fig 1**. For a given resonant frequency, the LC product is a constant:

$$X_L = 2\pi f_r L$$

$$X_C = \frac{1}{2\pi f_r C}$$

where at resonance $X_L = X_C$ and $X_L - X_C = 0$. Since the current (I) is common to both L and C, the resulting off-resonance reactance is $X_{Resultant} = X_L - X_C$.

3. The manner in which reactance varies with frequency is shown in **Fig 2**. Note that the resultant reactance curve is not the same above and below resonance. The resultant reactance is always smaller than the larger of the two individual reactances. The slope of the reactance curve is dependent on the LC combination.

Basic Principles of the Parallel-Resonant Circuit

The essential properties of a perfect parallel-resonant circuit with no losses are:

1. Similar to a series-resonant circuit, there is only one frequency for a given parallel LC combination where resonance occurs.

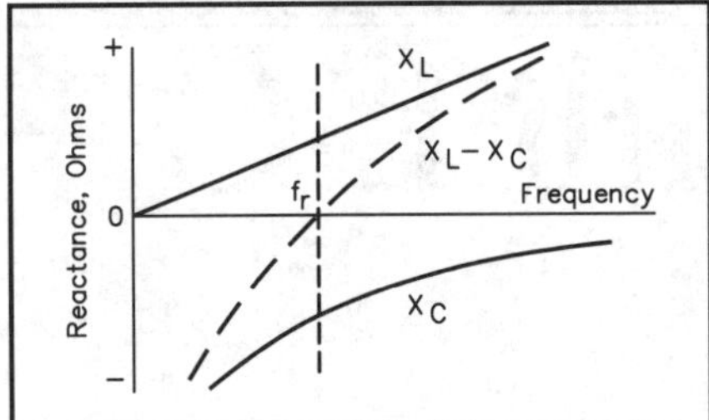

Fig 2—Curve of inductive and capacitive reactance change with frequency for individual elements in series-resonant circuit, and for combination (dashed line). The net reactance goes to zero at resonant frequency f_r.

Table 1
Feedpoint Impedance of 110 to 140 MHz VHF Antenna to be Matched Using Series-Resonant Network

f_x MHz	Impedance $R \pm jX\ \Omega$	Normalized Impedance $R/50 \pm jX/50\ \Omega$
110	43.5 + *j* 38.5	0.87 + *j* 0.77
115	75 + *j* 42.5	1.50 + *j* 0.85
120	110 + *j* 15	2.20 + *j* 0.30
125	100 – *j* 60	2.00 – *j* 1.20
130	57.5 – *j* 73	1.15 – *j* 1.45
135	34 – *j* 62.5	0.68 – *j* 1.25
140	19 – *j* 48.5	0.38 – *j* 0.97

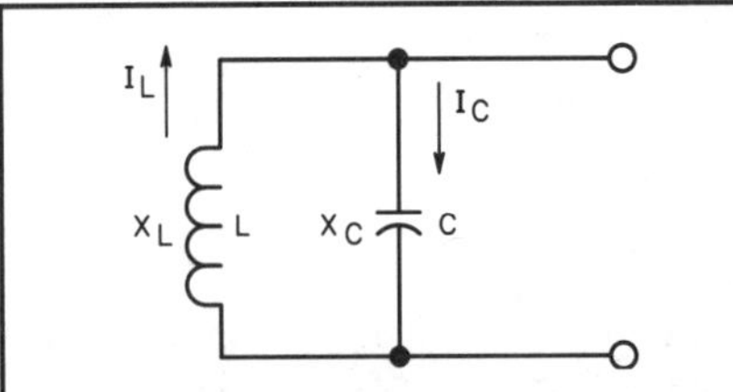

Fig 3—Lossless parallel-resonant circuit.

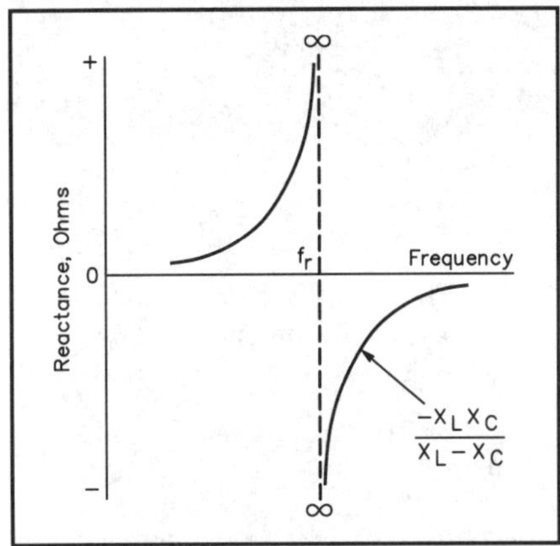

Fig 4—Curve of reactance of paralleled L and C versus frequency. At resonant frequency f_r the impedance across perfect parallel-resonant circuit is infinitely high.

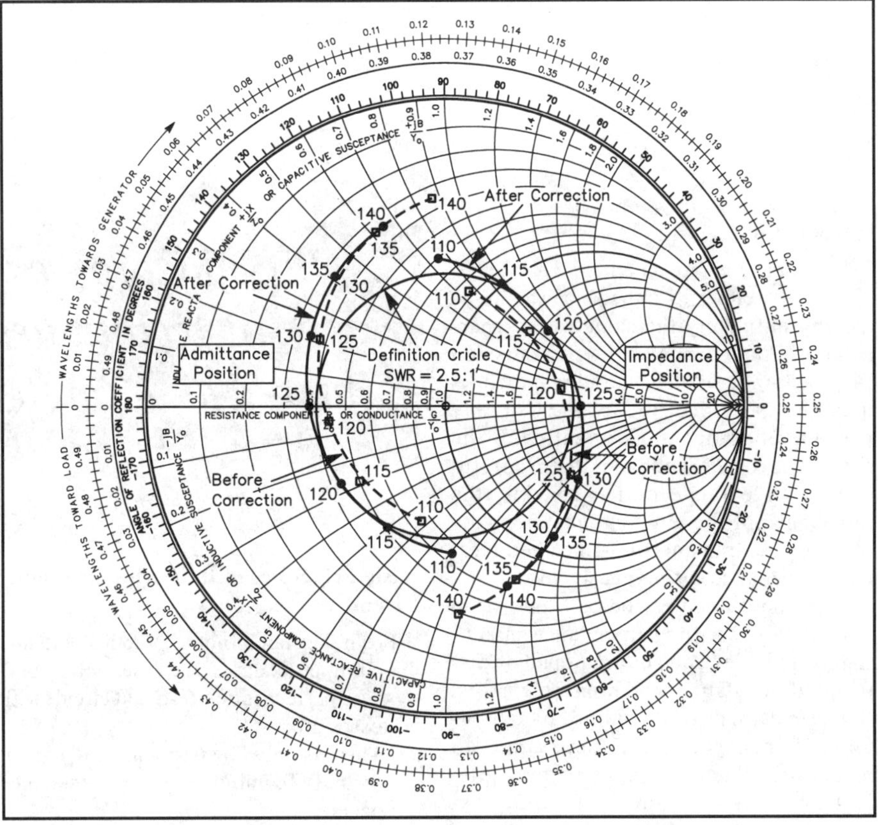

Fig 5—Smith Chart showing uncompensated impedance and admittance curves for VHF antenna (dashed lines) and compensated curves (solid lines) using a shunt coil of –0.18 chart siemens.

$$f_r = \frac{1}{2\pi\sqrt{LC}}$$

2. See **Fig 3**. For a given resonant frequency, the LC product is constant:

$$X_L = 2\pi f_r L$$

$$X_C = \frac{1}{2\pi f_r C}$$

At resonance $X_L = X_C$.

3. In a parallel-resonant circuit, the currents in the reactance are different; current lags in inductors and leads in capacitors. The total current (I) is equal to $I_L - I_C$, since the currents are 180° out of phase. Therefore, for a parallel-resonant circuit the total reactance is

$$X_{total} = \frac{-X_L X_C}{X_L - X_C}$$

and at resonance

$$\frac{-X_L X_C}{X_L - X_C} = \infty = \text{infinity}.$$

4. In the parallel-resonant circuit, the resultant reactance is positive (that is, inductive) if X_C is larger than X_L. The resultant is larger than the smaller of the two individual reactances. This is opposite the situation for reactances in series. The manner in which the net reactance of a parallel-resonant circuit varies with frequency are shown in **Fig 4**.

Example 1

The following procedure describes the sequence used to match a VHF antenna with a combination of matching elements, including a series-resonant circuit. The goal is to keep the SWR below 2.5:1 over the frequency range 110 to 140 MHz. The antenna exhibits the feedpoint imped-

Table 2
Admittance Versus Frequency, for Correction of –0.18 Chart Siemens at 140 MHz

f_x	*Correction Siemens*
110	–0.229
115	–0.219
120	–0.210
125	–0.202
130	–0.194
135	–0.187
140	–0.180

$$B_{f_x} = -j0.18 \times \frac{140}{f_x}, \text{ chart siemens}$$

Table 3
Corrected Admittance with Shunt Coil at Antenna Terminals

f_x	*Old Position*	*Correction*	*New Position*
110	0.60 – *j* 0.57	–*j* 0.229	0.60 – *j* 0.799
115	0.50 – *j* 0.28	–*j* 0.219	0.50 – *j* 0.499
120	0.44 – *j* 0.05	–*j* 0.210	0.44 – *j* 0.215
125	0.38 + *j* 0.22	–*j* 0.202	0.38 + *j* 0.018
130	0.34 + *j* 0.42	–*j* 0.194	0.34 + *j* 0.226
135	0.38 + *j* 0.63	–*j* 0.187	0.38 + *j* 0.443
140	0.35 + *j* 0.88	–*j* 0.180	0.35 + *j* 0.700

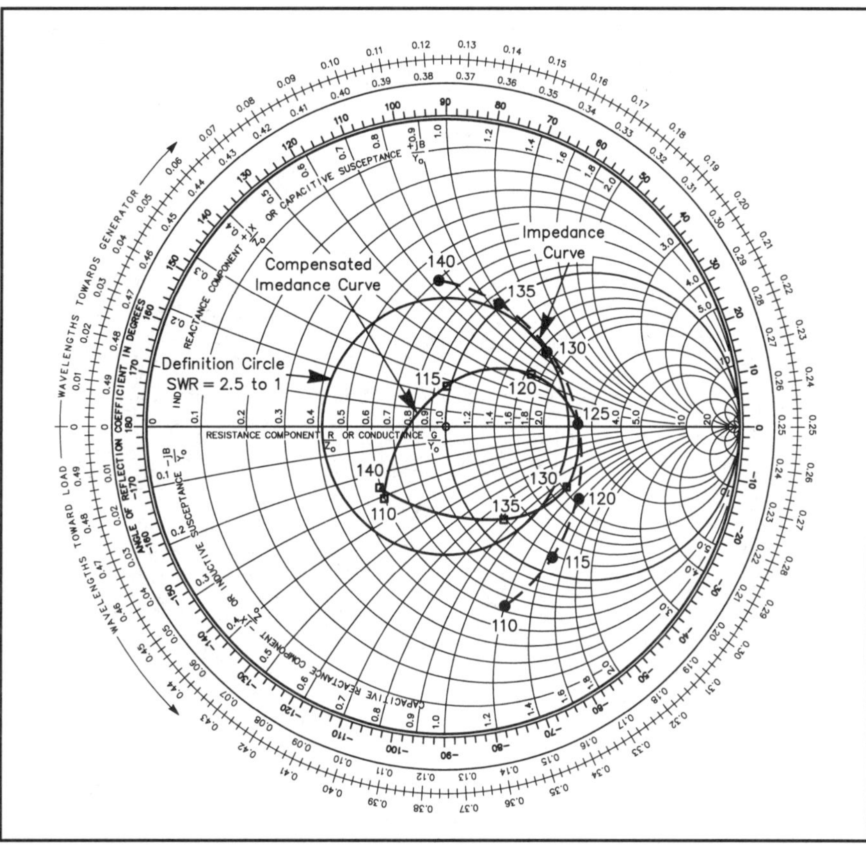

Fig 6—Smith Chart showing impedance curve of VHF antenna compensated with series-resonant circuit of 6 pF and 0.270 µH, which resonates at 125 MHz. The SWR falls outside the desired limit of 2.5:1 below about 125 MHz.

ances shown in **Table 1** over this frequency range.

Step 1

Plot the normalized impedance coordinates on a Smith Chart. See **Fig 5**. An examination of the impedance curve reveals that we have one problem preventing us from using a series-resonant circuit as the first matching element—the 140 MHz point. This point would follow the line of constant resistance and would not move into the 2.5:1 SWR definition circle. The curve must be moved to the right to place the 140-MHz point on a constant resistance line that passes through the definition circle. It appears that a shunt inductor is required at the antenna.

Step 2

When we use shunt elements, we must work with admittances on the Smith Chart instead of impedances. Admittance coordinates are diametrically opposite the impedance coordinates. See Fig 5 again. Obtain the admittance coordinates of the impedance. We need a correction of about –0.18 chart siemens at 140 MHz. Determine the amount of correction required at other frequencies (f_x). The amount of correction will be greatest at the lowest frequencies, as indicated in **Table 2**.

Step 3

Add correction to susceptance ± *j*B. See **Table 3**.

Step 4

Transfer the new admittance positions to their equivalent new impedance positions. See the Smith Chart in Fig 5 once more. Convert the normalized impedance to 50-Ω characteristic impedance. Enter the characteristic on a 50-Ω characteristic impedance Smith Chart. See **Fig 6**.

Step 5

This step is somewhat tedious because there are unlimited combinations of L and C that resonate at 125 MHz, the midpoint and self-resonant frequency. A combination of L and C is needed not only to provide resonance at 125 MHz, but also to provide the proper reactance at 110 MHz and 140 MHz too. After investigating numerous combinations, it was found that a capacitance of 6 pF and an inductance of 0.270 µH proved to be satisfactory.

For C = 6 pF, the capacitive reactance

$$X_{Cf_x} = \frac{1}{6.28 \times 125 \times 10^6 \times 6 \times 10^{-12}} = -212\ \Omega.$$

and

$$L = \frac{212}{6.28 \times 125 \times 10^6} = 0.270\ \mu\text{H}.$$

Step 6

Find the total reactance at other frequencies. See **Table 4**.

Step 7

Add resultant reactance (X_L–X_C) obtained above to the reactance obtained in Step 4. Draw a new impedance plot on the

Table 4
Net Reactance Versus Frequency of Series 6 pF Capacitor and 0.270 μH Inductor

f_x	X_L	X_C	X_L–X_C
110	187	–241	–54
115	195	–231	–36
120	203	–221	–18
125	212	–212	0
130	220	–204	+16
135	229	–197	+32
140	237	–190	+47

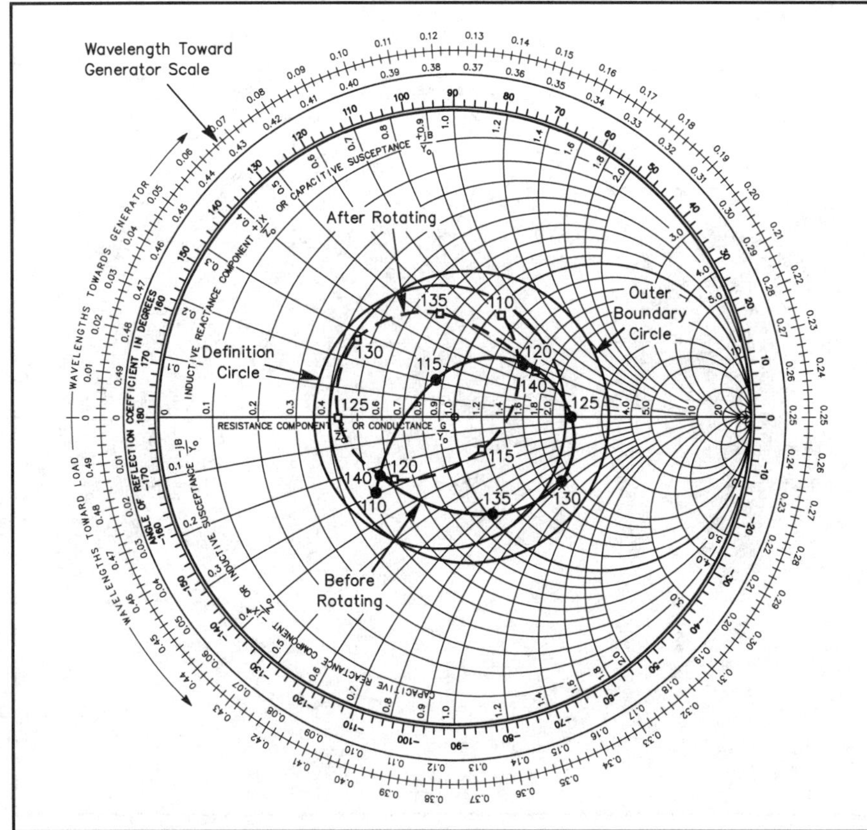

Fig 7—Smith Chart showing impedance curve of VHF antenna compensated with λ/4-transmission-line transformer for 125 MHz. The solid line shows results before rotating the curve, and the dashed line shows results after employing the transformer.

Table 5
New Impedance Position with Shunt Coil and Series Resonant Circuit

f_x	*Old Impedance* Ω	*Correction* Ω	*New Impedance* Ω
110	30 + *j* 37.5	–*j* 54	30 – *j* 16.5
115	49 + *j* 48.5	–*j* 36	49 + *j* 12.5
120	84 + *j* 50	–*j* 18	84 + *j* 32
125	130 + *j* 0	+*j* 0	130 +*j* 0
130	103 – *j* 68	+*j* 16	103 – *j* 52
135	58 – *j* 73	+*j* 32	58 – *j* 41
140	31 – *j* 60	+*j* 47	31 – *j* 13

Smith Chart. See Fig 6 again.

Step 8

With two matching elements, we have managed to move most of the antenna impedance curve into the SWR definition circle. Portions of the curve, from about 124 MHz to about 131 MHz, however, still lie outside the circle. From the nature of the impedance plot it appears that a line transformer might be effective.

Draw an outer boundary circle on the Smith Chart. The circle is drawn tangent to the definition circle at R = 20, with a radius arbitrarily chosen large enough so that the 125 and 130-MHz points are well within the circle. (For clarity, the definition circle is not shown in **Fig 7**. The high resistance R-intercept of the boundary circle is at R = 160. The characteristic impedance Z_0 of a line transformer having the boundary circle shown in Fig 7 is 56.6 Ω, the geometric mean between 20 and 160:

$$Z_0 = \sqrt{20 \times 160} = 56.6\ \Omega.$$

To determine the optimum length of the line transformer, we need to evaluate its effect at all frequencies. A line transformer one-quarter wavelength long at 125 MHz will be our first choice. The 140-MHz point is the most critical, since at that frequency the impedance curve is moving clockwise. The 140-MHz point may move outside the 2.5:1 definition circle.

Step 9

Determine the new position of all the points modified by the line transformer characteristic impedance Z_0 = 56.6 Ω. See **Table 5**. The definition circle is also modified by the line transformer, $20/56.6 = 17.5$ and $125/56.6 = 110$. Draw a new definition circle using the upper and lower boundaries determined above. See Fig 7.

Step 10

The line transformer rotates all of the impedance points in a clockwise direction on the Smith Chart. The amount of rotation is determined by the length of the transformer at various frequencies. A wavelength of 0.250 λ was chosen for 125 MHz. To determine the length at other frequencies:

$$0.250 \times \frac{f_x}{125}.$$

See **Table 6**.

Step 11

Determine the placement of each frequency point by drawing a line from the center of the Smith Chart, through the frequency point of interest, to the "Wavelength Toward Generator" scale. Write down the scale wavelength value as shown in **Table**

Table 6
Impedance Modified by Line Transformer, Z_0 = 56.6 Ω

f_x, MHz	Old Impedance Ω	New Impedance $R/56.6 \pm jX/56.6$ Ω
110	30 − *j* 16.5	26.5 − *j* 14.6
115	49 + *j* 12.5	43.3 + *j* 11
120	84 + *j* 32	74 + *j* 28.3
125	130 + *j* 0	115 + *j* 0
130	103 − *j* 52	91 − *j* 46
135	58 − *j* 41	51 − *j* 36
140	31 − *j* 13	27.4 − *j* 11.5

Table 7
Rotation in Wavelengths Versus Frequency for 0.25 λ Line at 125 MHz

f_x	Wavelength, λ
110	0.220
115	0.230
120	0.240
125	0.250
130	0.260
135	0.270
140	0.280

7. Add the line wavelength determined in Step 10 to the scale wavelength. This will provide the new impedance position, shown as a dashed line in Fig 7.

Step 12

The transformed impedance curve shown in Fig 7 is not the final impedance curve, although it shows that all the frequency points lie within the reference definition circle. From the transformed curve in Fig 7, determine the R and X values for each frequency and write them as shown in **Table 8**. To transform these values to the characteristic impedance of a 50-Ω line, multiply the new impedance values by the line transformer impedance (56.6 Ω), then divide by 50. $R \times 56.6 / 50$ and $\pm j X \times 56.6 / 50$. This is the final impedance plot, and no further matching efforts are required.

Likewise, the definition circle is transformed to the 50-Ω impedance by:

upper value: $11.5 \times 56.6 / 50 = 20$

lower value: $110.0 \times 56.6 / 50 = 125.$

Step 13

Draw the final impedance curve, shown in **Fig 8**, from **Table 9**.

Step 14

One step remains: the value of the shunt inductor used as the first matching section. It provided −*j*0.180 chart siemens at 140 MHz.

$$L = \frac{1}{2\pi f B_L Y_0}$$

where

B_L = 0.180 chart siemens

Y_0 = 0.02 siemens

f = 140 MHz.

Substituting

$$L = \frac{1}{6.28 \times 140 \times 10^6 \times 0.18 \times 0.02} = \frac{10^{-6}}{3.1651}$$

$L = 0.316 \times 10^{-6}$ henry or 0.316 μH.

Table 8
Summation of Scale Wavelength and Line Wavelength

f_x MHz	Scale Wavelength	Line Wavelength	New Position Wavelength
110	0.440	0.220	0.660 (0.160)
115	0.087	0.230	0.317
120	0.199	0.240	0.439
125	0.250	0.250	0.500
130	0.292	0.260	0.552 (0.052)
135	0.344	0.270	0.614 (0.114)
140	0.449	0.260	0.709 (0.209)

Note: numbers in parentheses are values minus 0.500 λ, since this goes completely around Smith Chart.

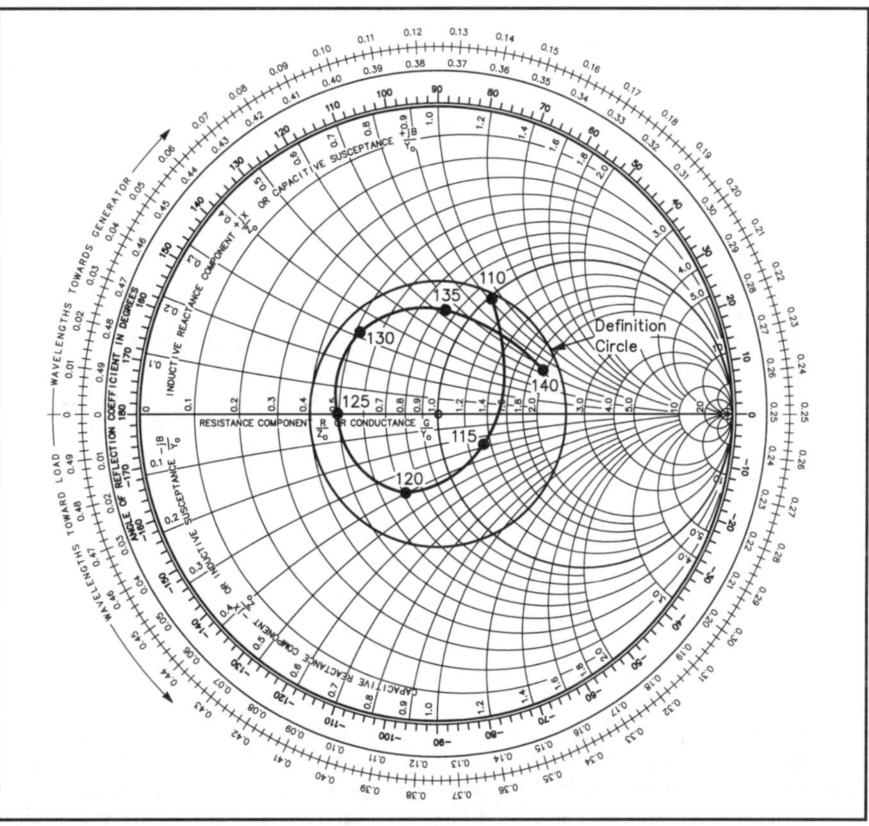

Fig 8—Final Smith Chart showing SWR across the entire frequency range with the complete matching network in place.

Table 9
Transformed and Denormalized Impedances from Smith Chart

f_x MHz	Transformed Impedance Ω	Final Impedance $R(56.6/50) \pm jX(56.6/50)$ Ω
110	52.5 + *j*40.5	50.4 + *j*45.8
115	58 – *j*12.5	65.7 – *j*14.2
120	31 – *j*13	35 – *j*14.7
125	21.5 + *j*0	24.3 + *j*0
130	22.5 + *j*14	25.5 +*j*15.8
135	37 + *j*28	41.9 + *j*31.7
140	82 + *j*29	92.8 + *j*32.8

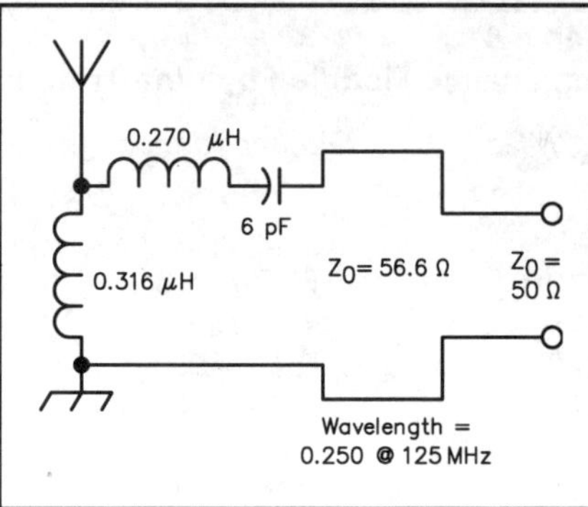

Fig 9—Final circuit showing use of series-resonant λ/4 transmission line with characteristic impedance Z_0 of 50 Ω, together with series-resonant circuit of 6 pF and 0.270 μH and a shunt coil across the antenna terminals of 0.316 μH. This matching system allows the antenna to be broadband, with an SWR of 2.5:1 or less from 110 to 140 MHz.

Table 10
Feed-Point Impedance of Short, Fat Monopole from 4.5 to 12.0 MHz

f_x	Antenna Impedance Ω	Impedance Normalized to 50 Ω
4.5	20 – *j*84	0.40 – *j*1.68
5.0	50 – *j*115	1.00 – *j*2.30
5.5	40 – *j*91.5	0.80 – *j*1.83
6.0	51 – *j*82.5	1.02 – *j*1.65
6.5	72.5 – *j*72.5	1.45 – *j*1.45
7.0	105 – *j*85	2.10 – *j*1.70
7.5	132.5 – *j*130	2.65 – *j*2.60
8.0	110 – *j*145	2.20 – *j*2.90
8.5	150 – *j*110	3.00 – *j*2.20
9.0	210 – *j*70	4.20 – *j*1.40
9.5	265 – *j*35	5.30 – *j*0.70
10.0	350 – *j*2.5	7.00 – *j*0.05

The final network is shown schematically in **Fig 9**.

Example 2

The following procedure describes the sequence used to match a broadband HF antenna with a combination of matching elements, including a parallel-resonant circuit. The antenna's feed-point impedance variation with frequency is shown in **Table 10**.

Step 1

Since we will be working with a parallel-resonant circuit across the antenna terminals, enter the normalized antenna impedance values on a normalized Smith Chart. See **Fig 10**. An examination of the impedance curve indicates that we must try to bunch up the curve and place all the impedance points on lines of constant resistance that enter the SWR definition circle. The frequencies between 4.5 MHz and about 8.0 MHz are on such lines, but those higher than 8.0 MHz are not. A parallel-resonant circuit may pull in the high frequency end, but we don't want to disturb the 4.5 MHz point, because it appears to be critical. A parallel-resonant circuit at 4.5 MHz appears to be well suited. We must now find a combination of L and C that not only resonates at 4.5 MHz, but that also provides the right amount of parallel reactance at higher frequencies.

Step 2

A parallel matching element requires that we work with admittances instead of impedances. Admittance coordinates on the Smith Chart are diametrically opposite the normalized impedance coordinates. Once the admittance coordinates have been plotted, you will find at 12.0 MHz that we need about +0.300 chart siemens to move the 12.0-MHz point along a line of constant resistance to fall inside the desired SWR definition circle. To accomplish this, we will need a second matching element.

Step 3

Find a combination of L and C that resonates at 4.5 MHz and that also provides a susceptance of +0.300 chart siemens at 12.0 MHz. I examined a number of possibilities and determined that a capacitance of 94 pF and an inductance of 13.3 μH fit the bill.

Step 4

Determine the amount of susceptance provided by the resonant circuit at other frequencies, f_x. The susceptance will be greatest at 12.0 MHz.

$$\text{Susceptance} = \frac{50}{\dfrac{-X_L X_C}{X_L - X_C}} \text{ chart siemens.}$$

See **Table 11**.

Step 5

Add the susceptance to the admittance coordinates and plot a new admittance curve. The resulting new admittance coordinates, are shown in **Table 12** and are plotted in **Fig 11**.

Step 6

Transfer the new admittance coordinates to their equivalent new impedance coordinates. This step is necessary because to move the new impedance coordinates into the SWR definition circle requires a series inductor.

Step 7

To assure that all of the new impedance coordinates move into the definition circle, the 12.0 MHz point will be moved as far as possible. This move requires +3.58 chart ohms. Determine the amount of chart ohms at other frequencies, f_x.

$$\text{Reactance} = +3.58\frac{f_x}{12.0} \text{ chart ohms.}$$

See **Table 13**.

Step 8

Add the normalized reactance to the old impedance coordinates of Table 13. Plot the final impedance curve, shown in Fig 11.

Step 9

The series inductance required is computed as follows:

$$L = \frac{3.58 \times 50}{6.28 \times 12.0} = 2.375\ \mu\text{H}.$$

Conclusion

The matching procedure presented here with the two examples provides the reader with practical knowledge of antenna impedance matching. The Smith Charts illustrate the impedance before and after a network is developed. They do not reflect the many hours (or days) needed to develop these steps. With many broadband antennas it is necessary to tailor the antenna by adjusting its height or length, its length-to-diameter ratio and even altering its feed-point geometry before we can achieve an impedance plot that will lead to a broadband match. As with many other endeavors, proficiency developing appropriate matching networks can only be achieved with practice.

Table 11

Susceptance of Parallel 94 pF and 13.3 H Coil Versus Frequency

f_x MHz	X_L Ω	X_C Ω	*Parallel Combination*	*Normalized to 50 Ω*
4.5	376	–376	∞	+*j*0
5.0	418	–339	–1794	+*j*0.028
5.5	460	308	–932	+*j*0.054
6.0	502	282	–643	+*j*0.078
6.5	544	261	–502	+*j*0.100
7.0	586	242	–412	+*j*0.121
7.5	627	226	–353	+*j*0.141
8.0	669	212	–310	+*j*0.161
8.5	711	199	–276	+*j*0.181
9.0	753	188	–251	+*j*0.200
9.5	820	178	–227	+*j*0.220
10.0	836	169	–212	+*j*0.236
10.5	878	161	–197	+*j*0.254
11.0	920	154	–185	+*j*0.270
11.5	962	147	–174	+*j*0.288
12.0	1003	141	–164	+*j*0.304

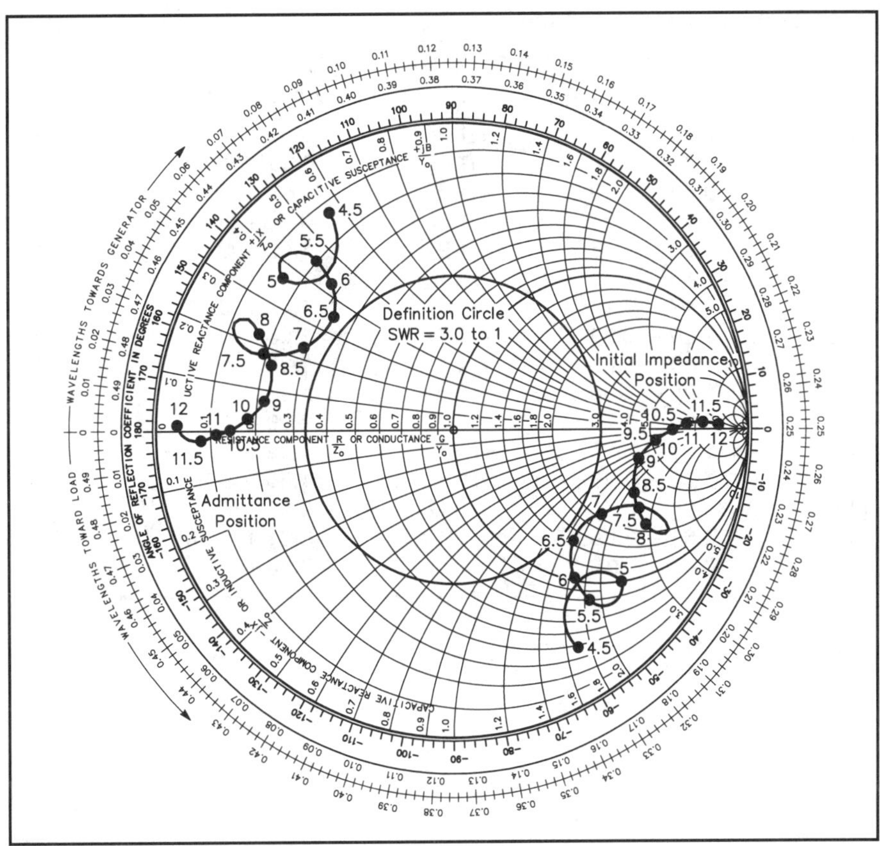

Fig 10—Impedance and admittance Smith Chart plots for broadband HF antenna with no compensation network.

Table 12
Adding Correction to Susceptance for New Smith Chart Curve

f_x *MHz*	*Old Admittance Siemens*	*Correction Siemens*	*New Admittance Siemens*
4.5	0.15 + *j*0.56	+*j*0.0	0.15 + *j*0.560
5.0	0.16 + *j*0.37	+*j*0.028	0.16 + *j*0.398
5.5	0.20 + *j*0.46	+*j*0.054	0.20 + *j*0.515
6.0	0.27 + *j*0.43	+*j*0.078	0.27 + *j*0.508
6.5	0.33 + *j*0.35	+*j*0.100	0.33 + *j*0.450
7.0	0.29 + *j0.23*	+*j*0.121	0.29 + *j*0.351
7.5	0.19 + *j*0.19	+*j*0.141	0.19 + *j*0.331
8.0	0.17 + *j*0.22	+*j*0.161	0.17 + *j*0.381
8.5	0.22 + *j*0.16	+*j*0.181	0.22 + *j*0.341
9.0	0.22 + *j*0.07	+*j*0.200	0.22 + *j*0.270
9.5	0.18 + *j*0.03	+*j*0.220	0.18 + *j*0.250
10.0	0.14 + *j*0.01	+*j*0.236	0.14 + *j*0.246
10.5	0.11 – *j*0.01	+*j*0.254	0.11 + *j*0.244
11.0	0.08 – *j*0.02	+*j*0.270	0.08 + *j*0.250
11.5	0.05 – *j*0.01	+*j*0.288	0.05 + *j*0.278
12.0	0.035 + *j*0.01	+*j*0.304	0.035 + *j*0.314

Table 13
Move 12.0 MHz Point by +*j* 3.58

f_x *MHz*	*Old Impedance* Ω	*Correction* Ω	*Final Impedance* Ω
4.5	0.40 – *j*1.69	+*j*1.34	0.40 – *j*0.35
5.0	0.85 – *j*2.20	+*j*1.49	0.85 – *j*0.71
5.5	0.68 – *j*1.70	+*j*1.64	0.68 – *j*0.06
6.0	0.80 – *j*1.56	+*j*1.79	0.80 + *j*0.23
6.5	1.08 – *j*1.46	+*j*1.94	1.08 + *j*0.48
7.0	1.40 – *j*1.70	+*j* 2.09	1.40 + *j*0.39
7.5	1.42 – *j*2.15	+*j* 2.24	1.42 + *j*0.09
8.0	1.05 – *j*2.39	+*j* 2.39	1.05 + *j*0.18
8.5	1.30 – *j*2.00	+*j* 2.54	1.30 + *j*0.54
9.0	1.95 – *j*2.30	+*j* 2.69	1.95 + *j*0.39
9.5	2.00 – *j*2.70	+*j* 2.83	2.00 + *j*0.13
10.0	1.90 – *j*3.20	+*j* 2.98	1.90 – *j*0.22
10.5	1.50 – *j*3.40	+*j* 3.13	1.50 – *j*0.27
11.0	0.97 – *j*3.40	+*j* 3.28	0.97 – *j*0.12
11.5	0.61 – *j*3.30	+*j* 3.43	0.61 + *j*0.13
12.0	0.40 – *j*3.15	+*j* 3.58	0.40 + *j*0.43

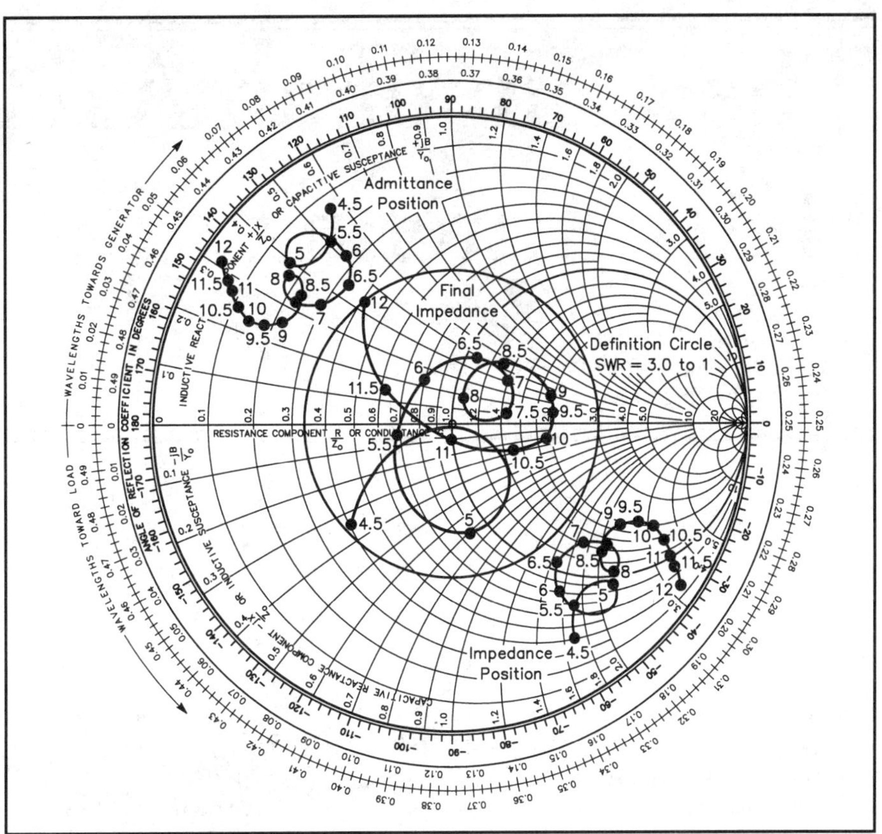

Fig 11—Final Smith Chart showing SWR across the frequency range of 4.5 to 12.0 MHz with parallel-resonant circuit and series inductor. SWR remains less than 3:1 across the range.

The Gain of an Axial-Mode Helix Antenna

By Darrel Emerson, AA7FV, G3SYS
3555 E. Thimble Peak Place
Tucson, AZ 85718

AA7FV had access to massive computing power—he used it to evaluate thoroughly the axial-mode helix antenna.

Introduction

The axial-mode helix antenna was first described by Kraus[1] in 1947, and now receives mention in nearly every antenna handbook. The antenna is inherently broadband, and provides a convenient way of generating circularly polarized radiation, with maximum gain along the axis of the helix.

There are conflicting claims for the gain of the antenna. Kraus originally gave the expression:

$$\text{Directivity} \approx 15\, C_\lambda^2 n S_\lambda \qquad \text{(Eq 1)}$$

where

C_λ = the circumference of the helix, in wavelengths

n = the number of turns

S_λ = the spacing between turns, in wavelengths

The equation is repeated in many antenna handbooks and textbooks.[2,3] The most extensive experimental study of the helix antenna to date is by King and Wong.[4,5] They derived a somewhat more complex empirical relationship for the gain, which they measured to be significantly lower than that given by Equation 1. In 1984 Lee and Wong[6] compared a theoretical model of the helix antenna with the empirical results, and with Eq 1 from Kraus. The theoretical gains were as much as 2 dB lower than those measured by King and Wong, but approximately 5 dB below Kraus. [In the second edition of *Antennas*, Kraus repeated the equation above and then stated: "A more realistic relation is Directivity $\approx 12 C_\lambda^2 n S_\lambda$. This is almost 1 dB less than his original formula.—*Ed.*]

Powerful antenna modelling programs are now readily available. In this article, a version of the classic *NEC2* program[7] is used to study systematically the gain of the helix antenna as a function of turn spacing, radius, total length, conductor diameter and resistivity, and the size of the groundplane. These extensive numerical results are then compared with published experimental data.

Modelling the Helix

Fig 1 is a sketch of the helix antenna used for the numerical modelling. The helix consists of n turns, with constant diameter D. The turn spacing S is constant for a given helix. The helix is mounted above a conducting groundplane, fed at the base of a short stub of length a. Other parameters are related to these quantities as follows, with the understanding that all are measured in units of wavelength.

Radius of helix	$R = D/2$
Circumference	$C = 2\pi R$
Pitch angle α	$\tan(\alpha) = S/C$
Total helix length	$L = n \times S$
Wire length per turn	$W = C/\cos(\alpha)$

Calculations were performed for different helical antennas of lengths from 2 to 7 λ. For a given model antenna the radius R and turn spacing S are held constant, but calculations were repeated for antennas with radii R ranging from R = 0.08 to 0.20 λ in steps of 0.005 λ, and with turn spacing S ranging from 0.08 to 0.20 λ, in steps of 0.01 λ or 0.005 λ. For most of the calculations, a "thin" conductor was used for the helix, of radius r = 0.00254 λ. However, the effect of varying the wire radius was also investigated. The wire was assumed to have the resistivity of aluminum. The antenna was modelled assuming an infinite groundplane, and also with finite, flat square groundplanes of different sizes. These finite groundplanes were modelled using a rectangular mesh of wires, with a mesh spacing of 0.05 or 0.1 λ.

The calculations were performed for a frequency of 299.8 MHz—that is, λ = 1 meter. Apart from the skin depth in the conductor, and hence its resistance per unit length, all dimensions will scale exactly with wavelength. Since the effect of resistance of the conductor is found to be minimal, almost identical results can be expected after scaling the antenna designs over a very wide range of wavelengths.

The program *NEC2* models an antenna as a number of short, straight segments of wire. The wire segments should be short compared

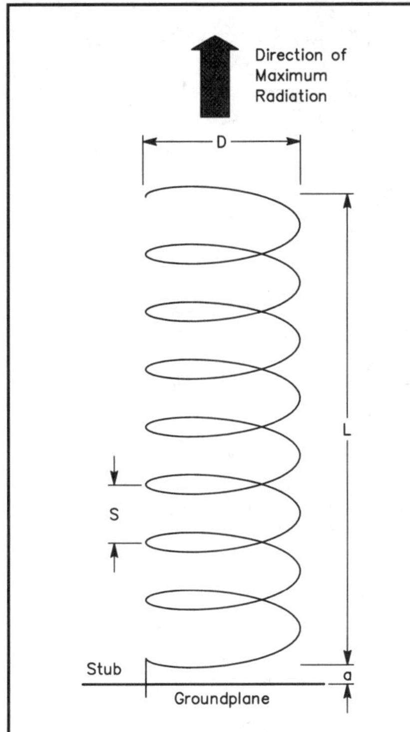

Fig 1—A sketch of the helix antenna modelled in this article, showing the turn spacing S, the diameter D and the total length L. Each helix antenna modelled has a constant diameter and turn spacing over its length. The helix is spaced a small distance "a" above a flat groundplane, and fed via a short wire stub.

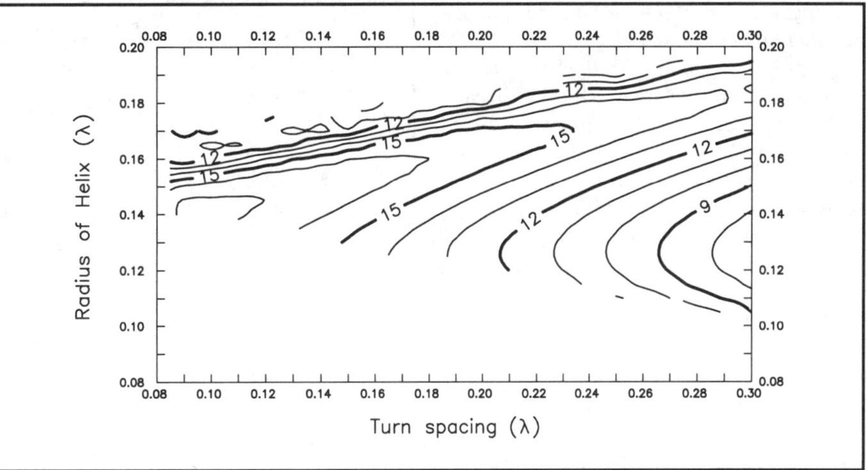

Fig 2—Contours showing how the gain of a helix antenna varies as a function of the turn spacing S, and of turn radius R (= D/2), for an antenna of length L = 6 λ. Gain contours are plotted every 1 dB, from 8 dB to 17 dB. In the region plotted the gain is fairly stable, showing only slow variations as a function of turn spacing and radius. Beyond this region the performance deteriorates rapidly and the gain becomes unstable.

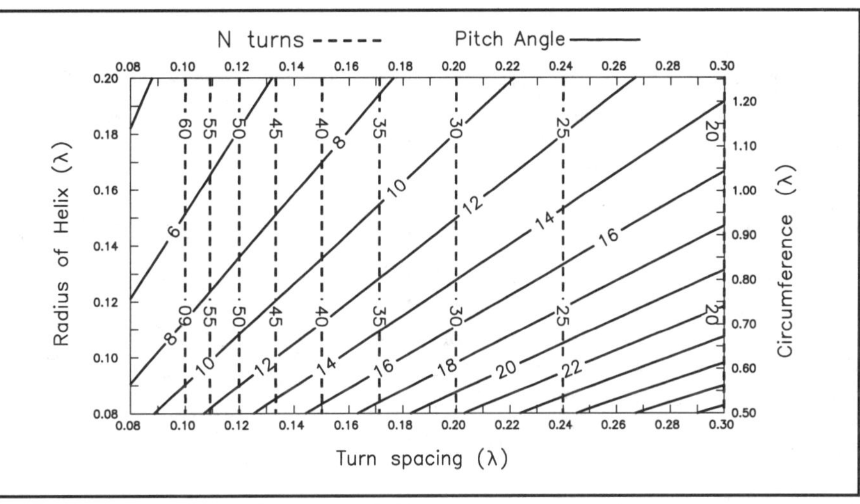

Fig 3—A plot showing the pitch angles in degrees (solid lines), and number of turns (dashed vertical lines) of a 6-λ helical antenna, for the same ranges of helix radius and turn spacing as Fig 2. The antenna circumference (C = 2πR) is shown on the right hand scale of this plot. The number of turns ranges from 20 (R = 0.3) to > 60 (R = 0.1), although for ≥60 turns the gains computed in Fig 2 will become less reliable because of the limited (600) number of straight wire segments used to model the antenna.

to λ in order to model the antenna accurately. Other important restrictions are discussed in Reference 7. I have tried to keep well within these known limitations of the program. Individual antennas were modelled using between 600 and 1400 segments. Some calculations were repeated, either doubling or halving the number of segments used to model the antenna, as a check on consistency of the results. Results were found to be very insensitive to the number of segments used, provided there were ≥ 10 segments per turn.

For the antennas modelled here, the errors in the computed gains should generally be much less than 1 dB. In the course of this project more than 10,000 separate antennas were modelled, requiring several thousand hours of (background task) parallel-processing computations on a network of workstation computers. Not every ham is fortunate enough to have access to such computing horsepower!

THE RESULTS

A 6-Wavelength Helix, with Infinite Groundplane

Fig 2 summarizes the gain of a 6-λ long helical antenna, as a function of turn spacing and radius. In this article, gain is measured relative to an isotropic radiator (dBi). This antenna was modelled over an infinite, perfectly conducting groundplane, but with aluminum wire of radius 0.00254 λ. The feed stub length was 0.025 λ.

For reference, **Fig 3**, drawn to the same scale, shows the number of turns, pitch angle and the circumference of a 6-λ antenna for the same ranges of radius and turn spacing. Gain contours in Fig 2 have only been drawn where the performance of the antenna is stable—that is, where moderate changes of turn spacing or radius only cause minor changes in gain. Beyond the region plotted, the gain and efficiency of the antenna rapidly deteriorate.

The gain is a very weak function of turn spacing. Most published helical antenna designs have recommended a turn spacing in the region of 0.24 λ; a peak gain of ~15 dB is obtained at this spacing. Slightly higher gains are suggested at closer spacings—up to ~17 dB by halving the spacing, or doubling the number of turns in the same length of antenna. However, the dimensions then become more critical—slight variations may produce large changes in gain and feed impedance. King and Wong had also noted that gain tended to increase as the turn spacing was decreased, but did not pursue the phenomenon in detail.

At the turn spacing S of 0.24 λ—that is, 25 turns—the peak gain of ~15.0 dB occurs with a turn radius R of ~0.174 λ, or a circumference C of ~1.09 λ. The gain drops off much more rapidly at larger radii than at smaller radii. In practice, with this length of helix, safe design parameters might be S = 0.24 λ and R = 0.15 to R = 0.16 λ (C = 0.94 to C = 1.01 λ). This would sacrifice 1 to 3 dB of potential gain, but allow very loose construction tolerances and give very wide bandwidth. If the turn spacing is decreased, then R should be decreased very slightly, following the peak contours of Fig 2. At a turn spacing of S = 0.14 λ, a peak gain of ~16.7 dB occurs at R = 0.15 λ.

Changing the frequency is equivalent to scaling the entire design. The pitch angle will remain constant, although all dimensions in terms of wavelengths will change proportionately. The gain of a helix antenna is a very weak function of length (see below), so a change in frequency is approximately equivalent to moving along a line of constant pitch angle. From Fig 3, a pitch angle of 13° is nearly parallel to the ridge of peak gain shown from the contours in Fig 2. That is, with a 13° pitch angle, the antenna will be *very* broadband; this is one of the important and well-known features of the helix antenna. Pitch angles from 11 to 14° are all nearly as good, although the optimum turn radius increases very slightly as the pitch angle is increased.

The basic features of the gain contours of Fig 2—that is, a ridge of peak gain, with optimum turn radius decreasing very slightly as the turn spacing is decreased, and with a sharper fall of gain with increasing rather than decreasing radius—are fundamental to this type of antenna. The absolute values of gain, and the precise position of the peak gain ridge, vary somewhat with other antenna parameters, such as the total length, but the basic features remain the same.

Fig 4 shows how the SWR at the feedpoint varies, as a function of helix turn radius and spacing. The SWR has arbitrarily been normalized to a resistive component of 230 Ω. Over the entire region plotted, the SWR is below 1.8:1, but for a substantial part of the region of interest the SWR remains below 1.1:1. A nominal value of 140 Ω is often quoted for the feedpoint of a helix antenna, rather than the 230 Ω chosen for normalization here. The precise value of the feedpoint is influenced by the stub length (see Fig 1), and the radius of the helix conductor, especially in the first quarter-turn. A common matching technique[8] is to increase the diameter of the conductor close to the feedpoint, in order to give a convenient match to 50 Ω. Either increasing the helix wire diameter, or reducing the feed stub length, reduces the feed impedance. The gain plot and renormalized SWR plots, such as Figs 2 and 4, are almost completely unaffected.

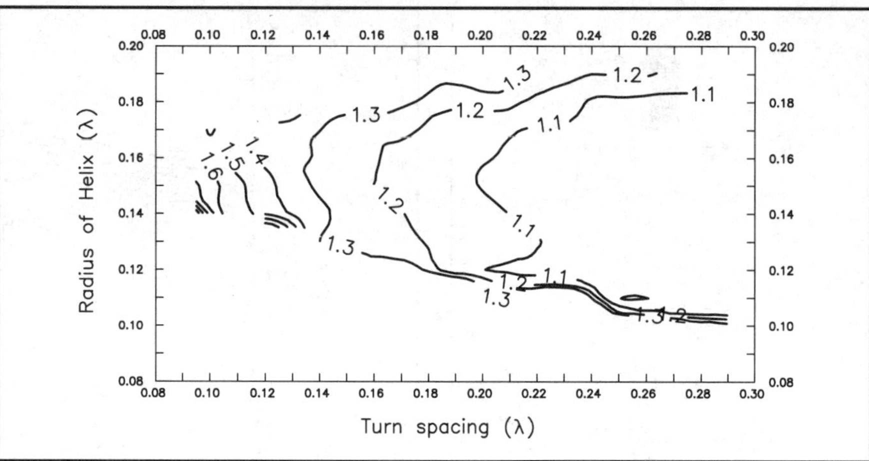

Fig 4—The SWR as a function of helix turn radius and spacing for the 6-λ helix in Fig 2 over an infinite conducting plane. The SWR has arbitrarily been normalized to a resistive term of 230 Ω; contours are plotted from 1.1, in steps of 0.1.

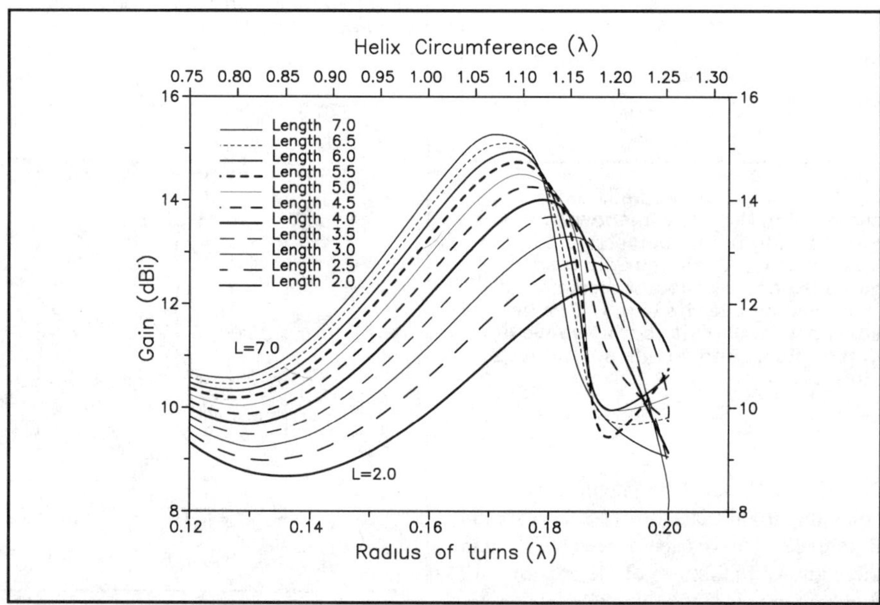

Fig 5—A plot of gain of a helix antenna as a function of length and turn radius, with a fixed turn spacing of 0.24 λ. Curves are plotted for helix lengths from 2 λ to 7 λ, every 0.5 λ. Note that the turn radius at which peak gain occurs decreases slightly as the helix length is increased. An increase in helix length from 2 to 3 λ increases the peak gain by ~1 dB, while an increase in length from 6 to 7 λ only improves the gain by less than 0.4 dB.

The Dependence of Antenna Gain on Total Helix Length

Calculations such as are displayed in Fig 2 were repeated for helix lengths from 2.0 to 7.0 λ. The same general characteristics of gain versus turn radius and spacing were found, although the radius at which peak gain occurs tends to increase a little as the helix is shortened. **Fig 5** summarizes the results. The gain is plotted as a function of turn radius, for a fixed turn spacing of 0.24 λ. Again, these calculations are for a helix over an infinite conducting plane, with wire of radius 0.00254 λ. As noted before, the gain falls more sharply with larger, rather than with smaller, turn radii. Although not plotted here, the polarization of the radiated signal from short helices becomes progressively elliptical, rather than purely circular.

Plotted in **Fig 6** is the peak gain G_{max} as a function of helix length, and also the turn radius R_{max} at which that peak gain occurs; both

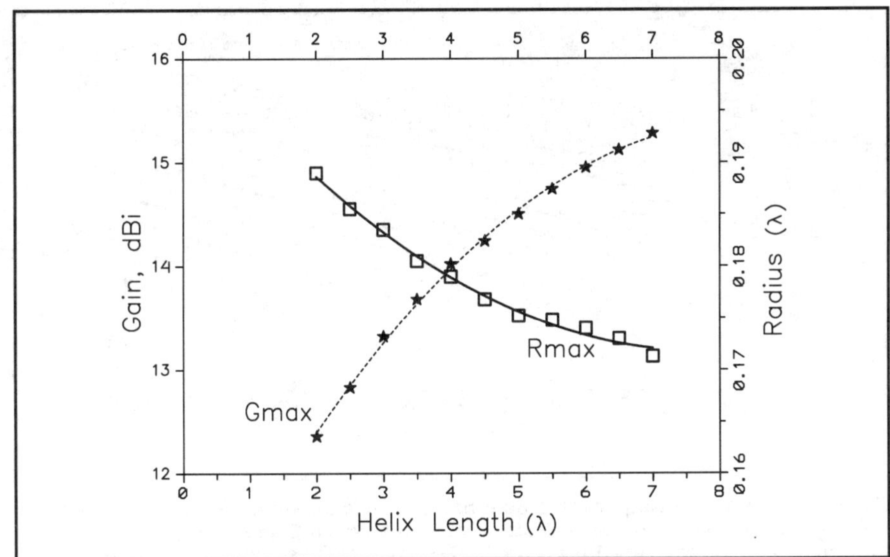

Fig 6—The peak gain (stars, left axis) from Fig 5, as a function of helix length, and the turn radius (squares, right axis) at which that peak gain occurs. As expected, the gain increases with helix length, but also the optimum radius decreases as the antenna is lengthened. The lines connecting the points are a smooth fit to the data; see text.

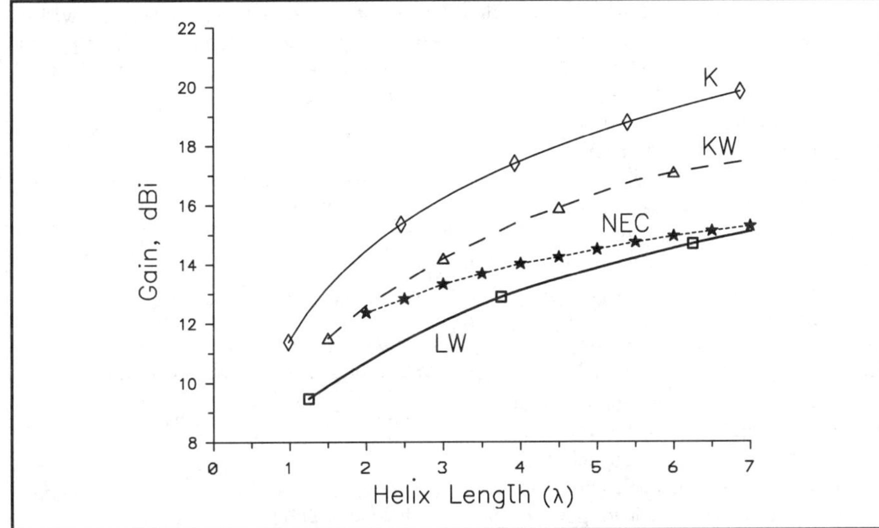

Fig 7—The peak gain of a helix antenna as a function of its length in λ. K is the gain given by the Kraus formula (Eq 1), KW (long dashes) is from the measurements of King and Wong.[4] NEC (short dashes) is from the numerical modelling in this article using *NEC2*, choosing the peak gain at a turn spacing S=0.24. LW is from the theory of Lee and Wong.[6]

quantities are derived from Fig 5. The lines joining the points are empirical fits to the data. For helical antennas of length L between 2 and 7 λ, with turn spacing S of 0.24 λ, and fed against a flat groundplane, the peak gain in dB and the turn radius in wavelengths for peak gain are given approximately by:

$$G_{max}\ (\text{dB}) \approx 10.25 + 1.22 \times L - 0.0726 \times L^2 \quad \text{(Eq 2)}$$

$$R_{max} \approx 0.2025 - 0.0079 \times L + 0.000515 \times L^2 \quad \text{(Eq 3)}$$

These expressions should not be applied outside the range of $2\lambda \leq L \geq 7\lambda$, or for turn spacings other than 0.24 λ.

Fig 7 compares the peak gains, taken from Fig 5, with Eq 1 from Kraus (K), the empirical results of King and Wong (KW),[4] and the theoretical results of Lee and Wong (LW).[6] The numerical modelling (*NEC*) of this article is in excellent agreement with the measurements (KW) at a length of 2 λ, but gives gains ~2 dB lower for a length of 7 λ. The *NEC* calculations give essentially perfect agreement with LW at 7 λ; this is consistent with the LW comment that their theory should be more accurate for longer antennas. The *NEC* calculations show a lower rate of increase of gain with antenna length.

The discrepancy between the numerical modelling and the measurements of KW is greater than the errors expected from the *NEC* calculations alone. However, there are some differences in the basic antennas: KW used a circular cavity at the feed end of the helix, whereas the modelling has assumed a flat groundplane. Furthermore, although it is unlikely to have any significant effect on gain, the KW construction included a metallic boom in the center of the antennas.

The formula from Kraus (Eq 1) gives values of gain that are much too optimistic, by nearly 5 dB for a 7-λ helix. However, Kraus derived his gain formula indirectly, deducing it from measured beamwidths rather than direct gain measurement, so the discrepancy is perhaps not surprising.

Effect of Helix Wire Thickness, and of Stub Length

Most of the *NEC* modelling used fairly thin wire for the helix: radius 0.00254 λ. Plots similar to Figs 2 and 4 were produced for an identical helix, but using wire of radius 0.01 λ. The gain curves were identical within a few tenths of a dB, with no significant displacement of the ridge of peak gain. The resistive part of the feed impedance was now close to 150 Ω. The SWR plot, after normalization to 150 Ω, was nearly identical to Fig 4, showing a similar slight variation as the dimensions of the helix are varied.

Varying the length of the short stub in series with the feed produces a similar effect. Reducing the stub length from 0.025 to 0.01 λ brought the resistive part of the original thin-wire helix from 230 Ω down to about 190 Ω, and as expected removed a few ohms of series inductive reactance from the feed impedance. The gain curves changed by less than 0.1 dB. The feed impedance is affected almost entirely by the ratio of wire diameter to spacing above the conducting plane for the first fraction of a turn of the helix. Shorter stub lengths would reduce the feed impedance further.

Effect of Wire Resistivity

The numerical modelling has included the effects of resistivity of the conductor, assuming the construction to be of solid aluminum wire. In the region of interest, where the antenna gain and impedance are fairly stable, power losses in the conductor are negligible. Additional calculations were made giving the wire a resistivity five times higher than that of aluminum, with the same wire radius of 0.00254 λ at 300 MHz. Even in this case, less than 1% of power was lost in the conductor.

Although decreasing skin depth will cause more power loss per meter of length in a given conductor as the frequency is raised, the loss *per wavelength* will actually decrease, if the wire diameter in mm is held constant. Similarly, the power wasted due to losses in a helical antenna of fixed length *measured in wavelengths*, will become even smaller at higher frequencies, for a given fixed diameter of wire.

Effect of the Groundplane

The numerical modelling shown so far has assumed an infinite conducting groundplane. Calculations were also made for a square groundplane with sides of either 0.5 or 1.0 λ. The square groundplane was first simulated by a square mesh of conducting wires, spaced 0.05 λ and connected at each crossing point. **Fig 8** shows the antenna gain for a 6-λ helix as a function of turn spacing and radius, for the 0.5-λ groundplane. This should be compared with Fig 2, which shows the gain of an antenna with an infinite groundplane. At a turn spacing of 0.24 λ, the gain with this square mesh is actually a few tenths of a dB higher than with the infinite groundplane, but the gain is a slightly sharper function of radius.

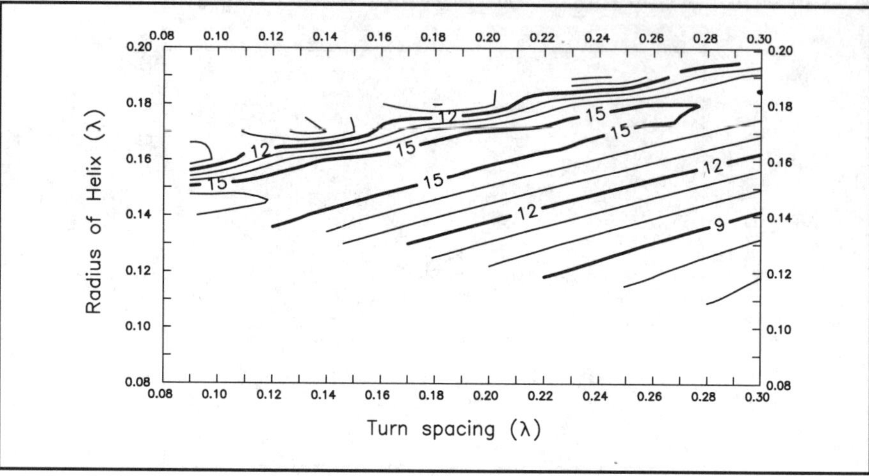

Fig 8—Contours showing how the gain of a 6-λ helix antenna, with a half-λ square mesh groundplane, varies as a function of turn spacing S and radius R. Gain contours are plotted every 1 dB, from 7 dB to 16 dB. These contours should be compared with Fig 2, a helix of the same length but with an infinite groundplane. Along the peak ridge of gain, there is now much less variation of gain as a function of turn spacing. Contours are omitted outside of the well-behaved region.

As the turn spacing is decreased, increasing the number of turns but keeping the helix length constant, the slight increase in peak gain is less pronounced than for the case of the infinite groundplane illustrated in Fig 2. A little less than 1 dB is won now by doubling the number of turns. Again, the precise dimensions of the helix become more critical. A plot of SWR at the feedpoint of this same 6-λ helix with 0.5-λ groundplane, as a function of turn radius and spacing, is essentially the same as for the infinite-groundplane case plotted in Fig 4, if normalized to a slightly higher resistive impedance of 250 Ω.

For a half-λ mesh with 0.1-λ wire spacing, the forward gain was about 0.5 dB lower than with 0.05-λ spacing. A 1-λ square mesh, with 0.05-λ wire spacing, yielded results closer to the infinite groundplane calculations.

Additional calculations were made using individual radial wires, instead of the wire mesh groundplane. With a system of 8 quarter-wave radials, each connected together at the feed point, spaced every 45° and isolated from each other, the average forward gain was ~3.5 dB lower than with the original half-λ, 0.05-λ mesh groundplane.

Conclusions

Extensive numerical modelling of the axial-mode helical antenna has been made using *NEC2*, giving results that are in reasonable agreement with previous experimental and theoretical data, although computed gains are somewhat lower than are usually quoted. With a large flat ground plane, slightly higher antenna gains may be achieved, for a given length of antenna, with a smaller turn spacing and hence more turns on the helix, although dimensions become somewhat more critical and the bandwidth will be narrowed.

At a given value of turn spacing, the optimum turn radius for peak gain decreases slightly as the helix is made longer. The gain is almost completely independent of wire diameter, or of the presence of a short feed stub, although the feed impedance changes with these parameters.

The resistance of the wire used to construct the helix, even if several times worse than aluminum, has little effect on the antenna efficiency. A half-λ square groundplane is nearly as good as an infinite groundplane, although the slight gain enhancement possible with smaller turn spacings is lost. The use of individual radials, rather than a continuous groundplane or a closely spaced mesh, gives a gain penalty of some 3.5 dB.

References

[1]J.D. Kraus, "Helical Beam Antenna," *Electronics*, vol 20, April 1947, pp 109-111.

[2]J.D. Kraus, *Antennas*, Second Edition (McGraw-Hill Book Company, 1988), p 284.

[3]H. Jasik (Editor), *Antenna Engineering Handbook*, First Edition (McGraw-Hill Book Company, 1961), p 7-5.

[4]H.E. King, & J. L. Wong, "Characteristics of 1 to 8 Wavelength Uniform Helical Antennas," *IEEE Transactions*, AP-28, March 1980, pp 291-296.

[5]C. J. Johnson, & H. Jasik, *Antenna Engineering Handbook*, Second Edition (McGraw-Hill Book Company, 1984), p 13-11.

[6]K. F. Lee, & P. F. Wong, "Directivities of Helical Antennas Radiating in the Axial Mode," *IEE Proceedings*, Vol 131, Pt H, No. 2, April 1984, p 121.

[7]G. J. Burke & A.J. Pogio, "NEC2, Numerical Electromagnetics Code Method of Moments," NOSC Technical Document 116, 1981. Distributed by NTIS, Springfield, VA 22161.

[8]J. D. Kraus, *IEEE Transactions*, AP-25, Nov 1977, p 913.

Programs for Antenna Analysis by the Method of Moments

By R. P. (Bob) Haviland, W4MB
1035 Green Acres Circle North
Daytona Beach, FL 32119

W4MB reviews the history of antenna modeling programs using the Method of Moments technique.

Overview

The essential equations describing antennas were well worked out nearly 100 years ago, not too long after the invention of the dipole and the loop antenna by Hertz. However, the equations were complex and "mathematically intractable," and the early solutions were for limited conditions and special cases.

The theories for very short dipoles and very small loops were the first to be developed. However, it took until the 1930s to get a good solution for dipoles with sinusoidal current distribution at resonance. The theory of arrays of antennas, based on the assumption that they could be regarded as point sources, came very early, and further advances were made in the 1930s. However, there was still a gap between calculations and measurements for most antenna types.

In the 1960s, a new method of obtaining solutions to the equations was developed. Instead of demanding that results were correct at all points on the antenna, exactly correct values were demanded only at selected points—the ends of the antenna and some specific intermediate points. It was recognized that the values away from these points could be in error, but the amount of error could be controlled by selection of the number of points, and by making some assumptions about the nature of variation between selected points.

Following mathematical nomenclature, the importance of the accumulated error is called its *moment* (as in the phrase, a momentous occasion). Consequently, the method of analysis came to be called the *Method of Moments*.

You should remember that the term "exact" is a mathematical fiction. The starting equations are exact, but they are difficult to solve. Approximations must be used to get numerical results and the accuracy of these must be watched. This use of approximations is found in all antenna solution methods. A goal of the originators of the programs has been to make the programs exact in the practical sense—hopefully, they are as accurate as any measurement made under conditions outside the laboratory.

For many years these techniques were exclusively used by specialists, often under limits of military security. Two factors made the techniques available to amateurs. One was the development of the small, powerful microcomputer, at a price within the amateur pocketbook. The second was the development and release of a simplified Method of Moments program specifically tailored for these smaller computers. Antenna analysis became relatively painless. It was possible to think of an antenna in terms of dimensions and connections, common amateur practice. The computer did all of the drudgery. As microcomputers became more powerful, even the time needed became negligible, going from minutes and hours to seconds and even fractional seconds today.

The first program to become widely available was called *MININEC*, standing for *Miniature Numerical Electromagnetic Code*. While the first version was used by a small number of amateurs, use did not become common until the third version became available, first as originally written, then with modifications intended to simplify use. The popularity of these programs can be judged by the number of articles based on *MININEC* results appearing in amateur publications.

MININEC was developed from the larger *NEC* program, which was designed for the large mainframe computers at government installations. *NEC* itself was later adapted to smaller computers, and is just becoming available to amateurs. It is more powerful than *MININEC,* and one purpose of this article is to show its capabilities. But these are not the only Method of Moments analysis programs available. In the following, other programs are also considered, together with the types of problems they are best suited to handle.

The Kraus-Newman Methods

In his book, *Antennas*, 2nd Edition,[1] Kraus, W8JK, has a good discussion of the Method of Moments, first as applied to charges on a wire and their electrostatic fields, and then as applied to the currents on a wire and the electromagnetic fields they produce to make the wire a radiator. Complete derivations of the equations involved are given, and a set of values is worked out using a typical "college" calculator. The techniques are later extended by Newman, to include self-impedance, radar cross-section and mutual impedance of short dipoles.

Although these are not programs in the computer sense, they are excellent explanations of the theory and methodology of the Method of Moments. The examples are also valuable to show the effect of variations assumed between the exactly calculated points—for example, you could compare a constant value to a sinusoidal variation. I strongly recommend that anyone really interested in antenna analysis work through the examples shown in Kraus, as well as some of the problems. This is particularly valuable if you want to make a change in one of the larger programs, since you will have a good feel for the processes. This is important to save time and to avoid mistakes. However, while the relations are easily rewritten into computer form, say as a BASIC program, this doesn't seem worthwhile. The same results can be obtained from the programs discussed below, with a lot less work and debugging!

The Ohio State Thin-Wire Program

In the 1970s, Richmond, of Ohio State University developed a Method of Moments antenna program for NASA.[2] Some of the capabilities and limitations of the program are:

- Uses sinusoidal current distribution on segments
- Can be used in lossy ambient media
- Near- and/or far-field calculations
- Can evaluate wire loss resistance
- Provision for a single or multiple voltage excitation
- Provision for excitation by an incident wave (radar scattering model).
- Provision for insulation around the conductor.
- Lumped loads allowed
- Limited to thin wires
- Limited to straight-wire segments

In addition, the program description gives hints for extensions which might be useful, such as curved wires. There is no provision for antennas above earth, but ideal-earth results can be calculated by doubling the number of elements, with the second set representing the below-ground mirror image of an antenna above earth.

The inclusion of insulation makes the program unique for amateur purposes. None of the other programs described will handle insulated wires. The lossy ambient capability means that the program is usable for antennas deep in sea water, or deeply buried. The program cannot handle two media, such as an antenna lying on the ocean's surface.

The scattering solutions give the radar echo area of the antenna, or of a structure simulated as a wire grid. They also give the pattern produced by scattering of a wave from any given direction.

As was common with early FORTRAN programs, input was designed for punched cards, and output for tape reels and lineprinters. Rewriting is required for other input-output. The alternate approach, used by the author, is to write a BASIC program producing a disk file to emulate punch cards. The changes for disk input and output are simple, and most compiler-computer combinations convert line output to PC printer formats. With these changes, the program is easy to use.

The author has used the results of this program to develop some approximations for designing insulated antennas.[3] The Ohio State program is also useful for comparisons. Since amateurs have little interest in buried antennas, or in radar scatter situations, the program is not used by amateurs, except rarely to serve as a check on the accuracy of *MININEC* analyses.

The Pozar Programs

Pozar has written and published a set of three BASIC programs using moment methods.[4] One is for dipoles, one for folded dipoles and the third for Yagis. An unusual feature is that the programs do not directly solve the field equations as part of the moment process. Instead, the mutual impedance between the segments are calculated directly, using relations developed by Hansen assuming sinusoidal currents on the segments. This makes for a short program, some 500 BASIC lines.

The programs are intended to give input impedance, efficiency, wire current and gain, but only for antennas in free space. Lumped loads may be included. The folded-dipole program outputs the transformation ratio. The Yagi program is of lower accuracy, since in order to make the program suitable for small computers it allows only a single calculation segment for each element. Forward gain accuracy, however, is quite good, F/B accuracy is acceptable, but drive impedance accuracy is estimated to be only ± 30%.

The Pozar programs are intended for small computers, and for education. They do not do anything that cannot be done as quickly and as easily as other programs, especially *MININEC*. However, if someone should want to write a fast program for a large number of wire segments, the impedance calculation algorithm should help keep down both the development and runtime needed.

MININEC

MININEC1 was developed at the Naval Ocean Systems Center specifically for antenna analysis on microcomputers. It was originally visualized as a cut-down version of the big *NEC* program discussed below, but considerable original work was added as the program developed. (All versions of *MININEC* were written in BASIC, although I understand a FORTRAN translation has been made. They could be compiled in the C language by using one of the BASIC-to-C translators now available.)

The first publicly available version was for the APPLE,[5] and has been translated for the IBM PC, the Commodore 64, the Amiga, and possibly to other small computers. It is still the only full-analysis program able to run on 8-bit, 64K early computers. It is slow on these—an hour of runtime is not unusual. Moreover, the fact that wire connections had to be explicitly entered to the program as data input made for many input errors. Other than these problems and the limit on antenna size imposed by the small memory of 64K computers, *MININEC1* is adequate for a lot of antenna work.

MININEC2 cleaned up some problems with the *MININEC1* code, and added a few features. However, although it was publicly available as early as 1986, it did not become popular with amateurs. *MININEC3* was the first program to become really popular.[6] One reason for this was a change to automatic calculation of wire connections, which eliminated a lot of input errors.

As usually used, *MININEC* calculates:

- Drive-point resistance and reactance
- Current distribution on elements
- Power at specified voltage input
- Far-field pattern and gain at specified E-Plane and H-Plane angles

These may be calculated for free space, or with a ground present. The ground directly under the antenna is always considered to be perfectly reflecting ideal earth. This means that drive impedance errors are appreciable for horizontally polarized antennas lower than about $\lambda/4$ above ground. This also affects the far-field pattern, in particular the depth of the nulls in the pattern. Errors are acceptably small for higher antennas and for completely vertically polarized antennas.

There is no specific provision for loss resistance in the original *MININEC*. This must be entered as a segment load, which is accurate only if a load is assumed at every segment. Some commercial versions have an option of doing this automatically.

One feature of *MININEC* that is often

overlooked is the ability to place the antenna on a hill or a ridge, or in a pit or a valley. The simulation is not perfect, since the standard program allows only five changes in elevation and ground characteristics. The ground is assumed to be level between elevation steps, meaning that the effect of sloping ground at the point of ground wave reflection is not correctly calculated. Except for hilltop locations, this is not a serious defect. When multiple ground media are specified, a ground screen may be specified for the one closest to the antenna. Since the program assumes perfect ground under the antenna, this feature is not very useful.

Another overlooked feature is the ability to handle elements with step changes in diameter, such as tapered telescoping elements used in Yagis. In fact, calibration data shows that *MININEC* is more accurate for tapering than is the larger *NEC3* program. For critical designs, it is better to use this feature instead of assuming constant element radius and applying a taper correction factor.

Mutual impedance between elements is sometimes needed. This is obtained by exciting each element in turn, and observing the currents in other elements. The exciting voltage divided by the element current is the mutual impedance.

MININEC includes a routine to calculate near-field intensities. This is normally considered to be the distance less than twice the square of the antenna dimension divided by the wavelength, or half a wavelength for a λ/2 dipole. However, the equations used in *MININEC* impose no limit on distance. Because of the way the equations are set up, the near field can be used to calculate field strength at a point or along a line, something not possible with the far-field calculation. The technique is useful out to the distance where Earth curvature becomes important.

One use of the near-field calculations is as a check of antenna performance by comparing actual measurements and *MININEC* calculations. A distance of several wavelengths is needed for even reasonably accurate far-field measurements. If this distance is not practical, you can calculate the near-field strength for a distance at which measurements are possible, and compare with your measurements. It is best to make the measurements with a very small pickup loop. This requires calculation of the magnetic field. The alternative is a very short dipole and the electric field strength. This is an area where some experimentation is needed.

Near-field calculations are likely to be more important in the future, as more emphasis is placed on the effect of radiated fields. For example, it could be important to show the field strengths of an amateur station and compare it to that from a local TV, Broadcast or Public Service station. This may help thwart future attempts to restrict Amateur Radio activity by those claiming to do so in the interest of public health.

MININEC has caused many changes in the design of amateur antennas, both home brew and commercial. For example, one of the larger commercial producers has redesigned his amateur line, improving performance over the designs based on antenna range measurements. Because of its relative simplicity of use, ready availability and good accuracy, it seems that this use will continue.

The Big Program—*NEC*

NEC stands for *Numerical Electromagnetic Code*. The first version of the program[7] was developed in the late 1970s, although it derived from an antenna-analysis program developed a few years earlier, about the same time as the Ohio State program listed above. *NEC* is now in its fourth revision, and is presently being validation tested. However, the last two revisions (*NEC3* and *NEC4*) are restricted to military use, so only the first two revisions are available to amateurs. (The third revision may be released for general use "any day now.") The following is based on *NEC2*, with a few indications of changes reported for later revisions.

One difference between *NEC* and the other programs above is the handling of assumed segment current variation. Instead of a single term, *NEC* uses a three-term relation, in the form: constant + sine term + cosine term. This means that *NEC* will give good results with fewer segments than needed for *MININEC*. This also means that large problems can run faster. In fact, because *NEC* provides automatic storage of data on tape when problems are too large for available memory, it can handle antennas far more complex than any of the programs above.

Another difference is that there is an alternative special routine to calculate current on the surface of a wire, rather than assuming that current is concentrated at the wire center. This means that *NEC* can be more accurate for fat wires.

The programs listed above have a single method for exciting an antenna by direct connection. *NEC* has three methods. One is a current source, and two are different methods of modeling voltage sources. There are three possible inputs for incident-wave excitation—linear (horizontal and vertical) and the two senses of circular polarization (right and left). There are some restrictions on use of sources; for example, no mixing of the three basic types, voltage, current or incident wave.

Other than these points, there is little difference between *MININEC* and *NEC* on a lot of amateur antenna problems. Results are essentially identical for, say, a 4-element Yagi located well above earth. There is no reason to abandon *MININEC* for a lot of work. A few tries with *NEC* will show that the simpler program is best for simple problems.

Where *NEC* really shows its value is in going beyond the limits of the other programs. For example, *NEC* will accept input describing wires bent in an arc, or even into helices. Internally, it handles these by simulation with straight segments. This can also be done with *MININEC*, but the process is automatic in *NEC*. Because of its handling of memory, *NEC* can solve large elements or even arrays of helices. (The spiral antenna used in printed circuit antennas is a helix of zero height—*NEC4* can handle logarithmic spirals.)

Another *NEC* extension is the ability to handle surfaces. These must be sections of a flat surface, joining other surfaces at the edge. Three- or four-sided surfaces are possible, singly or as divisions of a large plane surface. Wire-surface junctions can be made, for an accurate analysis of a 2-meter antenna mounted on an auto. This can be simulated with the other programs[8] by wire-grid models, but only crudely, due to memory size restrictions. Dish and horn antennas can be modeled as surfaces. *NEC* can handle very complex surfaces, either directly as solid sheets or as wire grids.

However, the most important feature of *NEC* for amateur use is the ability to accurately model antennas close to earth. Like smaller programs, NEC can model an antenna in free space or over ideal ground. In addition, it can use a reflection-coefficient approximation to ground. This, in essence, multiplies the radiation from the underground or "image" antenna by a factor to account for ground loss.

Finally, *NEC* has the ability to use the very accurate Sommerfeld-Norton ground model. This is based on the work of Sommerfeld and involves table-lookup of data prepared separately by another program. Use of this ground model slows down overall computation rather dramatically—an average of some four times slower than when simpler ground models are invoked. Any one table applies only to the ground condition and frequency specified, so studies of antenna-ground interactions at different frequencies and ground types can be even more time consuming as the special Sommerfeld-Norton ground files are prepared. The advantage, however, is that horizontal wires may be modeled that are very close to the ground—in fact, at heights equal to the diameter of the wire modeled.

Whatever ground model is selected, it applies to ground directly under the antenna as well as at the point of ground reflection. In addition, more ground conditions can be specified, but applying only to the far field. Far-field ground may be in circular or linear zones, with different elevations to simulate hills and valleys. The equations are valid for antennas close to the edge of a cliff. Ground screens can be specified, if desired, and both near and far fields can be calculated.

NEC includes a number of routines to simplify setting up antennas and structures. Symmetry can be used. For example, a rhombic can be specified by one wire and double symmetry. Quasi-circular structures are specified by one face and the number of faces. Arrays are easy, since any antenna or structure can be duplicated at one or more other locations. There are a few other time savers, such as the use of interaction-range approximations for well-separated elements.

Loads can be introduced into elements as series or parallel RCL circuits, as impedances, and/or as distributed wire-loss resistance. Transmission lines may connect point pairs on structures, and two-terminal networks may connect to any pair of structure points. True transmission-line relations are used, much more accurate than can be obtained with parallel wires in other programs. *NEC3* and *NEC4* include routines for insulated wires. *NEC4* can handle sagging wires directly, and includes detection of error-producing overlapping and intersecting wires.

The range of output data in *NEC* is large, huge in fact. The charge on a wire is available, as well as the current. Coupling between elements can be output. Far-field patterns can use an internally generated format, or one specified by the operator. Fixed frequency or stepped operation can be selected, with linear or constant percentage steps. Some intermediate results can be saved to shorten runtime on other similar problems.

The paper output of *NEC* can only be called *verbose*. It is divided into sections, with the input instructions printed first, then the pertinent conditions, followed by the actual output. The runtime of each section is shown, and usually is surprisingly small.

Overall, *NEC* is a powerful program, a considerable extension beyond *MININEC*. But providing this power, plus some other features makes *NEC* difficult to use. It is a huge program. A printout of the source code of programs and subroutines by pages is well over an inch thick. As is common with early mainframe programs, it is written in FORTRAN, and uses the punched card and tape records, and line printers of those computers. Then too, there are many inputs—some 36 card types are needed to cover all features, each have at least one input and some up to four integer and up to seven floating-point values. Setting up a problem is not easy!

One approach to simplifying the work is a program package called *NEEDS, Numerical Electromagnetic Engineering Design Station.*[9] This includes programs to work directly in FORTRAN card format, either new or stored, a program to handle plotting of data output, conversion between *NEC* and *MININEC* formats, plus some other utilities. (The package includes the latest available version of *NEC* and *MININEC*.)

Obtaining Programs

The Kraus-Newman programs are only available as equations and examples. Running through them once with a programmable calculator will save time on additional problems. As noted earlier, these programs are primarily for education.

The Ohio State program is a NASA contractor's report and will be found in depository libraries. It is also available from the National Technical Information Service, on paper and microfiche. The program is in FORTRAN source code form, and requires a FORTRAN compiler to make it usable. However, it is not an extremely long program, some 1000 lines, so the compiled code will run on quite small computers. The paper source copies are not the best, and it is very easy to make mistakes transcribing the material to the computer. The author has prepared a machine copy, currently running on the PC and Amiga. Its main value is handling insulated wires. If you have a real need for this, write to the author, including a formatted diskette and an SASE.

The Pozar programs are available in paper and as PC diskettes; see Note 4. There are a number of other, non moment method programs in the book and on the disk, making either a valuable addition to an antenna-oriented library.

The source for both *NEC* and *MININEC* is the Applied Computational Electromagnetic Society, ACES. This may be obtained as the *NEEDS* package, described above, or as individual executable programs, all with source code. You must become a member to obtain programs from this source. This also brings a newsletter and a journal, and makes many other programs available. Membership in ACES guarantees that you can obtain the latest released program versions.

Although much of the newsletter and journal material is based on complex mathematical techniques, there is also a wealth of practical material. This includes such items as hints on accurate modeling, design of small antenna ranges, graphs of earth characteristics, and reports of bugs and corrections to programs. The material is almost a necessity for any serious worker on antennas.

MININEC is now available commercially in several forms. Several packages have made significant improvements to the original code, allowing inputs from menus and generating on-screen graphs and tables. Several commercial packages include sample antenna files. The cost for these commercial versions varies almost directly with their complexity, but range from about $40 to $100. [Roy Lewallen, W7EL, sells *ELNEC*,[10] his highly customized version of *MININEC*, for $49.—*Ed.*]

There is some claim to have eliminated what seems to be a systematic error in *MININEC*, but the nature of this error is not clear, and the corrections and validation have not been published, to my knowledge.

Other than these factors, the programs are very much easier to use than raw *MININEC3*, a big factor in their popularity. Of course, most commercial programs do not include source code, so changes to customize the program are very difficult. One package is marketed based on keeping the original code, with the source code included. It provides a separate program to prepare the input data and another program to handle output plotting.

NEC is now available commercially. The ads indicate that data input is much easier than that needed by the original program, or even by the *NEEDS* package. Since this is a weakness of the original code, the new packages should help increase the popularity of *NEC*, making its great capability more widely available. [The *NEC for Wires*[11] from Brian Beezley, K6STI, is based on the code kernel from *NEC2*. To this date in the summer of 1994, the ability to handle surface patches has not been implemented. Beezley sells *NEC/Wires 1.5* for $100, or bundled together with *AO* (Antenna Optimizer) for $130.—*Ed.*]

See the ads and announcements in *QST*, other amateur magazines and in the technical journals and newsletters for future changes in availability. This computational field is still developing.

References

[1] John D. Kraus, *Antennas*, 2nd Ed (New York: McGraw-Hill, 1988), pp 384-408.

[2] J. H. Richmond, "Computer Program for Thin-Wire Structures in a Homogeneous Conduction Medium," Report No. NASA CR-2399, Ohio State University, Columbus. Available as N74-28708 from the National Technical Information Service, Springfield, VA 22161.

[3] R. P. Haviland, "Insulated Antennas," *Communications Quarterly*, Vol 3, Number 1, Winter, 1993.

[4] David Pozar, *Antenna Design Using Personal Computers* (Dedham, MA: Artech House, 1985), pp 35-59.

[5] A. J. Julian, J. C. Logan, J. W. Rockway, "*MININEC:* a Mini-Numerical Electromagnetic Code," Tech Doc 516, Naval Ocean Systems Center, San Diego, 1982.

[6] J. C. Logan, J. W. Rockway, "The New

MININEC (Version 3): A Mini-Numerical Electromagnetic Code," Tech Doc 938, Naval Ocean Systems Center, San Diego, 1986.

[7]G. J. Burke, et al, "Numerical Electromagnetic Code (NEC) Method of Moments, a User Oriented Code," Naval Ocean Systems Center, San Diego, 1977. Available as AD-A075 289 and AD-A0460 (Vol 1 and Vol 2), National Technical Information Service, Springfield, VA 22161. Also available from ACES. See Note 9.

[8]R. P. Haviland, "Antenna-Structure Interaction," *Communications Quarterly*, Vol 1, Number 4, Fall, 1991.

[9]*NEEDS*, The Numerical Electromagnetic Engineering Design System Ver 2, 1989, Distributed by The Applied Computational Society. For information, write Dr Richard Adler, Secretary, ACES, Naval Postgraduate School Code EC/AB, Monterey, CA 93943.

[10]*ELNEC*, Advanced Antenna Analysis Program, from Roy Lewallen, W7EL, PO Box 6658, Beaverton, OR 97007.

[11]*NEC/Wires 1.5*, from Brian Beezley, K6STI, 3532 Linda Vista Dr, San Marcos, CA 92069.

Quad Versus Yagi at Low Heights

By Carl Luetzelschwab, K9LA
1227 Pion Road
Ft Wayne, IN 46845

What determines the angle of skip—the antenna or the ionosphere? The answer may help you pick a new antenna.

I continue to read and hear about the performance advantage of a quad antenna over a comparable Yagi antenna when both are mounted at low height. To summarize briefly, the argument begins with the concept of a quad as two stacked dipoles and therefore it has a lower angle of radiation than the Yagi. This lower angle of radiation could result in fewer hops to get to a distant station and therefore less total attenuation.

But this argument, with the resulting conclusion, is not correct. I fell into this trap several years ago when making on-the-air comparisons of a 2-element quad and a 2-element Yagi.[1] After much thought and more analysis of patterns over ground, I realized the error of my ways. Thus the purpose of this article is to review patterns of the quad and Yagi at various heights, and to explain what the differences in the patterns really mean.

Quad and Yagi Design

The first order of business is to design a representative quad and Yagi. For simplicity, I used 2-element versions with identical boom lengths. To make an apples-to-apples comparison, the performance goals over the entire 20 meter band of both the quad and Yagi were taken as:

SWR: less than 2:1
Gain: within ± 0.6 dB referenced to gain at f_C (center frequency)
F/B: peaked at f_C

Both designs employ a driven element and a reflector, with the quad in a square configuration fed at the bottom center. The designs were iterated manually with *MN* Version 1.66.[2] Both designs use the equivalent of 20 segments per wavelength. The resulting free space performance at f_C = 14.175 MHz of the two designs is shown in **Table 1**.

Note that the quad, for the same boom length, has 0.86 dB free space gain over the Yagi. This is in excellent agreement with the 0.25 λ per side results in Table 4.4 of Chapter 4 of *Yagi Antenna Design.*[3]

Free Space Patterns

Fig 1 shows the free space E plane (azimuth) patterns of both the 2-element Yagi and the 2-element quad. **Fig 2** does the same for the H-plane (elevation or vertical) patterns.

Note that the main lobe E-plane patterns are quite identical. However, the quad has a narrower H-plane pattern. This narrowing provides the free space gain of the quad over the Yagi. (The net energy radiated remains the same. Therefore if the gain is slightly higher the pattern will be slightly narrower. *—Ed.*) It also results in pattern differences at low height, as we shall see.

Patterns at High Height

Fig 3 shows the H-plane patterns of both antennas mounted at 70 feet (the height of the quad is the height to the boom). On 20 meters, 70 feet is about 1 λ. Note that the quad has a higher gain than the Yagi at the lower major lobe. At this high height, as was

Table 1
Calculated Quad and Yagi Performance

Peak Antenna	*Gain, dBi*	*F/B dB*	*Max SWR Over 20m band*	*Boom Length*
2-el Yagi	6.16	11	2.0:1	7'5"
2-el quad	7.02	17	2.0:1	7'5"

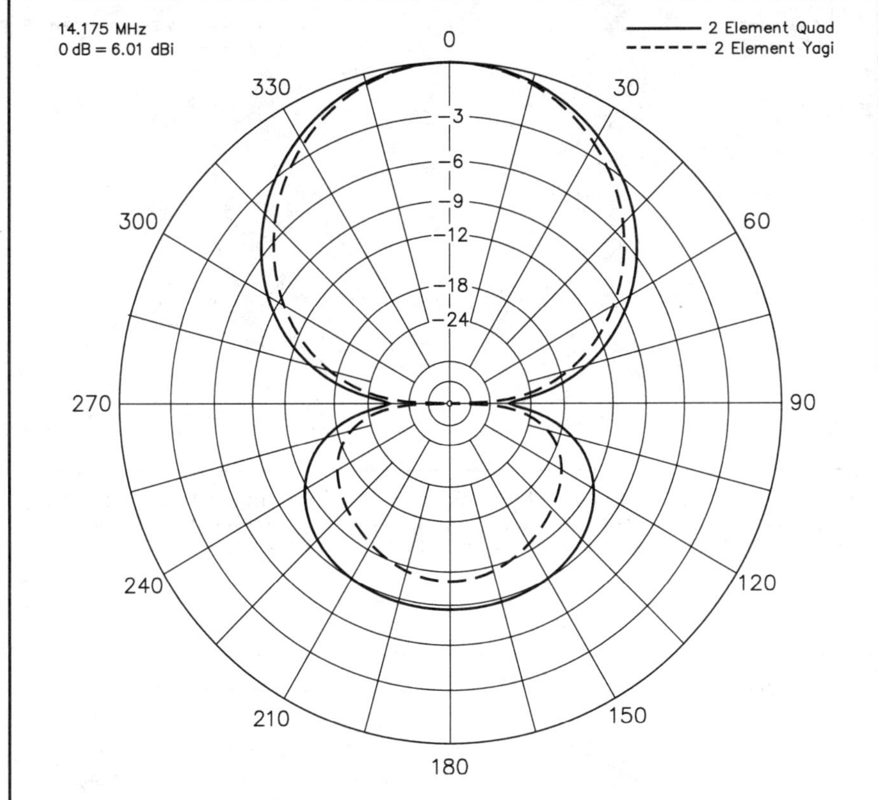

Fig 1—Comparison of quad (solid line) and Yagi (dashed line) E-plane (horizontal) patterns in free space at 14.175 MHz.

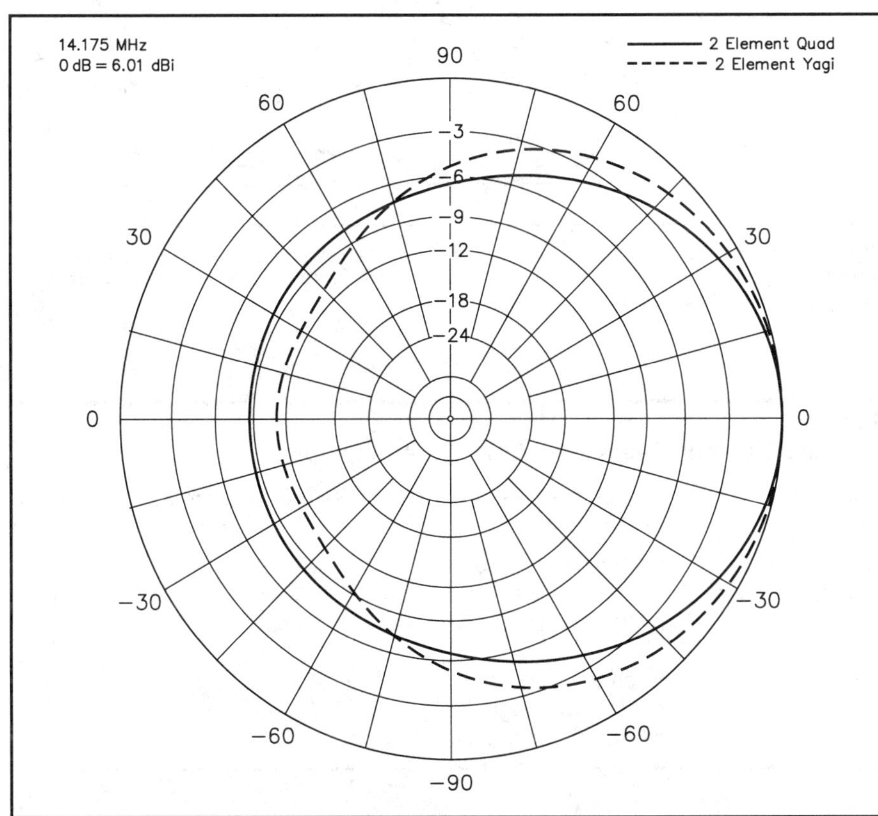

Fig 2—Comparison of quad (solid line) and Yagi (dashed line) H-plane (vertical) patterns in free space at 14.175 MHz.

the case in free space, the lobes peak at the same angles.

Patterns at Low Height

Fig 4 shows the H-plane patterns of both the 2-element Yagi and the 2-element quad when they're both mounted at 30 feet. On 20 meters, 30 feet is just under 0.5 λ.

The lobe of the Yagi occurs at about 30.0°, whereas the quad's lobe occurs at about 27.5°. This is where a basic misunderstanding leads to the erroneous conclusion cited in the beginning of this article. With simple Earth-ionosphere geometry, and assuming radiation only at these two angles, the quad's signal would appear to get to a distant station with fewer hops.

But this analysis is wrong. The above subtly implies (incorrectly) that it is the antenna patterns that determine the path through the ionosphere to a distant station. In other words, somehow, only the energy at 30.0° from the Yagi gets to the distant station, and only the energy at 27.5° from the quad gets to the distant station. From Fig 4 you can see that if the quad's 27.5° energy gets to a distant station, the Yagi's 27.5° energy also gets there. The reverse also holds. If the Yagi's 30.0° energy gets to a distant station, the quad's 30.0° energy also gets there.

What really determines the path through the ionosphere to a distant station is certainly not the antenna pattern, but rather only the distance involved and the state of the ionosphere. An antenna does not have a single angle of radiation. Fig 4 also shows that an antenna radiates over a range of angles. Some radiate over a large range (antennas at low height), while some radiate over a very restricted range of angles (a stack of Yagis, for example). But they all have the same thing in common—they *do not* determine the path through the ionosphere to the distant station.

Two things should now be evident—first the true comparison of two antennas is simply a comparison of their H-plane patterns at the elevation angle dictated by the ionosphere, and second the purpose of an antenna system is to put maximum energy at this angle.[4]

Now let's go back to Fig 4 again. Suppose the ionosphere says a 5° angle is required (or 2.5° or 7.5° or 10.0° or some other low angle—it doesn't matter). What's the difference between the Yagi and quad? Although the figure doesn't quite have the necessary resolution, a detailed evaluation shows that the gain difference varies from slightly less than 1 dB at 1° elevation to close to 0 dB at 32°.

The Correct Interpretation of Fig 4

Remember that in free space and at high heights the patterns of the Yagi and quad peak at the same elevation angles. However, at low heights this does not occur. If antenna A has X dB free space gain over antenna B, then at low elevation angles (the ones important to DX) antenna A will exhibit this X dB gain advantage over antenna B regardless of height. The only difference might be a small deviation due to the variation of apparent antenna impedance due to height. The gain relationship of two antennas at low elevation angles is maintained *regardless of height*. Nothing magical happens with a quad at low height. This relationship is summarized in words in **Table 2.**

This phenomenon necessarily comes out of the math when multiplying the free space H-plane pattern by the ground reflection factor to get the pattern over ground as seen in Fig 4. The narrower free space H-plane pattern of the quad (Fig 2) is why the pattern over ground peaks at a slightly lower angle.

This phenomena is not restricted to quads. It is a function of gain. To prove this, **Table 3** is a summary of the data of Table 5.2 and Table 5.3 of *Yagi Antenna Design*. It shows the peak of the main lobe of a 3-element Yagi versus a 6-element Yagi at various heights (in wavelengths) above ground.

The 6-element Yagi has 2.44 dB free space gain over the 3-element Yagi (10.71 dBi versus 8.27 dBi). At 0.50 λ, the 6-element Yagi's lobe peaks 4° lower than the 3-element Yagi. This makes sense compared to my 2-element results, as the higher free-space gain difference in this case necessitates a greater difference in the angle of the peak of the lobes at low elevation angles in order to maintain the free-space gain relationship. Up at 1.00 λ, the lobes are aligned as they should be.

I carried this concept one step further. I modified my 2-element Yagi (by shortening the reflector length) so it had the same free space gain as the 2-element quad (at the sacrifice of SWR bandwidth and F/B). When both antennas were modeled at 30 feet, the peaks of the lobes were very close. They would never be exactly the same, because although *gain* is the major contributor, the *shape* of the H-plane pattern is a second-order effect.

Summary

A higher gain antenna will necessarily have its lobe peak at a lower angle at low heights, but do not confuse this with the erroneous concept of an antenna's *angle of radiation*. An antenna does not have a single angle of radiation. When you see a plot comparing the angles of radiation of two antennas versus height, realize the divergence at low heights is simply what must happen to maintain the free-space gain relationship, and says nothing about the path through the ionosphere.

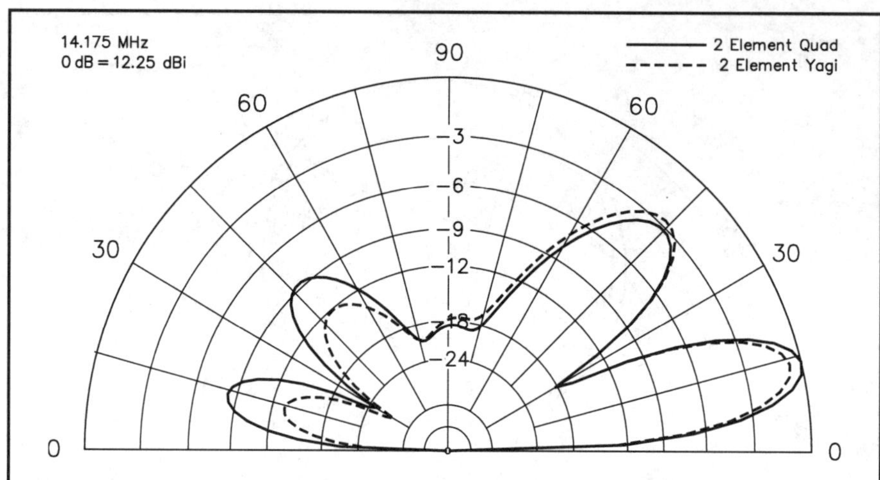

Fig 3—Elevation patterns at a 70-foot height over real ground on 14.175 MHz. The quad (solid line) has about 1 dB more gain than the Yagi (dashed line) but the peaks of both patterns occur at about the same elevation angle.

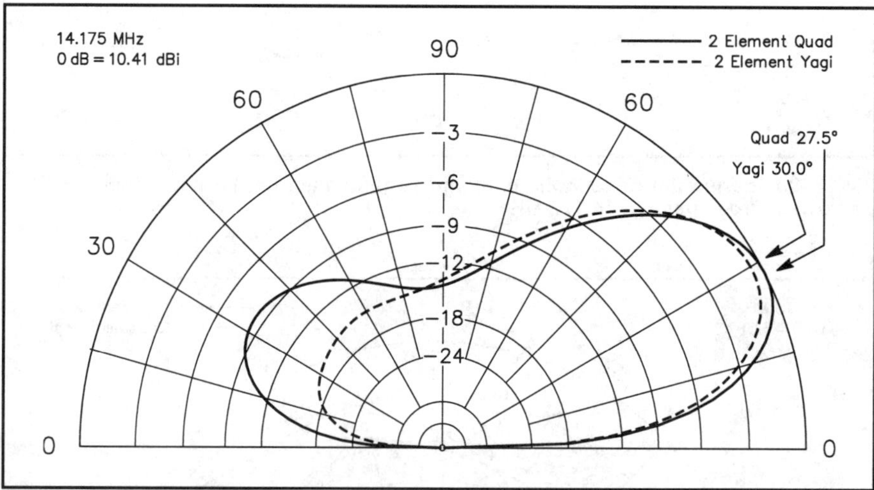

Fig 4—The discussion centers on this elevation pattern difference. At a 30-foot height the peak of the quad's response (solid line) is about 2.5° lower than the peak of the Yagi (dashed line).

Table 2
Antenna Gain and Beam Peak Angle Relationships

Antenna Height	*Angle of the Peak of the Beam*	*Gain Difference at Low Elevation Angles*
High (and free space)	Both antenna peak angles are the same (with X dB gain difference)	X dB
Low	Each antenna changes peak beam angle differently with height	X dB

Table 3
Comparison of 2 Yagi Designs at Varying Heights[5]

Height	*3-Element Yagi*		*6-Element Yagi*	
λ	*Gain, dBi*	*Angle*	*Gain, dBi*	*Angle*
0.10	9.82	58°	11.92	46°
0.25	11.08	41°	13.00	33°
0.50	13.19	27°	14.85	23°
0.75	13.84	19°	15.6	18°
1.00	14.04	14°	16.36	14°
Free space	8.27		10.71	

The purpose of this article was not to disparage quads. Quads are good antennas and they have their place in Amateur Radio. I believe a properly designed and implemented 2-element quad, can give a 2-element and 3-element Yagi (especially a trapped tribander) a run for their money. But don't put a quad up thinking that the laws of physics and mathematics will somehow be defied at low heights!

Notes

[1]R. C. Luetzelschwab, K9LA, "Antenna Angle of Radiation Considerations," *Communications Quarterly*, Summer 1991 and Fall 1991.

[2]These designs were done several years ago, and now can be done easier with *MN* versions that automatically optimize for specified criteria. Brian Beezley, K6STI, offers a set of such programs. Contact him at 3532 Linda Vista Drive, San Marcos, CA 92069.

[3]Dr. J. L. Lawson, W2PV, *Yagi Antenna Design* (Newington: ARRL, 1986).

[4]For a good analysis of the required angles versus time of day, season, sunspot cycle, and QTH, see R. D. Straw, N6BV, *All the Right Angles*, (New Bedford, PA: LTA, 1993).

[5]Very close to the ground (under 0.2 λ) some models are not totally accurate. See R. Lewallen, W7EL, "*MININEC*: The Other Edge of the Sword," *QST*, Feb 1991, pp 18-22.

An Improved Double Extended Zepp

By Rudy Severns, N6LF
32857 Fox Lane
PO Box 589
Cottage Grove, OR 97424

N6LF revisits the classic double extended Zepp to improve pattern and SWR bandwidth. He also offers some sage advice about computer modeling.

The sunspot cycle is headed down and the low bands are coming to life for DXing. The next several years will be the time to make those low-band contacts for DXCC. As always, the key to low-band operation is a good antenna. Unfortunately for 7 MHz and down, good antennas don't come in a box ready to assemble. Every location will have a unique set of limitations and opportunities.

One very important difference between this sunspot minima and all past ones is the availability of inexpensive, easy-to-use and powerful antenna modeling software.[1,2] This software allows you to design and optimize an antenna that exactly fits your situation and pocketbook. While cut-and-try experimentation is a very slow way to optimize antennas, modeling is so quick that a wide range of solutions can be investigated easily. The real problem with modeling is generating the will power to stop fooling with the variations and go out and build something!

The following article uses 40 and 80/75-meter double extended Zepps (DEZepp) as examples of what you can accomplish. By adding two small capacitors, made from short lengths of RG-8, in just the right place, the pattern can be improved and the driving-point impedance changed from reactive and narrowband to resistive and wideband. This allows the antenna to be used without a tuner and with an SWR < 1.5:1 over the entire 40-meter band, or with SWR < 2:1 over the entire 75/80-meter band.

A Look at the Classical DEZepp

The classical DEZepp is simply a piece of wire 1.25 λ long, fed at the center, usually with open-wire transmission line and a tuner at the transmitter. The DEZepp displays a useful amount of gain over a dipole of approximately 3 dB. The radiation pattern for a DEZepp designed for 7.15 MHz and suspended 80 feet above ground is shown in **Fig 1**, along with the pattern for a λ/2 dipole at the same height for comparison. The elevation angle is 26°, the peak of the main lobe. The current distribution along the antenna is shown in **Fig 2**.

The DEZepp does indeed provide gain over the dipole, but only over the relatively small angle of approximately 40°. The beamwidth between 3 dB points is 35°. Unless the antenna is pointed directly toward the receiving station, the gain is not usable due to the narrow beam width. In addition to the narrow main lobe, there are significant sidelobes. These are not big enough to be helpful in those directions, but they will also certainly pick up noise and interference. The impedance of the antenna is very reactive, and even when matched at midband does not allow the entire band to be covered without retuning.

For this reason, the DEZepp has traditionally been used with an antenna tuner. This is

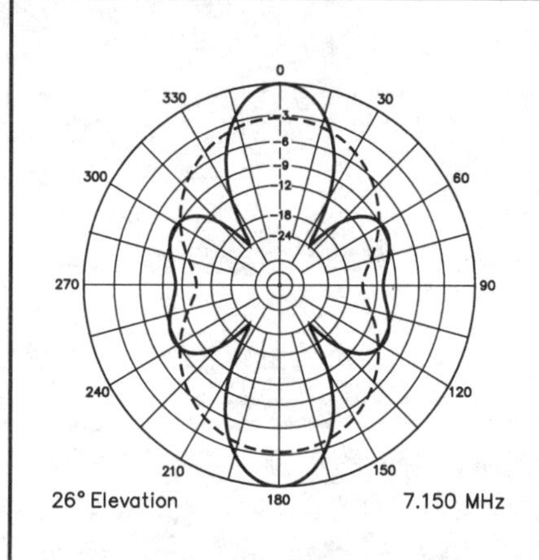

Fig 1—Azimuth pattern of classic double extended Zepp (solid line) at 7.15 MHz, compared with standard dipole (dashed line), both 80 feet high over average ground. Patterns are shown at 26° elevation, where the gain is maximum. The wire runs along the 270° to 90° axis on the graph. Note significant sidelobes for DEZepp.

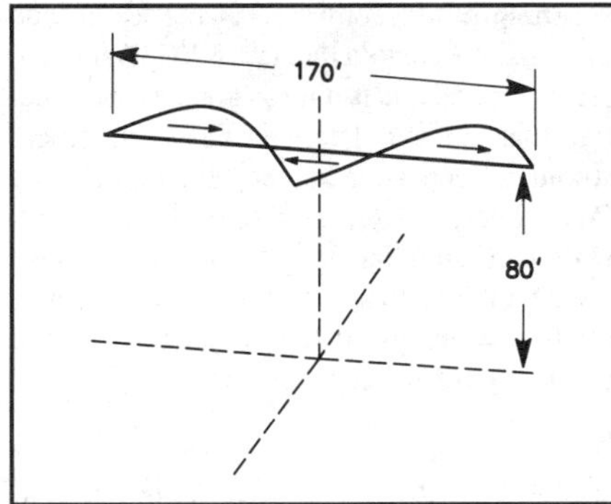

Fig 2—Schematic for classic DEZepp, showing current distribution along antenna. The "bulging out" of the current in the opposite direction near the center of the antenna is responsible for the sidelobes seen in Fig 1.

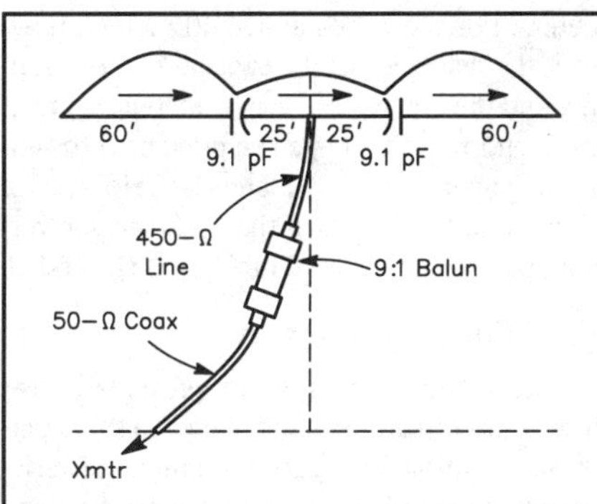

Fig 3—Schematic for modified N6LF DEZepp, with new current distribution. Overall length is 170 feet, with 9.1 pF capacitors placed 25 feet each side of center. Now current distribution doesn't create sidelobes.

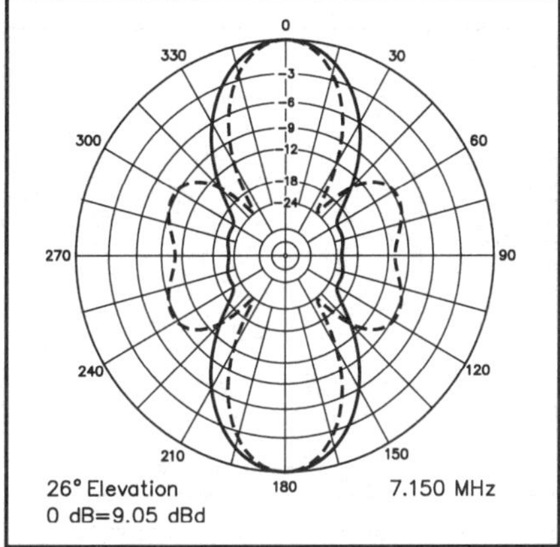

Fig 4—Azimuth pattern for N6LF DEZepp (solid line), compared to classic DEZepp (dashed line). The main lobe for the modified antenna is slightly broader than that of the classic model, and the sidelobes are suppressed better.

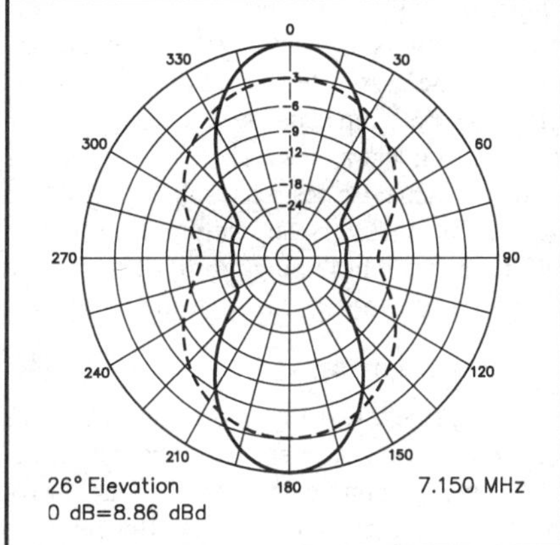

Fig 5—Azimuth pattern for N6LF DEZepp (solid line), compared to dipole (dashed line) at the same height.

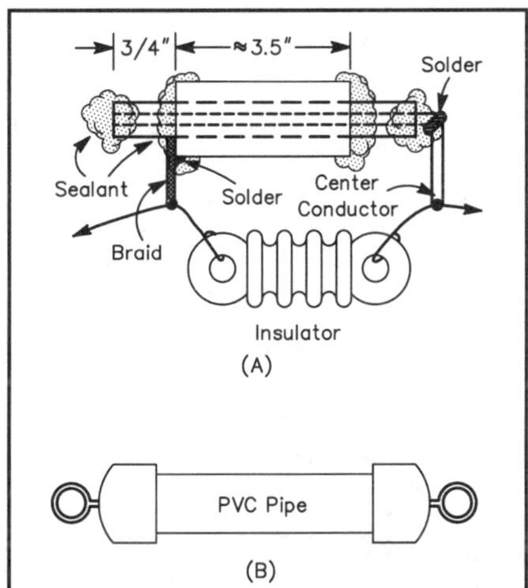

Fig 6—Construction details for series capacitor made from RG-213 coaxial cable. At A, the method used by N6LF is illustrated. At B, a suggested method to seal capacitor better against weather is shown, using a section of PVC pipe with end caps.

not a terrible hardship but it would be nice if the tuner could be eliminated, at least on one band, and a low SWR presented to the transmitter over a whole band.

The gain displayed by the DEZepp is due to the separation between the two current maxima. The small inverted current in the center section subtracts a little from the main lobe and contributes to the sidelobes. The DEZepp is essentially two end-fed collinear dipoles. The transmission line and the center portion of the antenna are the feed system.

It would be very beneficial to suppress the sidelobes and put that energy into a broader main lobe, retaining most of the gain if possible.

A Modified DEZepp

The key to modifying the radiation pattern is to modify the current distribution. One of the simplest ways to do this is to insert a reactance(s) in series with the wire. This could either be an inductor(s) or a capacitor(s). In general, a series capacitor will have a higher Q and therefore less loss. With either choice it is desirable to use as few components as possible.

As an initial trial I decided to use only two capacitors, one on each side of the antenna. I varied the value and position of the capacitors to see what would happen. It quickly became clear that I could tune out the reactance at the feedpoint by adjusting the capacitor value, making the antenna look like a resistor over the entire band. The value of the feed-point resistance could be varied from less than 150 Ω to over 1500 Ω by changing the location of the capacitors and adjusting their values to resonate the antenna. The *AO 6* (Antenna Optimizer) software[1] has the nice feature that it will automatically adjust a variable to tune out reactance. Simultaneously, the pattern was also changing in useful ways.

A number of interesting combinations were created. The one I elected to use is shown in **Fig 3**. The antenna is 170 feet in length. That is a couple of feet shorter than the classic DEZepp, but that also just happens to be all the distance I had between my supporting trees! Two 9.1 pF capacitors are located 25 feet out each side of the center. The antenna is fed with 450-Ω transmission line and a 9:1 three-core Guanella balun[3] used at the transmitter to convert to 50 Ω. The transmission line can be any convenient length and it operates with a very low SWR.

That's all there is to it. The radiation pattern, overlaid with that for a standard DEZepp for comparison, is shown in **Fig 4**. A comparison to a standard dipole is shown in **Fig 5**. The sidelobes are now reduced to below 20 dB. The main lobe is now 43° wide at the 3-dB points, as opposed to 35° for the original DEZepp. The antenna has gain over a dipole for > 50° now. The gain of the main lobe has dropped only 0.2 dB below the original DEZepp.

The reason for the pattern change can be seen in Fig 3, showing the modified current distribution. The main current maxima are still pretty much in the same place, but the current in the center of the antenna now flows in the opposite direction. The resulting pattern is much cleaner.

Experimental Results

I managed to pry myself away from the computer and actually build the antenna. It was made from #14 wire and the capacitors were made from 3.5-inch sections of RG-213, shown in **Fig 6A**. Note that great care should be taken to seal out moisture in these capacitors. The voltage across the capacitor for 1.5 kW will be about 2000 V so any corona will quickly destroy the capacitor. One of the nice features of modeling software is that it gives the current amplitude along the antenna, making it easy to determine the stresses on any series reactances.

I used silicon sealant and then covered both ends with coax seal, finally wrapping it with plastic tape. The solder balls indicated on the drawing are to prevent wicking of moisture through the braid and the stranded center conductor. This is a small but important point if long service out in the weather is expected. An even better way to protect the capacitor would be to enclose it in a short piece of PVC pipe with end caps, as shown in Fig 6B.

Note that all RG-8 type cables do not have exactly the same capacitance per foot and there will also be some end effect adding to the capacitance. I trimmed the capacitor with a capacitance meter. It isn't necessary to be too exact—I checked the effect of varying the capacitance ±10% and the antenna still works fine.

The results proved to be close to those predicted by the computer model. **Fig 7** shows the measured value for SWR across the band. These measurements were made with a Bird directional wattmeter. The worst SWR is 1.35:1 at the low end of the band! With a little adjustment of the antenna length this could have been lowered a bit more, but I figured why bother?

My antenna was oriented to work into Europe. Prior to putting up this antenna I had been using a dipole. I could hear a few Europeans but was unable to work them. Three dB may not seem like much gain but after putting up this antenna I immediately heard many more signals and have been regularly working into Europe with 56/57 reports.

Dick Ives, W7ISV, was sufficiently impressed by the success of the 40-meter version of this antenna to ask me to design a 75-meter version for him. In his location one end of the antenna could only be 60 feet high (< 0.25 λ), and I was concerned about the accuracy of the modeling program, because *MININEC*-based programs are known to be inaccurate for gain and feed-point impedance at low heights. Fortunately, Brian Beezley, K6STI, has a *NEC*-based program called *NEC Wires.*[1] This does model ground accurately and is just the ticket for low antennas. Using this program I designed a new antenna for W7ISV.

Despite the temperatures in mid-December, Dick erected the antenna as shown in **Fig 8**. The series capacitors are 17 pF, and since he isn't interested in CW, Dick adjusted the length for the lowest SWR at the high end of the band. The antenna could have been tuned somewhat lower in frequency and would then provide an SWR < 2:1 over the entire band, as indicated by the dashed line in Fig 8.

This antenna provides wide bandwidth and moderate gain over the entire 75/80-meter band. Not many antennas will give you that with a simple wire structure.

Multiband Operation

When operated with an antenna tuner, one of the advantages of the classical DEZepp is that it is a multiband antenna. Typically a 40-meter DEZepp behaves like a dipole on 75/80 meters and like a long wire on the higher frequency bands. Adding the two series capacitors decouples the ends of the wires on 75/80 meters and a rather poor antenna results. It behaves more like a 30-meter dipole being used on 75/80 meters. For the bands above 40 meters, however, the reactance of the capacitors drops rapidly and the behavior is very much the same as for the normal DEZepp. The price paid for improving operation on 40 meters is the loss of 75/80 meters. Similarly, in the 75/80-meter version, performance on 160 meters is sacrificed.

Some Final Thoughts

The antenna shown here represents a very simple modification of an old idea to suit a particular situation. There are any number of variations that could have produced similar results. Two important lessons were learned during this effort. First, the modeling software is pretty accurate, particularly now that *NEC*-based software is available. The results obtained were very close to that predicted—and this is not the first time I have seen this. Second, the modeling process is a great teacher. It helps you to learn how antennas really work and cuts through many misconceptions. By viewing the current distributions, the associated radiation patterns and driving-point impedances, it becomes much easier to understand which way to modify a design to achieve a desired result. Being able to get results quickly is very helpful also.

There is a whole new world of low-band antennas out there waiting to be created!

Acknowledgments

I want to express my appreciation to Dick Ives for braving the wet and cold to verify the 75/80-meter version of this antenna in time for me to add it to this article. Brian Beezley was very helpful with comments on the design, and of course he created the new *NEC Wires* program. An antenna is only as good as the materials that go into it. I used the beautiful "silky" stranded copper-clad steel wire supplied by Press Jones, N8UG, and also some of his 450-Ω transmission line. It was wonderful to have a wire antenna that didn't try to envelop you (and every nearby bush) like a boa constrictor due to a built-in set in the wire.

Notes

[1] *AO 6.1* and *NEC Wires*, by Brian Beezley, K6STI, 3532 Linda Vista Dr, San Marcos, CA 92069.

[2] *ELNEC 3.0*, by Roy Lewallen, W7EL, PO Box 6658, Beaverton, OR 97007.

[3] Jerry Sevick, W2FMI, *Transmission Line Transformers*, Second Edition (ARRL, Newington, CT, 1990), p 9-28.

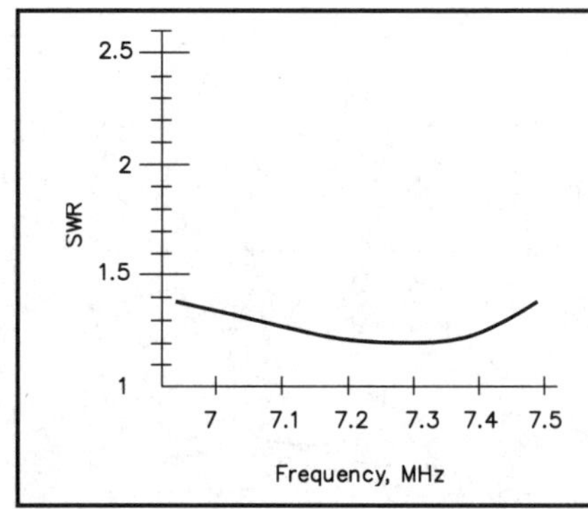

Fig 7—Measured SWR curve across 40-meter band for N6LF DEZepp.

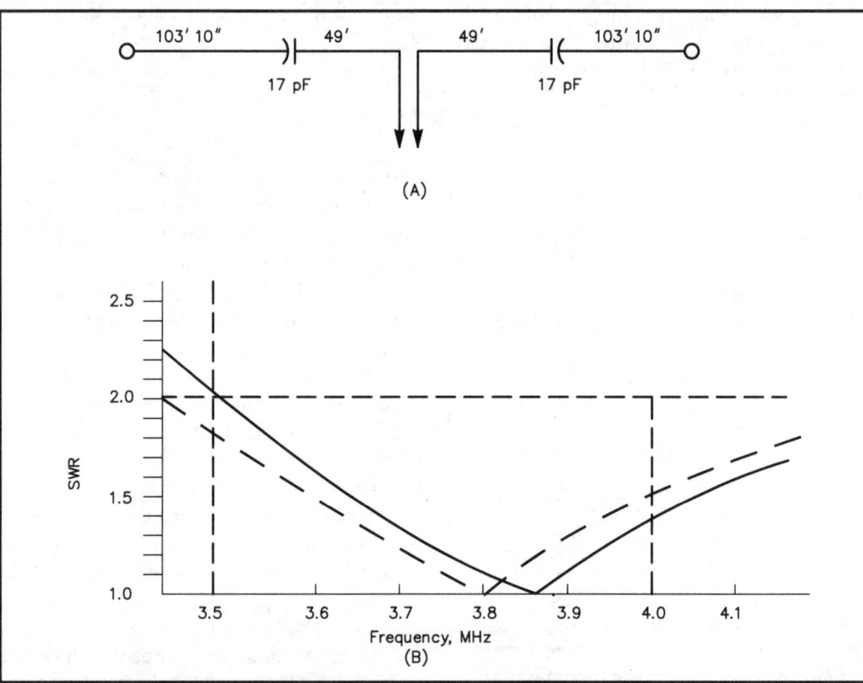

Fig 8—75/80-meter N6LF DEZepp, designed using *NEC Wires*. At A, a schematic is shown for antenna. At B, SWR curve is shown across 75/80-meter band. Solid line shows measured curve for W7ISV antenna, which was pruned to place SWR minimum higher in the band. The dashed curve shows the computed response when SWR minimum is set to 3.8 MHz.

Mobile Antennas

Short Coil-Loaded HF Mobile Antennas: An Update and Calculated Radiation Patterns

By John S. Belrose, VE2CV, ARRL TA
17 Tadoussac Drive
Aylmer, QC
J9J 1G1, Canada

There is more to consider than appearance and convenience when installing your HF mobile antenna, since the frame and the body of the vehicle are a part of the radiating system.

Introduction

It has been known for some time that the efficiency of an electrically short whip is improved when the tuning coil is moved from the monopole base (*base fed*) and located in series with the monopole itself. The coil is then called a *loading coil*. The earliest paper on inductively loaded HF mobile antennas is by Belrose,[1] who analyzed the antenna as an opened-out transmission line. Although this analysis is only approximate, the trends are correct. The results obtained agree rather well with the more rigorous analysis by Hansen,[2,3] who analyzed the inductively loaded antenna by numerical moment-method techniques. Both authors showed that maximum efficiency for a monopole of height h occurs at a load point approximately 0.4 h from the feed.

The author's early (simple) analysis assumed a linear current distribution on the antenna, an assumption that limits the applicability of the method. Later this restriction was removed (see Note 4 and Annex A) by assuming a sinusoidal current distribution on the antenna. The modified analysis can be used to calculate the radiation resistance with reasonable accuracy for heights up to about 0.2 λ.

However, the analyses described above neglected the effect of the mounting structure on antenna performance. From personal experience, the author has recognized that received signal strengths using a rear-bumper-mounted HF mobile whip on the 20 meter band can be up to two S-units (about 10 dB) stronger when the vehicle is heading in the direction of propagation. It is clear that currents on the frame and body of the vehicle have a profound influence on the performance of the antenna.

In this article I summarize results of recent measurements, and of a detailed numerical modeling study employing *MININEC*,[5] and *NEC-2*,[6] for an HF center-loaded mobile antenna bumper mounted on a GMC Jimmy truck. The study begins with the calculated performance for a ground-mounted HF center-loaded monopole. These results can be compared with the earlier simple method of analysis.

Ground Mounted

Three types of electrically short monopoles are sketched in **Fig 1**. Here we are concerned with the inductively loaded monopole in Fig 1B. The monopole height is $h = h_1 + h_2$ and d_1 and d_2 are the average diameters of each section. These antenna dimensions can be measured in feet, inches, meters or millimeters. The author uses mm in his detailed numerical models.

The first case study is for a 110-inch center-loaded mobile monopole. Here, h = 2794 mm; $h_1 = h_2 = 1397$ mm; $d_1 = 5.55$ mm; and $d_2 = 15.9$ mm. The inductance L_0 is adjusted so that the base input impedance of the center-loaded whip is purely resistive; that is, the antenna is resonant. To do this by numerical modeling using *MININEC* (*ELNEC* or *MN* version) proceed as follows:

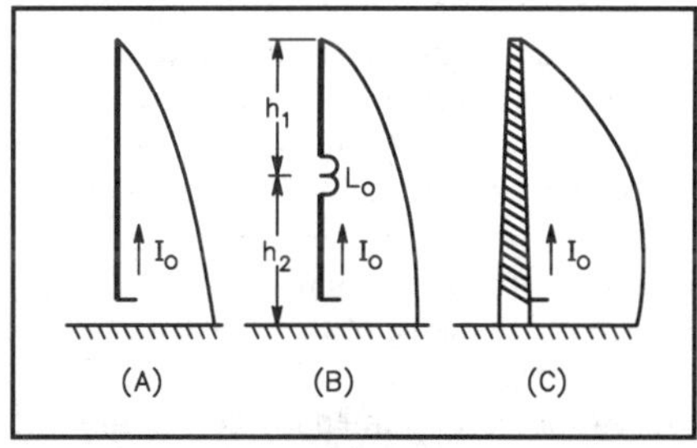

Fig 1—Electrically short monopoles: at A, base-loaded; at B, inductively loaded; and at C, continuously loaded.

Table 1

A Comparison between the Simple Analysis (Sinusoidal Current Distribution) and *ELNEC* for the Case Study Antenna Ground Mounted

	Frequency MHz	R_r Ω	X_C Ω
Simple Analysis	3.75	1.15	*j* 2922
Ground Mounted	7.15	4.22	*j* 1598
	14.15	15.40	*j* 807
ELNEC	3.75	1.20	*j* 3020
Ground Mounted	7.15	4.19	*j* 1520
	14.15	14.80	*j* 631

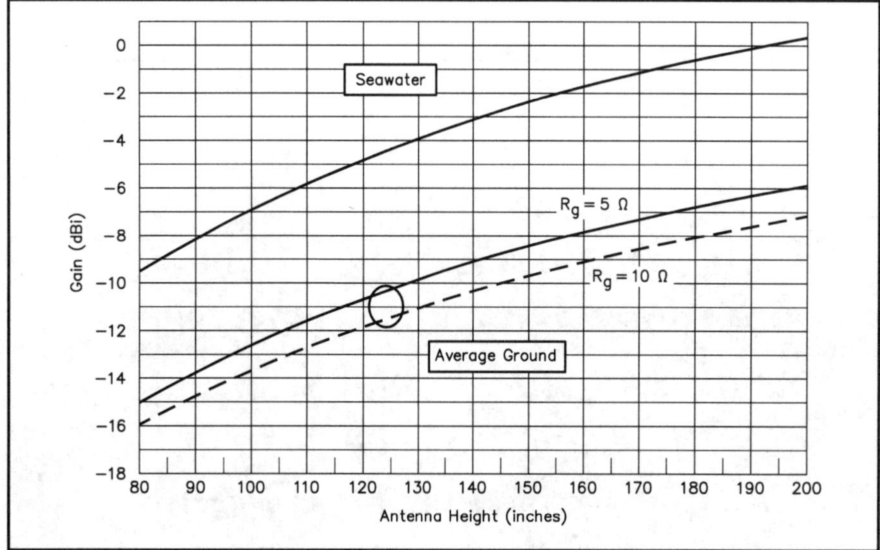

Fig 2—Gain in dBi of a coil-loaded HF mobile antenna versus antenna height, where the antenna is over seawater and then over ground of average conductivity. The frequency is 3.75 MHz (80 meters), and h_2/h = 0.4.

1) Place the load and the source at the junction between h_1 and h_2, and set the load impedance to zero (zero reactance and zero resistance).
2) Calculate the source impedance. For our mobile antenna at 3.75 MHz, $X_s = -j$ 3055 Ω. This is the reactance to be canceled by the inductive reactance of the loading coil L_0.
3) Set the load reactance to the required value (+ *j* 3055 Ω).
4) Place the source generator at the base of the antenna, and calculate the source impedance for zero loss resistance. Since all loss resistances (coil loss resistance R_c, ground loss resistance R_g and conductor loss resistance) are zero, the source resistance is the radiation resistance R_r referenced to the base of the antenna (for our case study R_r = 1.2 Ω).
5) Now include assumed losses. For the coil loss assume a Q-factor = 300; for the ground loss resistance assume for the present example R_g = 12 Ω, see below; and for the conductor loss enter the appropriate conductor material (aluminum assumed).
6) Calculate input impedance (R_a = 26 Ω). Radiation efficiency, antenna gain and bandwidth can be calculated.

The radiation efficiency of our 110-inch center-loaded whip at 3.75 MHz is

$$\eta = \frac{R_r}{R_r + R_c + R_g}$$

$$= \frac{R_r}{R_a} = \frac{1.2}{26} = 0.046 = 4.6\%$$

The calculated values compared with the analysis assuming a sinusoidal current distribution are tabulated in **Table 1**. For estimation of radiation efficiency and gain, a ground loss resistance of 12 Ω corresponds to measured values for an 80-meter HF mobile antenna on an automobile over average ground. For a ground-mounted antenna, this might correspond to a stake ground. For 7.15 and 14.15 MHz, the ground loss resistances for this case study are 9 Ω and 29 Ω; see below.

Effect of Real Ground on Antenna Performance

The gain of a monopole antenna over real ground is dependent on the conductivity of the ground beneath the antenna (which affects the radiation efficiency of the antenna), and in front of the antenna (which affects the vertical plane pattern of the antenna). When the conductivity of the ground is finite the antenna's performance is seriously affected for elevation angles less than 15°, angles which are comparable with the most probable arrival angle for HF skywaves from distant stations.

In the discussion above we implicitly assumed that the radiation efficiency for an electrically short monopole at a low height over real ground is determined by the radiation resistance divided by the total antenna resistance. The total antenna resistance is the sum of the radiation resistance (R_r), the ground loss resistance (R_g), the coil loss resistance (R_c) and the antenna's conductor loss resistance. Except for electrically small loop antennas, the conductor loss resistance is small with respect to the other resistances. The ground beneath the antenna introduces a loss resistance, which is a function of the frequency, the height of the antenna above the ground and the ground conductivity. The smaller the antenna the higher its capacitive reactance, the larger the tuning coil required for resonance, and the greater the loss associated with the coil-loss resistance. The gain of a short antenna therefore decreases as the frequency, height or length of the antenna decrease.

In **Fig 2** we have plotted the maximum skywave gain for an electrically short coil loaded antenna, computed by *ELNEC*, as a function of antenna height or length of the antenna. This is for antennas over seawater and over average ground (conductivity, σ = 3 mS/m, dielectric constant, ε = 13). The frequency is 3.75 MHz (80 meter), and h_2/h = 0.4. We have further assumed that R_g = 0 for the case of an antenna over seawater. Clearly, while the ground loss resistance is an important parameter, a change of ground loss resistance from 5 Ω to 10 Ω results in a decrease in gain by only 1 dB. A most important parameter is the finite conductivity of the ground in front of the antenna, which affects the ability of the antenna to launch skywaves.

Computer programs like *NEC* and *MININEC* express antenna gain with respect to an isotropic radiator (dBi). For electrically short vertically polarized antennas, a better reference might be either radiation efficiency (since power radiated can be calculated) or gain with respect to a well-grounded quarter-wave monopole. Both the antenna under study and the reference antenna are equally influenced by the conductivity of the ground in front of the antenna.

Vehicular Antenna—Ground Loss Resistance

Ground loss resistance is a function of vehicle size (electrical size in wavelengths), placement of antenna on vehicle (according to Brown[7]), and conductivity of the ground over which the vehicle is traveling. Newer, smaller vehicles with front-wheel drive do not have a complete frame. Undoubtedly they provide an inferior ground plane compared to older, full-size vehicles with rear-wheel drive. More than 40 years ago the author [previously referenced 1953 *QST* article] estimated that the ground loss resistance R_g for a 3.8 MHz whip on his 1940 Dodge was about 12.5 Ω.

The author has recently measured the ground loss resistance for a bumper-mounted Swan Model 45 HF mobile whip (one that had never been used) mounted on his 1992 GMC Jimmy truck. That is, the antenna was "new" and the truck was not rusted. A home-brew low-capacity bumper-mount base insulator was used. The antenna's input resistance at resonance was measured using modern instrumentation—a Hewlett-Packard Impedance/Gain/Phase Analyzer Model 4194A, configured to measure balanced impedance using an HP supplied balun. One terminal of the balanced input was connected to the antenna; the other terminal to the vehicle ground. The effect of the balun and the short lead length to reach the antenna were calibrated out, so the antenna's input impedance was measured with the vehicle in effect isolated from instrument and power ground.

To estimate ground loss resistance the antenna's input impedance at resonance was calculated using *ELNEC* and a wire grid model to simulate the vehicle (see below), for an assumed coil Q-factor of 300, and no ground loss resistance. The Swan Model 45 whip employs a large air-wound coil. The difference between the measured value of R_a (at resonance), which includes ground loss, and the calculated value assuming $R_g = 0$, gives the estimated value of ground loss resistance. With the whip's band selector switch set at the 3.8 MHz tap, the resonant frequency was 3.77 MHz, and the impedance at resonance $R_a = 22$ Ω. The calculated antenna resistance $R_r + R_c = 9.7$ Ω. Hence $R_g = 12.3$ Ω.

For 7.077 MHz (the antenna's resonant frequency on the 40-meter band) the antenna's resistance R_a measured is 16.4 Ω. The calculated antenna resistance $R_r + R_c = 7.55$ Ω. Hence $R_g = 8.8$ Ω.

For 14.79 MHz (the antenna's resonant frequency on the 20-meter tap position) the antenna's resistance R_a measured is 46.6 Ω. The calculated antenna resistance $R_r + R_c = 14.5$ Ω. Hence $R_g = 32$ Ω.

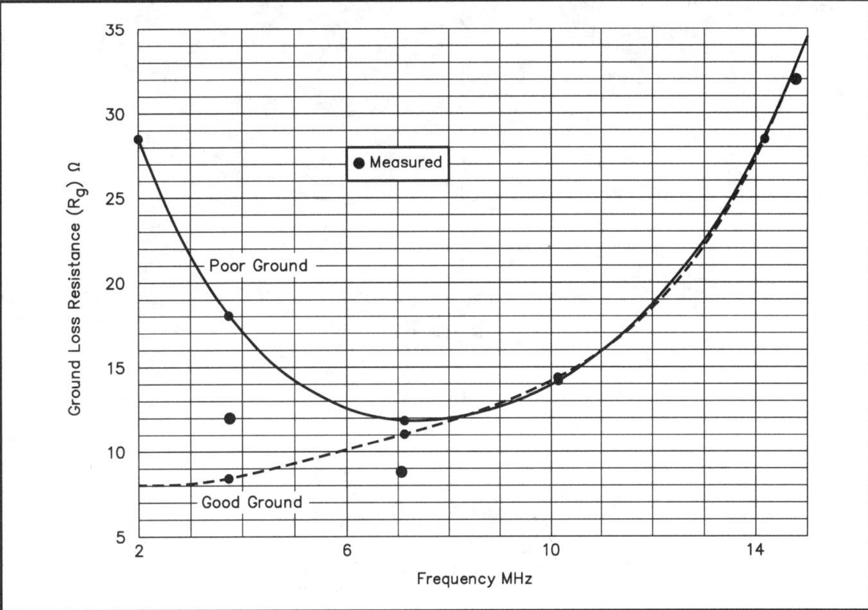

Fig 3—Ground loss resistance (R_g) measured (see text), and calculated using *NEC-2* for an electrically short HF mobile antenna on the basic frame of the vehicle, for two ground conductivities.

For a whip mounted on the basic frame of our vehicle (see below) we have calculated the ground loss resistance using *NEC/Wires 1.5* for good (σ = 5 mS/m, ε = 13) and for poor (σ = 1 mS/m, ε = 5) ground. The results of this analysis, and the measured values are plotted in **Fig 3**. This figure shows two interesting features: (1) There is a rather good agreement between the measured and calculated values for the ground loss resistance; and (2) The type of ground has a strong influence on the ground loss resistance for frequencies below about 7 MHz. For frequencies above about 7 MHz, the type of ground (good or poor) has little or no effect.

Vehicular Antenna—Radiation Pattern

First we will consider an inductively loaded antenna fed against the frame of the vehicle (see **Fig 4A**). The reason for doing this is two-fold. First, no bumper bracket is needed—a bumper bracket is a very short wire and this may give trouble with the *MININEC* model. Second, the return currents that flow on the basic frame are orthogonal to the current on the antenna. Radiated fields due to these return currents therefore cannot destructively interfere with radiation due to current on the whip. We will further discuss this below.

The currents on the wire model are shown in **Fig 5A**. The radiation patterns, elevation pattern for azimuth angle φ = 337° (the azimuth of maximum gain), and the principle plane azimuth pattern (elevation angle of 31° for maximum gain) for 3.75 MHz are

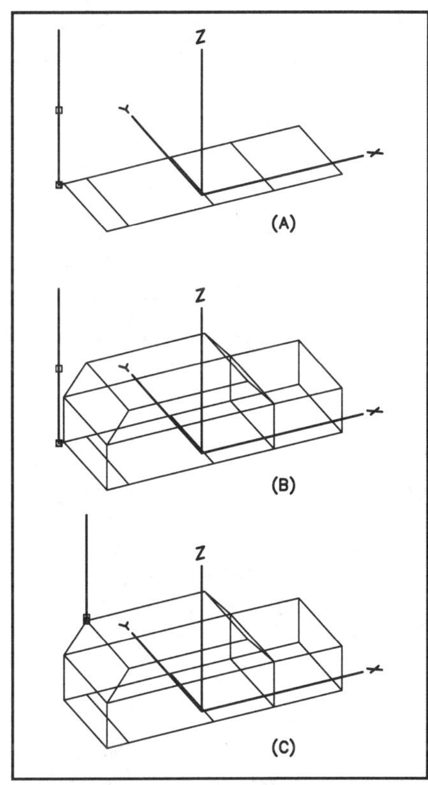

Fig 4—Wire models of a center-loaded HF mobile antenna: at A, rear left-bumper mount on the basic frame of a motor vehicle; at B, on a GMC Jimmy truck; and at C, on the left-rear corner of the roof (whip tip height the same as for B).

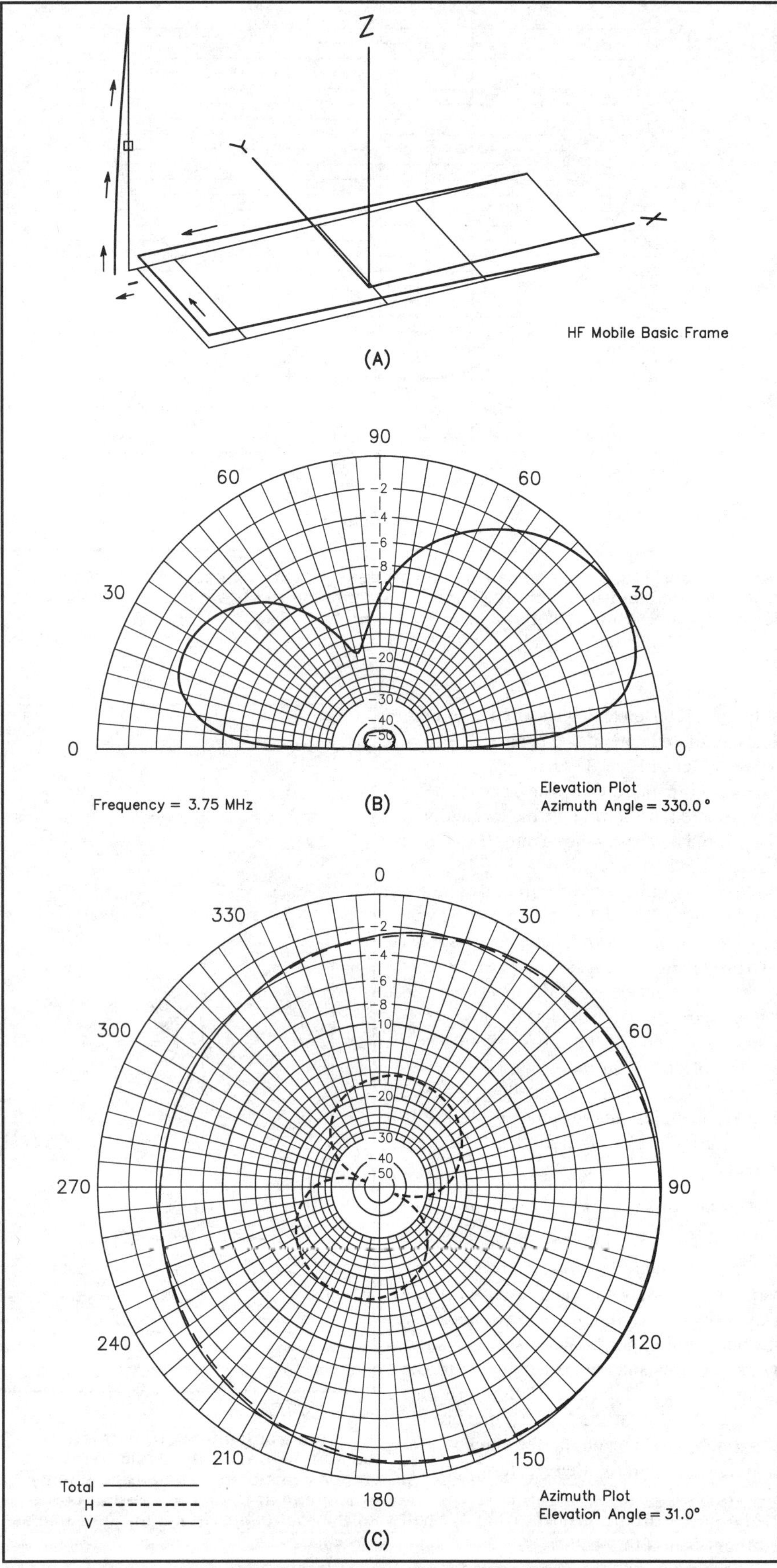

Fig 5—Center-loaded 110-inch mobile whip on the basic frame of a motor vehicle: at A, wire model and currents for 3.75 MHz; at B, elevation pattern at azimuth of maximum gain; and at C, principal plane azimuth pattern at elevation angle of maximum gain. Note the vehicle is in the 0° to 180° plane facing the 0° azimuth direction.

given in Fig 5B and 5C. Notice that the maximum gain is in the direction of the long axis of the vehicular frame. For a left-rear-corner-mounted antenna, maximum gain is in the diagonally opposite direction.

We calculated the radiation resistance (for loss resistances zero) $R_r = 1.3\ \Omega$, the loading coil for resonance and the gain for a loss resistance $R_g = 12\ \Omega$, with and without a short bumper bracket. A bumper bracket is needed for the Jimmy vehicle model; see below. The reactance of the loading coil for resonance and the gain agreed almost exactly, thus validating the model with a bumper bracket.

Let me comment on the model, for those radio amateurs interested in modeling. The segment length on all wires simulating the vehicle and the frame of the vehicle are made approximately the same length (number of segments proportional to wire length). Four segments are used on the bracket on which the whip is mounted, and the segment tapering option (*ELNEC* program) was used, tapering to this segment length at the source and load ends of the antenna wires.

Now let us look at a bumper-mounted antenna on a Jimmy truck; for the wire-grid model see Fig 4B. Here, return currents on the conductors of our wire-grid model of the vehicle can be oppositely directed to the current on the antenna, see **Fig 6A**. The resultant radiation field associated with these currents is in part canceled. This results in a lower effective value of the radiation resistance. For this antenna at 3.75 MHz, $R_r = 0.6\ \Omega$. Compare with the value for the antenna mounted on the frame, where $R_r = 1.3\ \Omega$. The whip and the vehicle body are closely coupled—a change of current on the antenna is reflected by a change of current on the body of the vehicle in a complicated way, and the wire-grid model may not exactly simulate the metal surface of the vehicle. Nevertheless, the author considers the results obtained to be reasonable, and the radiation patterns (Fig 6B and 6C) are in accord with expectation. Compare these with Fig 5B and 5C.

Let us now consider further the reduced value of the radiation resistance due to the oppositely directed return currents on the body of the vehicle. The calculated radiation resistance for an antenna having the same tip height, mounted on the left-rear corner of the roof, is $R_r = 0.53\ \Omega$. See Fig 4C. This is only marginally less than the value for the bumper-mounted whip. Indeed, fields due to current flow on the lower part of the mobile antenna and the vehicle almost cancel.

Finally, we have modeled the vehicle with a roof-mounted whip, which is not a very practical installation for passing under low bridges and beneath overhanging trees.

This lifts the antenna and reduces to a minimum the effect of oppositely directed return currents. The antenna is more or less symmetrical with respect to the support structure. The calculated radiation patterns for 3.75 MHz are given in **Fig 7**.

Radiation Pattern for 40 and 20 meters

For the amateur interested in mobile operating at high frequencies, the patterns in Fig 4B for 40 and 20 meters are plotted on **Fig 8** and **Fig 9**. The tuning inductances are j 1559 Ω and j 647 Ω respectively. Clearly, the basic azimuth pattern found for the 75/80-meter band is retained, and the efficiency and directivity increase as frequency increases.

Predicted Gain

The predicted gains for 80, 40 and 20 meters for a 110-inch whip on the GMC Jimmy truck are –13.5 dB, –6.2 dB and –3.8 dB with respect to a well-grounded quarter-wave monopole respectively. The front-to-back ratio is about 9 dB at 14.15 MHz, which is in agreement with operational experience.

HF Mobile for NVIS

The traditional HF antenna used with land vehicles is a vertical whip, which produces little radiation straight up, making it poorly suited for near vertical incidence skywave (NVIS) communications, a typical requirement for mobile operators using the 80 and 40-meter bands. For 20 meters and up a low launch angle is desired. The approach to overcome this problem for military NVIS tactical communications involves the use of a tilted whip. Wallace[8] described the use of a whip-tilt adapter that raised the whip 2 feet over the vehicle's top, with the top part of the whip pointed forward. He claims that this works well for NVIS communications. The author modeled such an antenna system, see **Fig 10A**. There does not seem to be much gained (see Fig 10B) over a bumper-mounted mobile whip. Even though the antenna is much longer (20 feet) the maximum gain according to *ELNEC* is less, and the gain at high elevation angles is not improved. But the pattern is omnidirectional (Fig 10C). Perhaps it is this difference that leads military communicators to believe that the tilted whip resolved the NVIS communications problem.

If the whip can be tilted, an alternative arrangement is to direct the whip rearward, at an angle to the ground, although this is hardly a practical solution for a vehicle in motion. This tilted-back arrangement provides high-angle radiation, since the whip and the rubber-tired vehicle radiate somewhat like a funny-looking dipole, the vehicle being one "arm" of the dipole. However, it should be noted that military whips

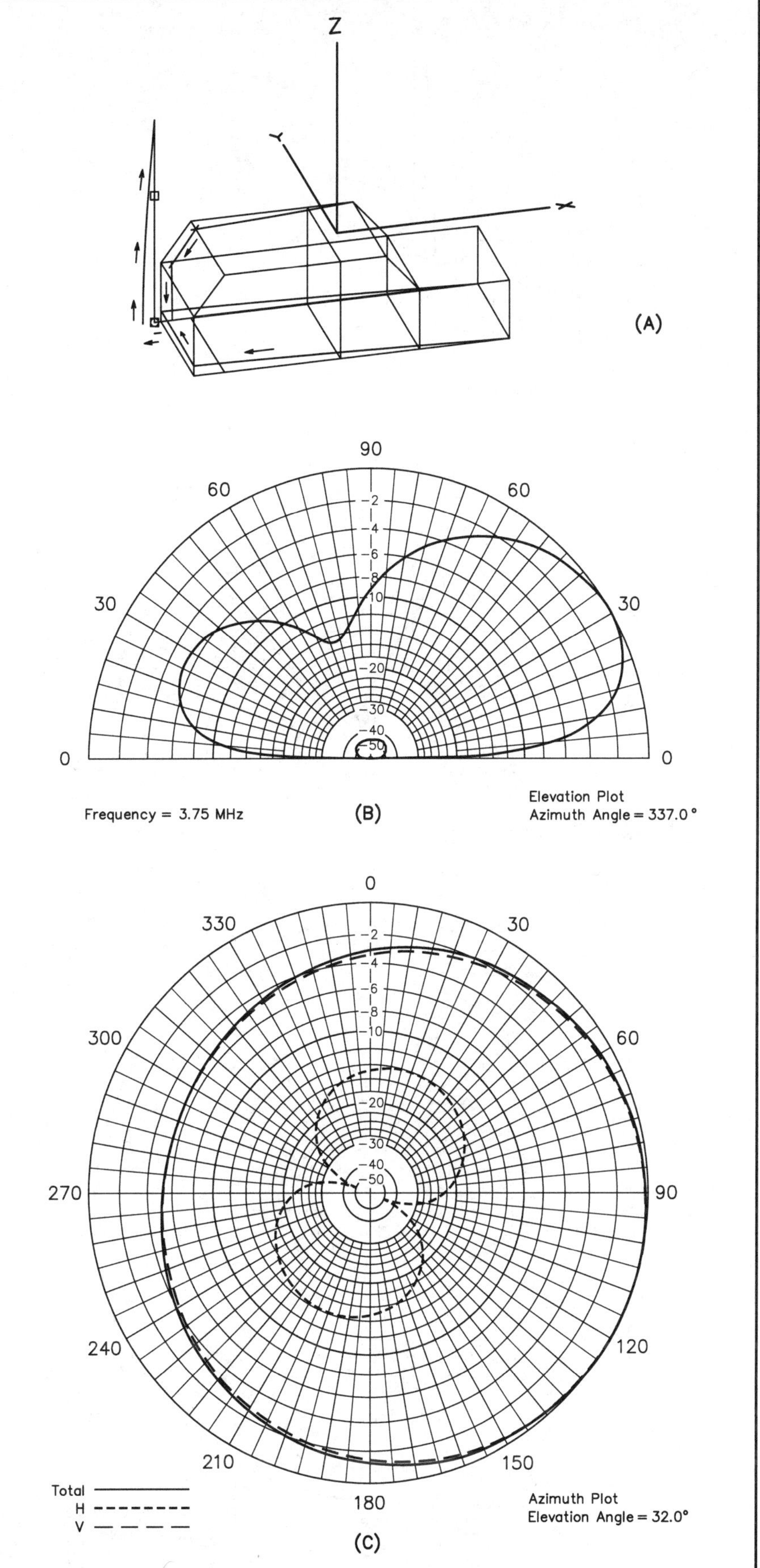

Fig 6—As for Fig 5, but the whip is bumper-mounted on a GMC Jimmy truck (Fig 4B), frequency 3.75 MHz.

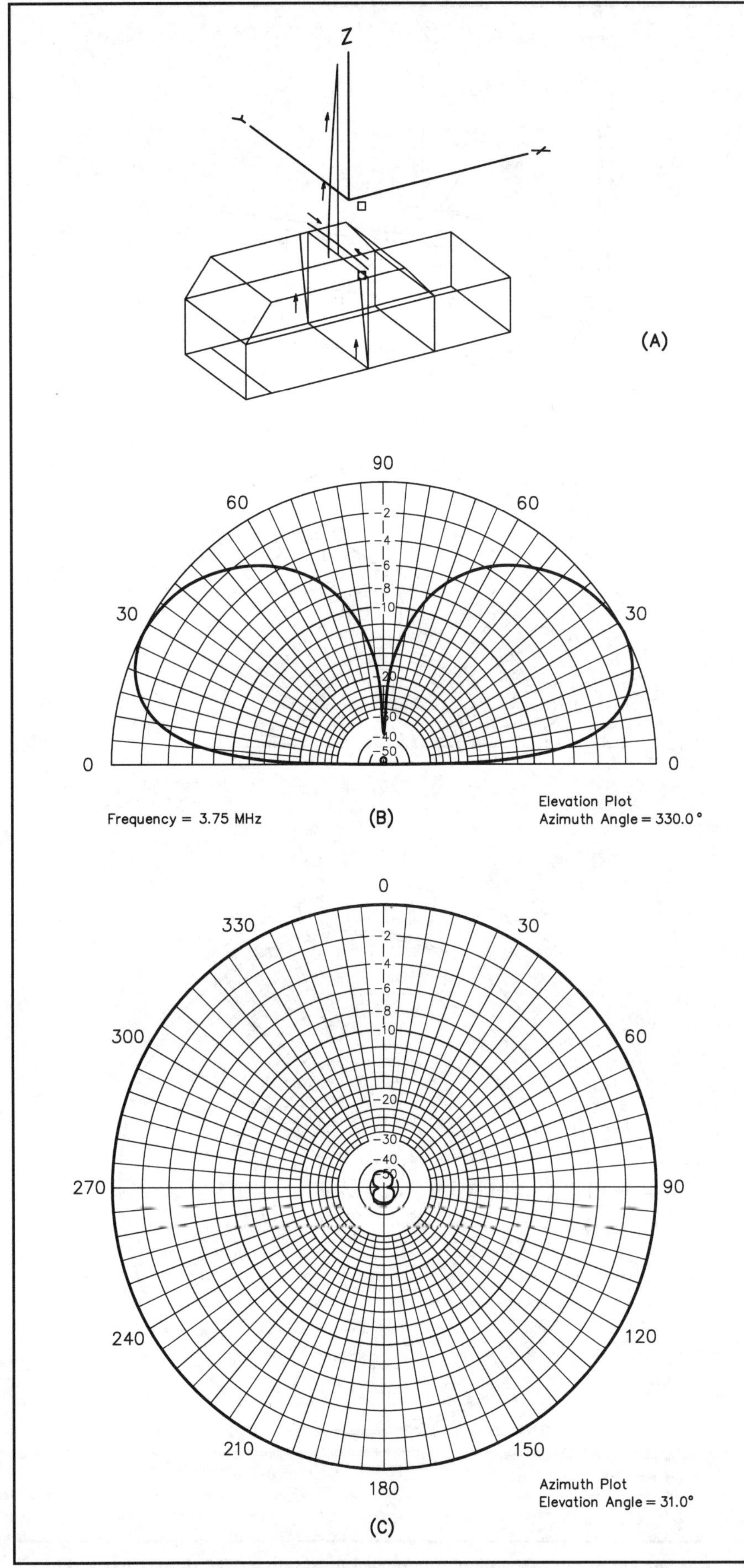

Fig 7—Roof-mounted 110-inch center-loaded whip, frequency 3.75 MHz.

are 16 feet long, which is quite a different length compared with the 9 foot (110-inch) whip described here. In fact, the military whips can be extended all the way to 32 feet. It is no wonder that only the lower half of the whip is used, and the end of the whip is tied down when traveling.

Practical Considerations: Tuning/Matching and Installing HF Mobile Antennas

HF mobiling continues to be interesting to the HF Amateur Radio community, as is perhaps evident by two articles published in *QST* in 1993. Jeff Gold, AC4HF,[9] briefly described his experience with a number of commercially available HF mobile antennas. Roger Burch, WF4N,[10] describes how he made his own HF mobile antenna and mount. A number of the author's fellow amateurs are duplicating D. K. Johnson, W6AAQ's design (see below).

In his own 1953 *QST* article, the author addressed the subject of tuning and matching the HF mobile antenna. He recommended resonating the antenna using base or center loading, but specified the resonator must be a part of the antenna, external to the vehicle. The reason for this is readdressed below. Since a resonant antenna's impedance is not usually a match for 50-Ω coax, author Belrose described circuitry to achieve an exact 50-Ω match, using an L-section network. But this additional matching circuitry is an inconvenience for the mobiler who wants to operate on several bands. Besides, most modern transceivers come with a very useful automatic antenna system tuning unit (ASTU), providing the SWR is not too high. Therefore, it is really only necessary to resonate the antenna for minimum SWR with the ASTU switched off, since the automatic ASTU can then correct for the resistance mismatch on the 50-Ω coaxial feed line.

Further, a number of commercially available automatic ASTUs are available. These can be installed close to the antenna feedpoint (say, in the trunk of the automobile). But do not use such a unit to resonate an untuned electrically short mobile whip.[11]

The capacitance C_a of a 110-inch whip at 3.75 MHz is 33 pF. Since the feed-through capacitance C_{bi} of a typical base insulator can be 3-15 pF, an appreciable non-radiating current can flow through it (see **Fig 11A**) if the tuning coil is inside the car. If the antenna is resonated "outside" the metal vehicle, by a base- or center-loading inductor (see Fig 11B), the capacitance of the feedthrough becomes unimportant. Certainly a coaxial cable, no matter how short, should not be used to connect an untuned mobile whip to the tuner. Furthermore, the Q of the inductances used in a compact automatic ASTU will be low with respect to the Q for large diameter air-spaced resonator coils.

The loading coil is an important consideration, since a high-Q is wanted, particularly for 80 meters. The inductance required is very near to the self-resonant frequency for a coil with no space between turns. An air-wound coil with a space between turns should be used, to reduce the self capacity of the coil and to increase its self-resonant frequency. Ideally, a separate coil should be used for each band. An interesting alternative to using a separate coil for each band is described by D. K. Johnson, W6AAQ [reference his application notes for the "Big DK^3 HF Mobile Antenna," or his new book on "HF Mobileering," printed by Worldradio, 1993]. This antenna uses a large lower antenna mast (2-inch diameter), and the series tuning coil moves down inside the lower mast as the frequency is increased—the coil is fully extended at the lowest frequency for the antenna. Integrity is assured by installing fingerstock at the top end of the lower mast section. Although coil turns are in effect eliminated as they go into the pipe mast—there are in effect no shorted turns—they are not turns anymore, since current flows on the top section whip, on the exposed part of the inductor, and on the outside surface of the lower mast section.

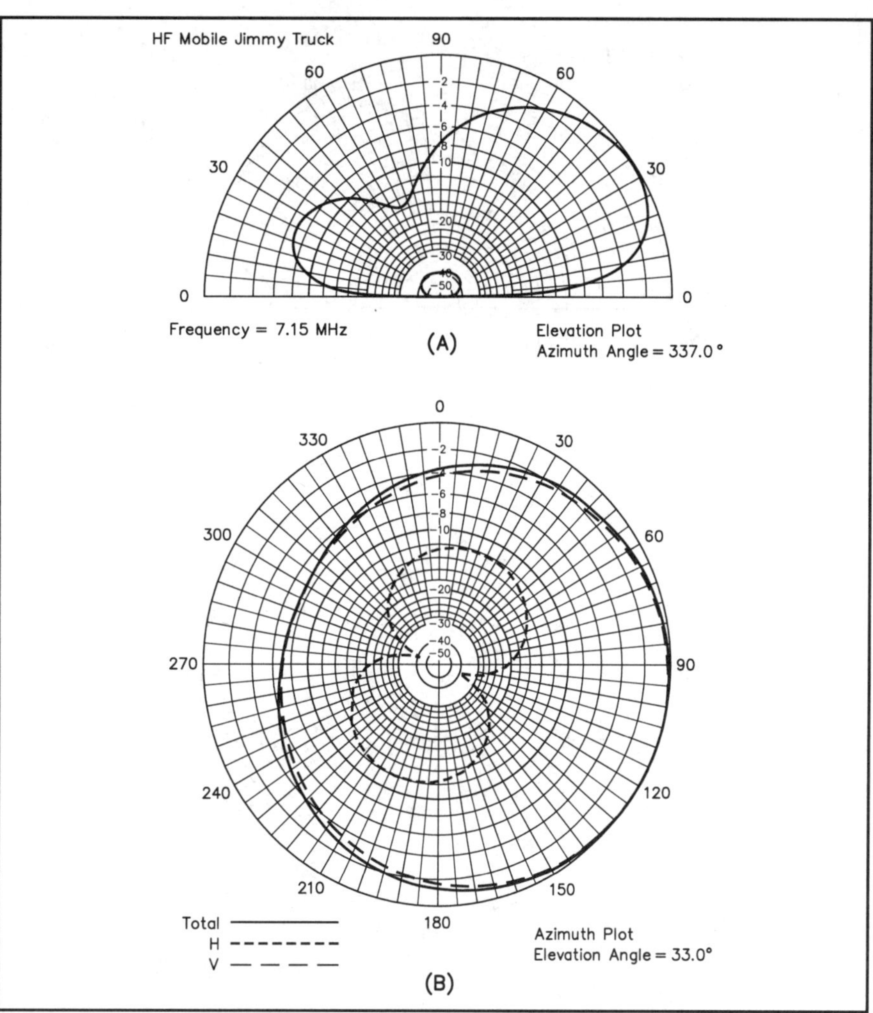

Fig 8—Bumper-mounted whip on GMC Jimmy truck, as for Fig 6, frequency 7.15 MHz.

Concluding Remarks

Mobile antennas are strongly affected by the structure on which they are mounted. This study has shown the patterns to be expected with typical land vehicular mobile antennas. In particular we have emphasized the effect of the close proximity of a bumper-mounted antenna to the back of a 4×4 type vehicle. Clearly, a bumper-mounted mobile antenna on a 4-door sedan with a deep low-profile trunk, or on a pick-up truck, would give better performance, but the difference would only amount to about 2 dB.

The directional azimuth pattern when employing a rear-bumper-mounted HF coil-loaded whip on a rubber-tired vehicle can be used to advantage for communications in a crowded band, but only providing the vehicle is stationary with the desired azimuth orientation, or that it is driven in the direction of propagation! When the vehicle is parked, advantage can be taken of local specifically chosen terrain (for example, a mountaintop or a seaside location with the ocean in front of the antenna). An insulated wire approximately $\lambda/4$ long could be laid on the ground to enhance the directivity in the desired direction.

For NVIS communications using HF antennas on rubber-tired vehicles, an electrically small (compact) loop could be employed.[12] Whatever the mobile platform, rubber-tired or track vehicles, or marine vessels, a compact loop is a better antenna for NVIS.

Acknowledgment

The author works for the Communications Research Centre, Shirleys Bay, ON. He acknowledges the opportunity to use instrumentation and computing facilities not normally available to the average radio amateur.

Notes and References

[1]J. S. Belrose, "Short antennas for mobile operation," *QST*, Sep 1953, pp 30-35, p 108.

[2]R. C. Hansen, "Optimum inductive loading of short whip antennas," *IEEE Trans. Veh. Tech., VT-24*, 1975, p 21.

[3]R. C. Hansen, "Efficiency and matching trade-offs for inductively loaded short antennas," *IEEE Trans. on Comm., COM-23*, April 1975, pp 430-435.

[4]J. S. Belrose, "VLF, LF and MF antennas," in *The Handbook of Antenna Design*, ed. Rudge, Milne, Olver and Knight (London: Peter Peregrinus, 1983), pp 627-630, 633.

[5]The author uses the *ELNEC* version, by Roy Lewallen, PO Box 6685, Beaverton, OR 97007.

[6]The author uses the *NEC/Wires* version, by Brian Beezley, 3532 Linda Vista Dr, San Marcos, CA 92069.

[7]B. F. Brown, "Optimum Design of Short Coil-Loaded High-Frequency Mobile Antennas," *ARRL Antenna Compendium, Vol 1*, 1985, pp 108-115.

[8]M. A. Wallace, "HF Radio in Southwest Asia (during Desert Storm)," *IEEE Communications Magazine*, Oct 1991, pp 58-61.

[9]J. Gold, "HF Mobiling—Taking it to the Streets," *QST*, Dec 1993, pp 67-69.

[10]R. Burch, "You Can Operate HF Mobile!" *QST*, Feb 1993, pp 29-30.

[11]J. S. Belrose, "Automatic Antenna Tuners for Wire Antennas," Technical Correspondence, *QST*, Mar 1994.

[12]J. S. Belrose, "An Up-Date on Compact Loops," *QST*, Nov 1993, pp 37-40.

APPENDIX A

Impedance of a Short Monopole: Simple Analysis

Belrose[1,4] analyzes inductively loaded whips by an equivalent transmission line

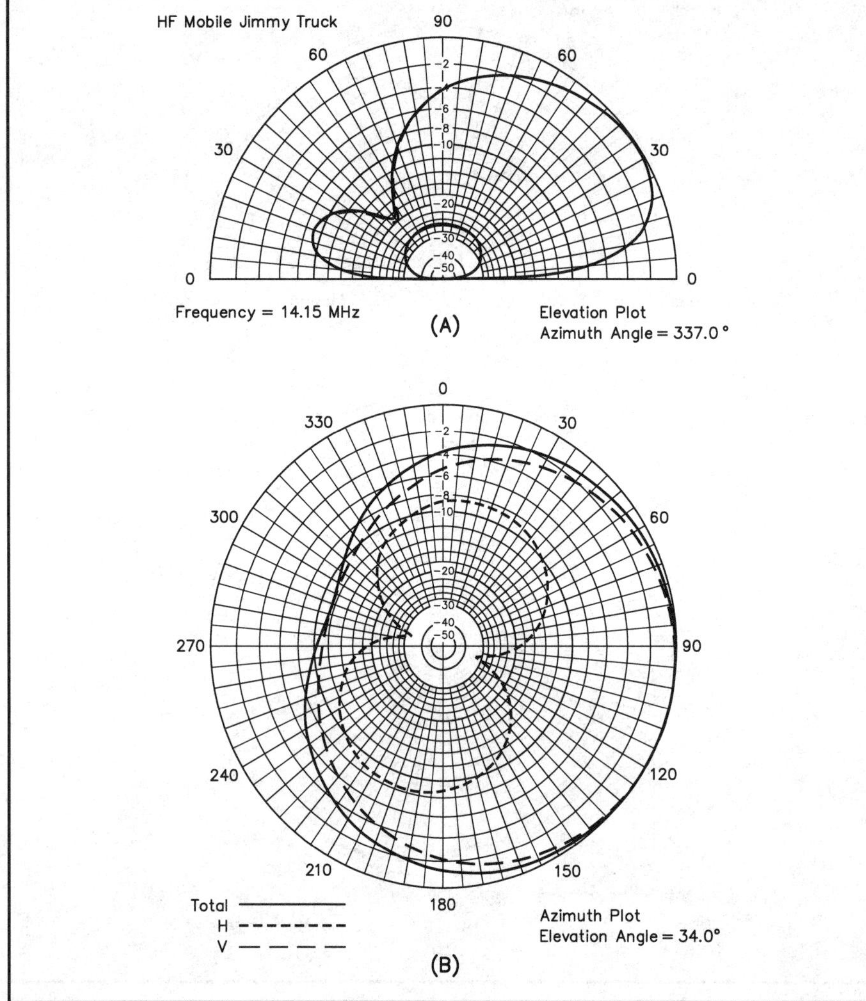

Fig 9—Bumper-mounted whip on GMC Jimmy truck, as for Fig 6, frequency 14.15 MHz.

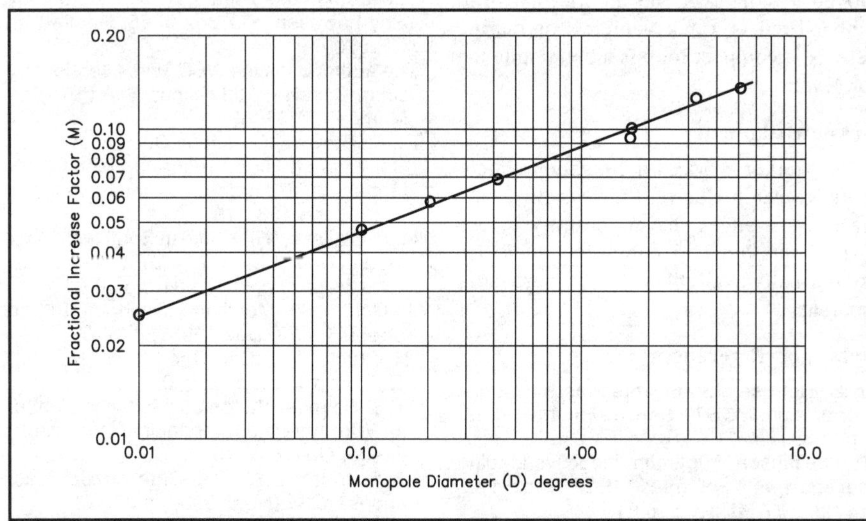

Fig A1—Graph of Fractional Increase Factor (M) versus monopole diameter D in degrees.

method. The earlier method assumed linear current on the antenna. In the later analysis the author assumed a sinusoidal current distribution on the antenna.

The following is a sample calculation for a center-loaded whip for 3.75 MHz. The height of the monopole:

$h = h_1 + h_2 = 2.794$ meters, and the average diameter d = 8 mm.

The inductance L_0 is adjusted so that the base reactance is purely resistive, that is, the antenna is resonant. The lower section of the antenna is considered to be an opened-out transmission line loaded by the capacitive reactance of the top section and the inductive reactance of the loading coil in series. For resonance

$$X_{Lo} = j\,Z_0(\cot G_1 - \cot G_2)$$

where G_1 and G_2 are the electrical heights of h_1 and h_2, and Z_0 is the average characteristic impedance of the monopole of height h.

$$Z_0 = 60\ (\ln\frac{2h}{d} - 1)$$

For $h_1 = h_2 = 1.397$ m

$$G_1 = G_2 = (1+M)\ \frac{h}{\lambda}\ (360°)$$

$$= \frac{1.035\ (1.397)\ (360°)}{80} = 6.5°$$

The factor (1 + M) = 1.035 is the reciprocal of the usual antenna factor k. Information on the relation between M and the electrical diameter in degrees (D) of the whip is shown in **Fig A1**. For our mobile whip, the electrical diameter D:

$$D = \frac{d}{\lambda}\ (360°) + \frac{8(360°)}{80 \times 10^3} = 0.036°$$

Hence M = 0.035. The characteristic impedance

$$Z_0 = 60\ (\ln\frac{2(2.794)}{8} - 1) = 333\ \Omega$$

and $X_{Lo} = j333\,(\cot 6.5° - \tan 6.5°) = j2922\,\Omega$

The radiation resistance can be calculated by adding the current areas (degree-amperes) on the two parts of the antenna, since the currents are in phase.

$$A_1 = \frac{180}{\pi}\left[\frac{1-\cos G_1}{\sin G_2}\right]\cos G_1$$

= 3.23 degree-amperes, and

$$A_2 = \frac{180}{\pi}\sin G_2$$

= 6.49

Hence $R_r = 0.01215\,(A_1 + A_2)^2 = 1.15\ \Omega$. These values can be compared with those calculated by *ELNEC* in Table 1.

Fig 10—A long mobile whip (20 feet) tilted over the top of the vehicle: at A, wire model and currents for a frequency of 3.75 MHz; at B, elevation pattern for 0° azimuth; and at C, principal plane azimuth pattern.

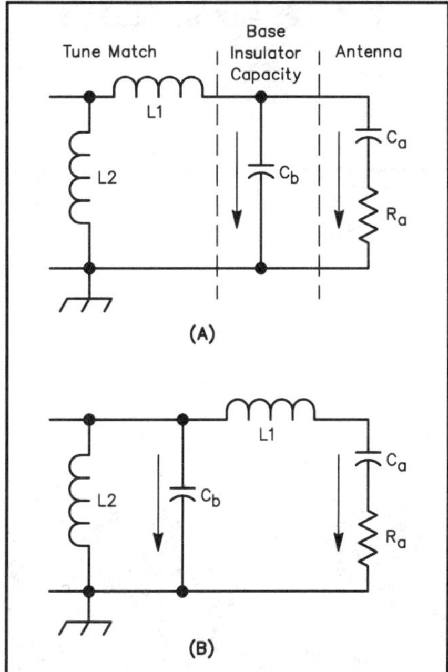

Fig 11—Equivalent circuit illustrating incorrect (at A) and correct method (at B) of tuning an electrically short monopole, where a base insulator feedthrough to the inside of a metal vehicle or an ASTU is employed.

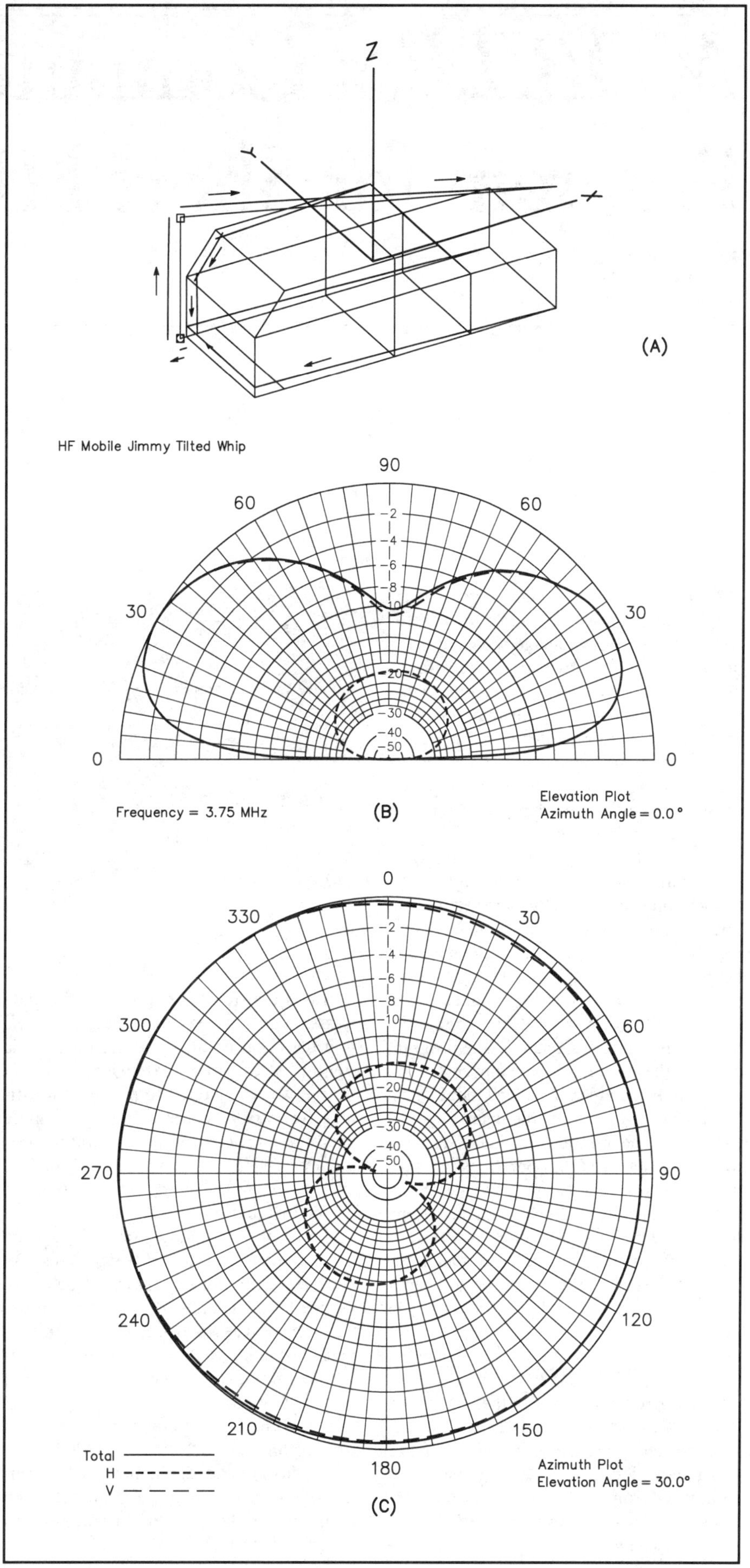

MOBILE, a Computer Program for Short HF Verticals

By Leon Braskamp, AA6GL
1459 Via Cortez
Placentia, CA 92670

AA6GL describes an interactive computer program for designing and evaluating mobile whips. The program is on diskette bundled with this book.

The design of a coil-loaded vertical antenna involves a number of compromises. Every change seems to bring on the need for two or three others. *MOBILE.EXE* is an interactive program that will provide some help. It displays all characteristics on a single screen, and important values may be changed at will. Each change results in an immediate screen update, showing the effect on other parameters and on overall performance.

Do you want to compare base-loading with center-loading?—simply lengthen the top section and shorten the base section. Do you want to change coil diameter?—Enter the change and it will display the new coil length, turn count, Q estimate, and overall performance. For physical changes to the elements, it will always display the inductance required for resonance. The program changes the coil length (and turns count) to continually provide the needed inductance.

Appendix A contains the details needed to fully use the program. A good first step is to simply run it! Most operations are supported by on-screen prompts and messages, and a short trial will give you a sense of the look and feel of it. Before doing any serious work, the Appendix should be read carefully. It describes the more obscure functions, which might include just what you need.

Program Development

Bruce Brown's article in the first *ARRL Antenna Compendium*[1] provided all of the math related to the antenna itself. To make a complete picture, this was combined with formulas from other ARRL publications.[2,3,4] This left two missing links—ground loss and coil Q.

Brown's article[5] contained several data points on ground loss resistance, and is the only data that could be found on vehicle ground losses. As an extension, a simple routine was developed to provide an estimate for various combinations of vehicle size, antenna location and operating frequency. There is a provision for manual entry to allow use with fixed station and marine installations.

The Q of high-quality coils cannot be directly measured at HF frequencies. By chance, some representative Q values were obtained[6] and a series of mathematical expressions was developed to provide an estimate. The math used has no technical foundation; it was developed by trial and error to produce reasonable results at several baseline combinations of coil form-factor and turn spacing. Wire "skin effect" RF resistance is also included. It appears to be accurate over a wide range, and an accuracy warning will show on the last line of the screen if that range is exceeded.

How It Works

Physical dimensions are used to determine the coil inductance needed for resonance. How this coil is built can have a big effect on performance.

The initial coil configuration is determined by the computer. It is not supposed to be a perfect coil, but it should be practical. Up to 125 μH, it is something that may be purchased at reasonable cost. You may want to change the wire size, coil diameter, and turn spacing. For each change, a new turns count and coil length will be found producing the same inductance. An estimate of coil Q is also made, together with the calculation and display of the resulting performance.

Error handling is in two phases. For each input, a trial calculation is made. If any parameters would go outside program limits an error message is shown, together with alternate ways to recover. A warning message describes the cause if the results are within program limits but with possibly degraded accuracy.

The last line of the screen displays helpful information to the operator. It may show a short-cut alternate command, an accuracy warning, or it may describe the use of the Escape key to back out of an error.

A Sample Application

Fig 1 shows the working display for a representative 75-meter mobile antenna; an 8 foot center-loaded whip, using a 3.00 inch coil made from standard coil stock. The base section length of 4.7 feet allows for the spring and mount. A mid-level mounting location was selected for reduced ground loss.

Pressing "E" (for Extended display) results in the screen shown in **Fig 2**. This uses the working area of the screen to show data that is not frequently used. The antenna represented in this display will be used as a standard for comparison with two alternate configurations.

A Trade-Off of Length and Mounting Location

A minimum of experience with the program will demonstrate that antenna length is the major factor determining performance. For this example, the antenna will be moved to a bumper mount (even though ground loss will increase) and lengthened so that it is the same over-all height. Fig 2 shows a top height of 12.7 feet, so that will become the limit in the following examples.

The results are shown in **Fig 3**. A base length of 4.5 feet (plus the 0.7 foot spring and mount allowance) was chosen since threaded commercial base sections are available in this size (Hustler MO-3). The top section is an 8 foot CB whip cut down as necessary to 5.4 feet. Note that performance has been improved by 1.7 dB despite a 47.5% increase in ground loss.

Winding Your Own Coil

If you are capable of making your own air-wound (or equivalent) coils, further improvement may be made by increasing coil Q. The Up and Down (arrow) keys permit scanning the effect of larger or smaller coil diameters in quarter-inch increments. With #14 wire and 8 turns-per-inch pitch, the optimum is quickly found at 5.25 inches.

The final antenna is shown in **Fig 4**. Top section length has been increased another 0.4 foot, made possible by the shorter coil. The combined changes result in a 3.2 dB improvement. That may not seem like much since it is still 8.5 dB below perfection. However, it has a greater effect than doubling transmitter power, is less expensive than an amplifier and is much easier on the battery!

This interactive program allows you to exercise your imagination without having to build anything. Various combinations of available components may be evaluated, changed, and discarded without cost. Like any good computer game, it provides a scorekeeper in the form of the continuous display of performance information.

APPENDIX A

Reference information for running *MOBILE.EXE.*

Equipment Required

MOBILE.EXE will run on any MS-DOS computer. The display is text mode graphics, using standard IBM characters. It checks for a color monitor, and if EGA or VGA is installed it provides a full color screen. If not, a high-contrast monochrome display is used.

For laptop users—Some LCD laptops provide a monochrome VGA display using shades of gray to simulate colors. To secure better contrast you may force a true monochrome display by entering the program as "MOBILE <space> B" or "MOBILE / B".

Program Limits

- Frequency—1.8 to 30 MHz.
- Antenna length—One-quarter wave maximum
- Coil length—Varies with frequency; from 39-inches maximum at 1.8 MHz to

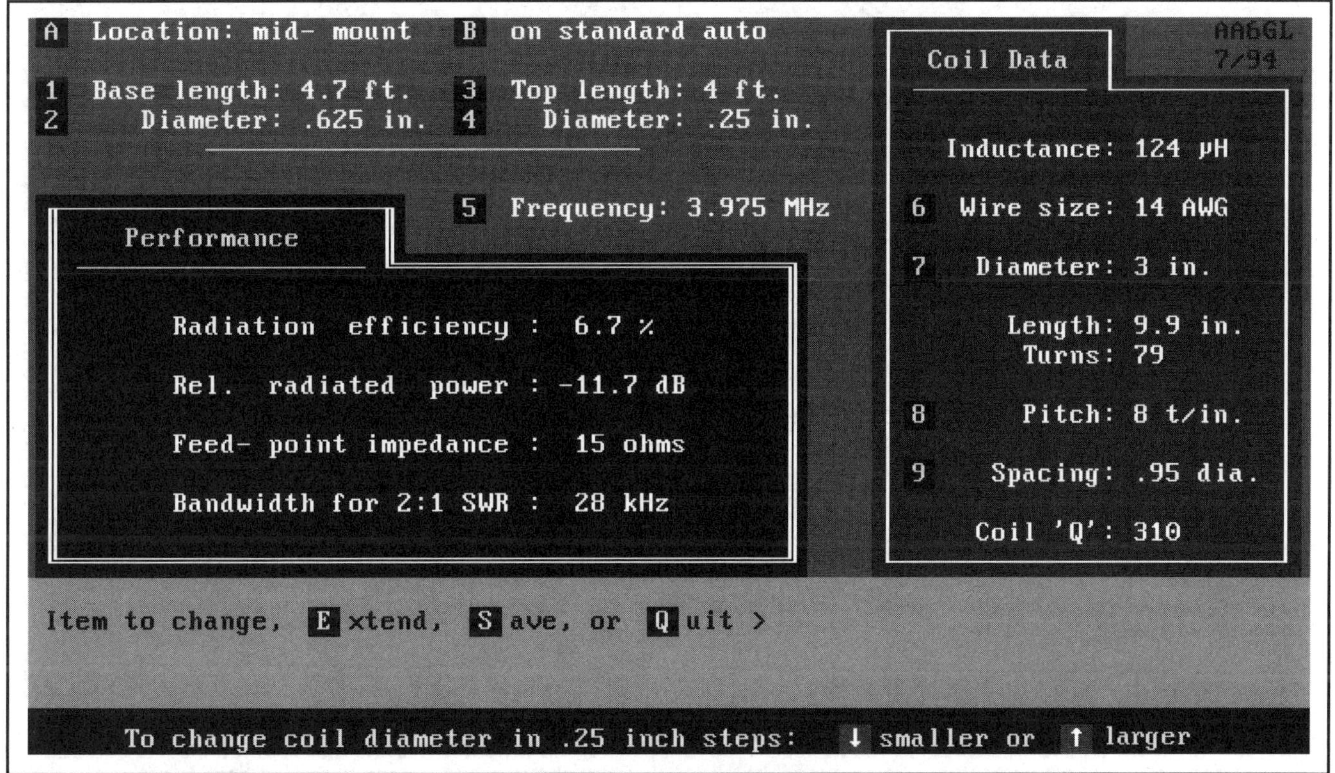

Fig 1—Screen print from *MOBILE.EXE*, showing nominal 8 foot whip mounted on rear trunk lid of standard auto. The base diameter is 0.625 inch, and the average diameter of tip is 0.25 inch. A loading coil with 124 μH inductance and a Q of 310 is assumed. The efficiency is 6.7% for this configuration, down 11.7 dB from an ideal λ/4 vertical installation.

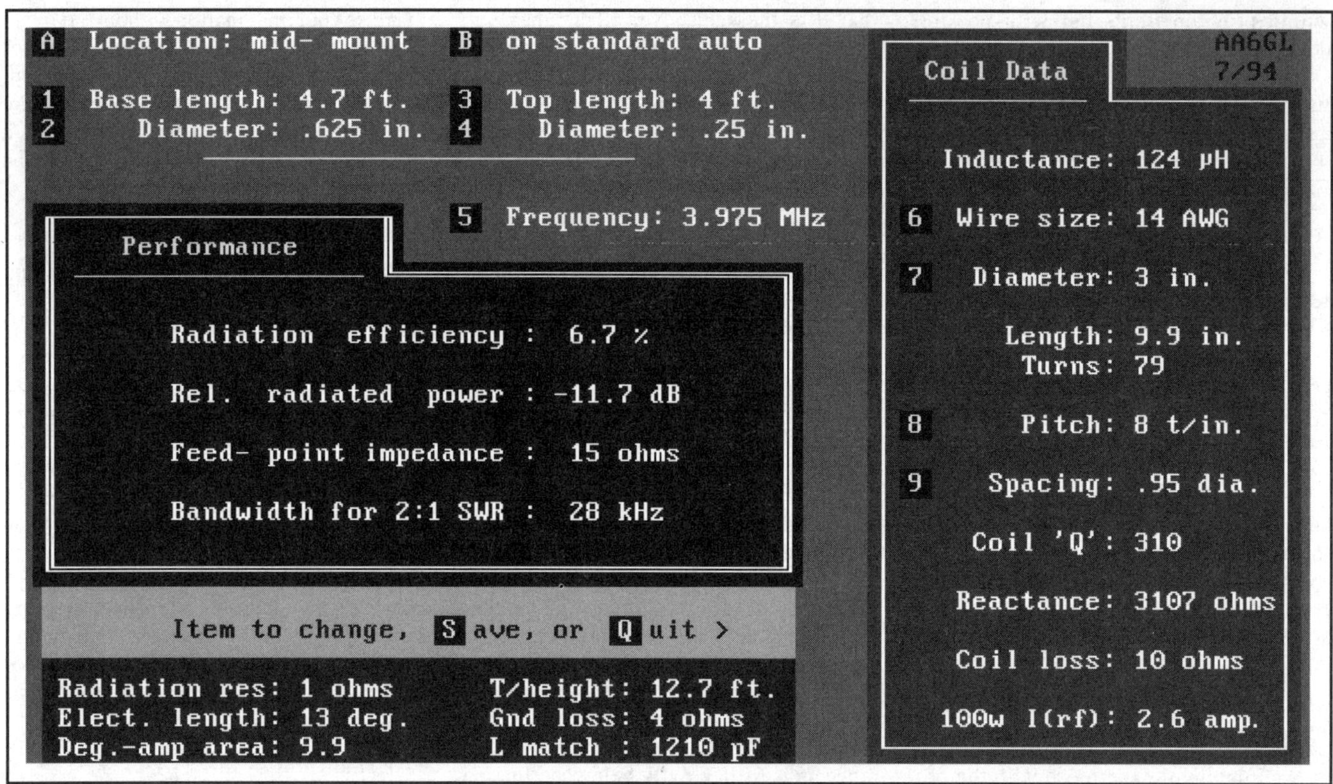

Fig 2—Screen print for same conditions as in Fig 1, except that "E" (Extended screen) has been pushed by operator. Overall height of whip is 12.7 feet above ground. The ground loss is estimated at 4 Ω.

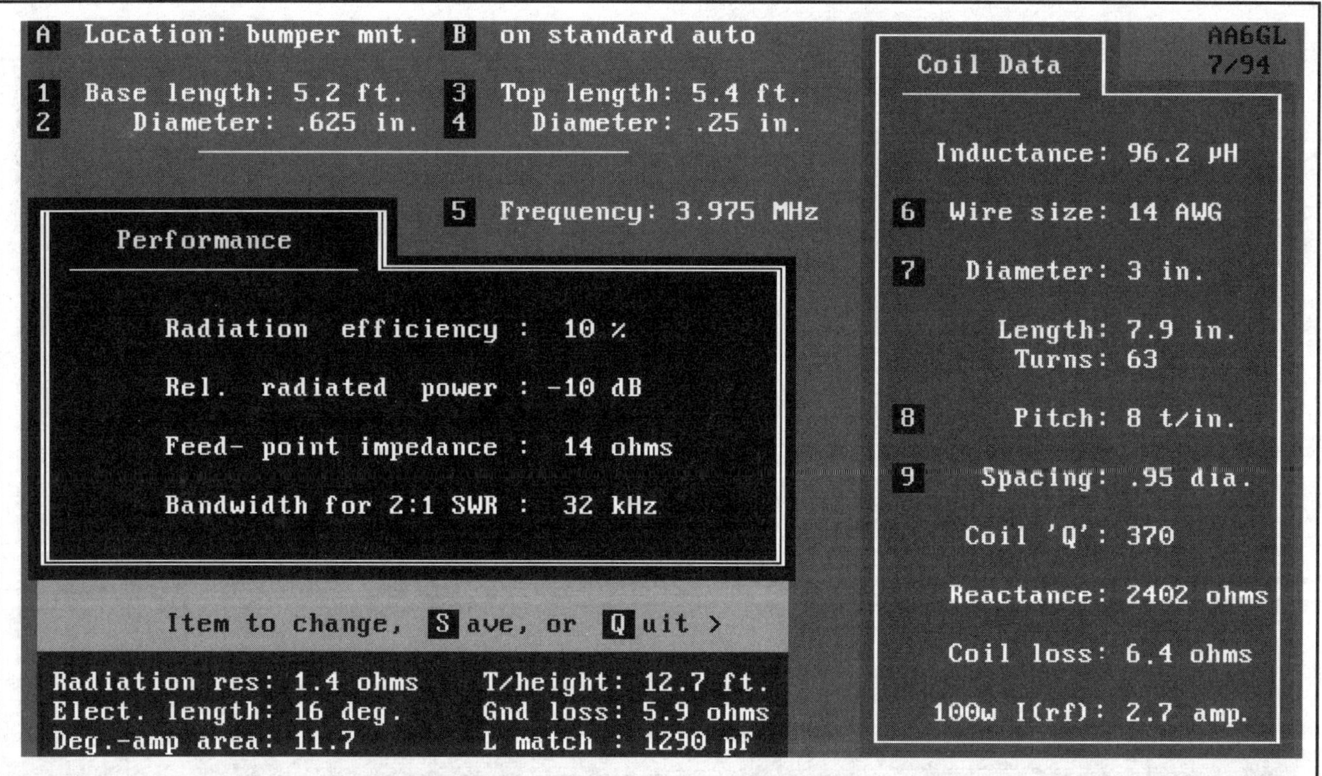

Fig 3—Screen print from *MOBILE.EXE*, showing effect of moving base of the whip down to the bumper on the same car, and extending top portion of whip to achieve same overall height of 12.7 feet above ground. Despite the increase in ground resistance, the efficiency has improved in this configuration by 1.7 dB, because the longer radiator has increased radiation resistance faster than ground loss.

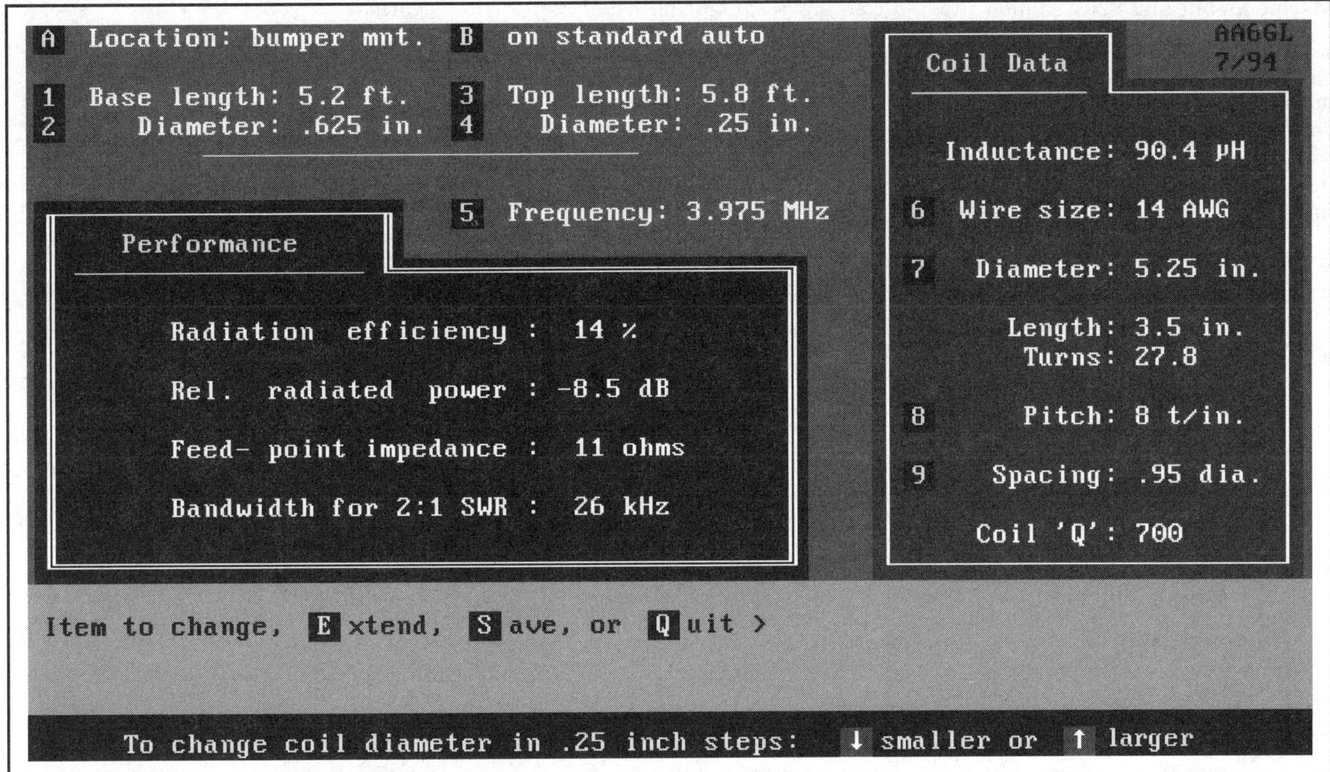

Fig 4—Screen print showing that increasing diameter of loading coil, while shortening the coil length and increasing the whip's length slightly to achieve same 12.7 feet overall height, has increased system efficiency by another 1.5 dB. These changes yield a total of 3.2 dB overall improvement over the case in Fig 1 and 2.

2-inches maximum at 30 MHz

- Spacing factor—1.5 maximum; "close-wound" minimum (any outside entry results in an input at the limit)
- Coil pitch—The limit varies to keep spacing factor in range

Program Accuracy

The accuracy of performance numbers is hard to evaluate, but the relationship of physical measurements to antenna resonance is easy to measure. Results have ranged from zero error to coils a few percent too long. Always allow for the unknown; wind a few extra turns and then trim for resonance.

Base-Loaded Antennas

Although developed for center-loaded antennas, MOBILE may also be used for base loading. Enter the sum of spring height and the length of the lead-in wire (to the coupler) as "base section length."

Wire Size Entries

Coil wire size, Item [6], is normally entered by AWG number. Even numbered sizes from #8 to #16 are allowed. For odd sizes or small tubing, wire size may also be entered by diameter. Any entry of 0.250 inch or less results in a display change to "Wire Diameter" and the specified value is used in the calculations.

Pitch and Spacing Factor

[8] Pitch (turns-per-inch), and

[9] Spacing Factor (turn spacing as a fraction of wire diameter)

These are two ways to specify the same thing. Enter either one and the other will be calculated during screen update. (The program actually works from spacing factor; pitch has been included to permit the entry of commercial coils directly from data sheets.)

Close-Wound Coils

Although not very efficient, a close-wound coil may be entered. Select item [9] (Spacing Factor), and enter either "C" or "0" (zero). The display will change to "Spacing: close" and the calculated pitch should closely agree with published wire tables.

Understanding the Extended Display

The area on the right is an extension of the coil data area. The item labeled "100 W I(rf)" is an abbreviation for the RF current in the coil with 100 W applied at the feedpoint. These antennas work at a low impedance, and planning for a fixed station running high power should include a good look at the wire or tubing size used for the coil (for power over 100 W the current may be directly scaled up).

The extended display area at the bottom of the screen contains data applicable to the antenna itself. The three items in the left column are best understood by a review of the original math.[1] The term "T/height" is an abbreviation for "top height," or overall height above ground.

To operate a low-impedance antenna without a tuner, the easiest way to get up to 50 Ω is an L-match consisting of a slight addition to the loading coil and a shunt capacitance at the base.[2] The "L-match" shown in the extended display is the correct value for this capacitor. Install the capacitor before trimming the coil and you will resonate and match in a single operation.

The Performance Data Table

- Radiation efficiency—The RF radiation, expressed as a percent of that from a zero-loss quarter-wave vertical working against a perfect ground.
- Relative radiated power—The RF radiation, expressed as dB down from the zero-loss antenna.
- Feed-point impedance—The impedance as seen at the base of the antenna.
- Bandwidth—The 2:1 SWR bandwidth of the antenna itself, not including any matching circuitry.

Marine Mobile and Fixed Station Use

For these applications, mount height and ground loss resistance must be determined separately and directly entered. As you enter the program, from the "Vehicle Size" menu on the second screen, select item [6] and prompts will follow for these items.

Estimating Coil Q

The Q-estimating routine is based on the use of a single layer of air-wound copper with no taps, shorted turns or close proximity to metal objects. The following parameters are used to establish coil Q:

- Coil length / diameter ratio
- Turn spacing
- Wire resistance, including RF "skin effect"

Perfectly proportioned wire coils will vary from about 600 to 750 depending on size and frequency. Q values may run as high as 850 with 1/4-inch tubing. The minimum estimated Q is 100; any lower estimate is displayed as "< 100" and the value of 100 is used in performance calculations.

Notes

[1]Bruce Brown, W6TWW, "Optimum Design of Short Coil-Loaded High-Frequency Mobile Antennas," *ARRL Antenna Compendium, Vol 1,* pp 108-115. All the math for antenna performance and required coil inductance was derived from this article.

[2]*The 1994 ARRL Handbook,* p 2-17 (coil winding), p 33-21 (L-section match).

[3]*The ARRL Antenna Book,* 17th Edition, (ARRL, 1994), p 26-29 (bandwidth).

[4]Steve Trapp, N4DG, "A Ham's Guide to Antenna Modeling," *ARRL Antenna Compendium, Vol. 2,* p 157 (RF wire resistance).

[5]Bruce Brown, (see Note 1)

[6]Jerry Sevick, W2FMI, private correspondence.

Performance Comparison Between the Use of Coil-Loaded Mobile Whips and Antenna Couplers

By Jack Kuecken, KE2QJ
2 Round Trail Drive
Pittsford, NY 14534

You might not need to use any loading coil if you match your mobile whip with a properly connected antenna coupler.

Over the years many hams have been fascinated by mobile operation on the lower bands. Automobile HF antennas, however, have some unfortunate characteristics. They are notoriously lossy, in many cases quite ungainly and usually very narrowband. This article describes several variations on mobile whip antennas and couplers with a goal of providing efficient, multi-band operation and the ability to change bands while in motion.

Perhaps because of my professional background in designing military HF and VHF antenna couplers I believe any good land mobile antenna system should be weather proof, roadworthy and tunable to any desirable frequency with the vehicle in motion. The typical military land-mobile HF antenna coupler will tune 2 to 30 MHz. The antenna will withstand some level of abuse while striking fixed objects.

Height limitations imposed by roadway conditions are a fundamental problem in mobile operations. Interstate highways and most electrified railroad crossings generally give 12.5 feet of clearance to permit tractor-trailer rigs to pass. Having an antenna hit something solid at even 25 mph is a startling experience; having the antenna contact a 1500-V catenary of an electrified railroad could be disastrous!

My experience shows a maximum height of 11 feet, 4 inches above the roadway will clear about 95% of highway obstructions. I must avoid the two 10.5-foot high bridges within 5 miles of my home QTH. Further, most drive-throughs and porticos are lower than 10 feet, and most garages only allow an antenna about 7 feet high. Thus, any automobile antenna over 7 feet above the pavement will need a quick and convenient method of laying it down.

If we accept the 11-feet, 4-inches height limitation for general open-highway use and subtract the distance from the bottom of the antenna to ground, it means the active portion of our antenna can be no more than 9 or 10 feet long. For any frequency below 24 MHz the antenna is electrically short. In fact, at 3.5 MHz, a 10-foot antenna is only 0.0356 wavelength long!

The Electrically Short Antenna

A simple "buggy whip" antenna is usually considerably shorter than a quarter wave. It behaves like a small resistance in series with a capacitor. For electrically small radiators without any top loading, the radiation resistance is related to the part of the power actually radiated and is given by Kraus[1] as:

$$R_r = 40\left(\frac{\pi H}{\lambda}\right)^2 \qquad \text{(Eq 1)}$$

where

R_r = radiation resistance, in ohms
λ = wavelength, in meters
H = radiator height, in meters

Eq 1 assumes the current in the radiator decreases linearly from a maximum at the radiator base to zero at the top. If the radiator is heavily top-loaded the current distribution tends to be uniform and the radiation resistance is effectively twice as great. A formula for a fairly accurate value of capacitance at frequencies below resonance is given by:[2]

$$C_A = \frac{17L}{\left[\left(\ln\frac{24L}{D}\right)-1\right]\times\left[1-\left(\frac{fL}{246}\right)^2\right]} \qquad \text{(Eq 2)}$$

where

C_A = capacitance, in pF
L = radiator length, in feet
D = radiator diameter, in inches
f = frequency, in MHz

The radiation resistance of the antenna is not the only source of resistance in the cir-

Fig 1—Photo showing antenna coupler and test antenna mounted on the author's car. The insulated lower section braces the antenna. A 6-inch cord and hook is used to hold the antenna in the lowered position.

Fig 2—One of the adjustable loading coils used for the tests was taken from a military surplus transmitter. It is fastened to an insulated section that was alternately inserted into the whip as a base- and center-loading coil.

cuit. Loading-coil losses are present as a loss resistance, and there is always a certain amount of ground-loss resistance present as well. In practical cases, when the antenna is electrically quite short, the loading coil and ground losses far exceed the radiation resistance. The BASIC language computer program ANTCAP.BAS on the accompanying disk solves the equations for capacitance, radiation resistance and coil losses.

The Mobile Coupler

All electrically small monopoles have a high reactance. This reactance must be tuned out to allow any appreciable current to flow in the antenna. Quite often this is done with a loading coil. Conventional ham wisdom says a center-loaded antenna is more efficient than a base-loaded antenna. We shall investigate this premise from a theoretical and experimental standpoint.

In practice a loading coil of any size in the center of a military land-mobile antenna is mechanically unacceptable because it would be subject to snagging in branches and destruction on contact with a fixed object at any significant speed. Workable military land-mobile antennas consist of spring stainless or fiberglass whips tuned with a coupler at the base. In many cases the tuning range of the coupler is continuous and it will handle the maritime mobile and commercial frequencies as well as the ham bands.

The mechanical features of these units are interesting. **Fig 1** shows one of my remote couplers, mounted on a car with a plastic bumper unsuited for direct mounting of the antenna. The coupler mounts on the lower lip of the trunk. I drilled through the plastic bumper, and mounted two reinforcing plates beneath it. Each aluminum bracing plate has three 10-32 threaded holes. The outer two are used to keep the brace in place, and the central hole holds the antenna coupler to the car. *A piece of half-inch braid grounds each plate to the car body.* When the trunk lid is closed, this hardware is hidden, securing the coupler from possible theft.

The whip base is held by a brace attached to one of the bolts on the bumper shock absorber. A horizontal brace runs to the unibody beneath the trunk. This unibody attachment brace is designed to shear off in the event of a collision to avoid defeating the 5 mph bumper. The lowest section of the antenna is an insulating extender. The end of this section is about 23 inches off the pavement. This piece also clamps to the antenna coupler. An 18-inch length of 1-inch diameter aluminum tubing leads up to the 4-inch long spring. All of the pieces attach together with 3/8-24 threaded stock mating into aluminum plugs driven in the ends.

I normally use a 7.5-foot stainless steel whip above the spring. This whip tapers from 1/4-inch diameter at the base to 3/32-inches at the tip. On the very top I mount a 3/8-inch diameter corona ball. It also serves to keep people from poking an eye out when the whip is tied down. A metal tie-down clip fits between the rear window and the door. A 6-inch length of cord holds a hook straddling a 3/8-inch diameter aluminum cylinder. The cylinder is driven onto the whip about 18 inches below the top. The whip base is fitted with a stainless fixture with a 3/8-24 thread. With the antenna erect, the tip is 11.25 feet off the pavement.

I run at highway speeds without any bracing or guys on the antenna. If a branch is clipped, the whip simply deflects. When tied down, the maximum height in the bow of the whip is about 6 feet and the tip of the antenna is just behind the front door. The elevated placement of the spring permits the trunk to be opened and closed. Passengers can enter and leave through any door with the whip tied down.

Center- Versus Base-Loading

I wanted to determine how much was being lost by the use of base-loading, rather than center-loading. This led me to construct a special antenna to permit the use of either center- or base-loading of the same radiator. The top section of this radiator is a 4-foot, 3-inch length of 3/8-inch diameter aluminum tubing, with a stainless fitting on the bottom. I made up a 6-inch long insulator of G-10 fiberglass tubing, with threaded aluminum fittings in the ends. A 22-inch length of one-inch aluminum tubing finished the antenna, giving a total radiator length of 9 feet. The radiator tip is 10-feet, 11-inches above the pavement. With this special setup, the insulator can be placed either in the center of the 9-foot radiator or at the base.

A pair of mounting brackets were fashioned to hold a roller coil on the insulator. The coil used for most of the measurements was the antenna loading coil from the T-22 ARC-5 radio. This unit is shown in **Fig 2** fitted with a knob and mounted on the 6-inch insulator. The coil has a measured Q of 144 at most settings. For the 3.5-MHz tests, a larger coil with a sliding short was used. This coil had a Q on the order of 150.

One of the points made in favor of a center-loaded antenna is the assumption of a more uniform current throughout the system. This is supposed to double the radiation resistance. It does, but only for the lower half of the antenna. The radiation resistance in the upper section is unchanged so the total radiation resistance does not double.

Furthermore, radiation resistance is only a small fraction of the total circuit resistance in an electrically small antenna. The inductive reactance required to resonate the center-loaded antenna is nearly twice as great as the value required by the base-loaded antenna. If both coils have the same Q, the loss resistance in the center-loaded antenna coil will be nearly twice as great as the loss resistance in the base-loaded antennas.

One point not in favor of the base-loaded antenna is the sensitivity of the antenna to stray capacitance. Any current flowing through the stray capacitance associated with the whip base and the loading coil contributes to the losses, but not to the radiated power of the antenna. For the sake of efficiency the connection between any coupler used and the radiator must be a short open wire, with as little capacitance as possible. The antenna mounting base should be designed for minimum capacitance to ground.

If you try to use as little as a foot and a

half of RG-58 coax for the connection between a coupler output and the whip base, you will introduce about the same capacitance as the antenna and therefore only half of the current passing through the coil in the coupler (and its loss resistance) will flow in the antenna. Since the losses in the coil vary as the square of the current, doubling the current will quadruple the loss. Similarly, halving the current in the antenna will reduce the radiated power by a factor of four. Never use coaxial cable to connect an antenna coupler to the base of an electrically small antenna! This restriction does not apply to a center-loaded antenna where the reactance has been canceled.

Impedance Measurements

The 9-foot test antenna was mounted on the car using the same mounting facilities used for the coupler. The car was parked on an asphalt driveway two lanes wide. A Hewlett-Packard 8656B signal generator was used for the signal source. A Tektronix 2710 Spectrum Analyzer was used as the detector and the measurements were made with a General Radio 1606B impedance bridge. The roller coil and insulator assembly were placed first in the center and later at the base of the antenna. At each frequency the roller coil was adjusted to give a minimum SWR and the impedance was then measured through a short length of RG-58C/U (0.84-m length).

The cable length was mathematically "rotated out" of the data with a computer program.[3] As a check on the validity of the theoretical program, a comparison was made between the computed and the experimentally derived inductance data. **Figs 3** and **4** show very good agreement between the theoretical and measured values for the center-loaded and the base-loaded inductors. **Table 1** summarizes the resistive portion of the data. R_{MEAS} is the actual measured resistive component. At each frequency the coil inductance and Q at the measured setting were taken and the coil loss was computed for the R_{COIL} column.

Radiation resistance is not directly measurable in the ordinary sense of the word. In outer space if one could measure the radiated power over each square meter of a large sphere (10 wavelengths or so in radius) and divide the sum by the square of the antenna current you would have the radiation resistance in ohms. Above lossy earth the measurements would be a little more difficult to do. A large-scale pattern measurement range would be required and it would be difficult to measure the energy radiated into and absorbed by the earth.

On a theoretical basis, Eq 1 for radiation resistance stands on relatively firm ground, being honored with longevity if nothing else. On the other hand, the formula for cal-

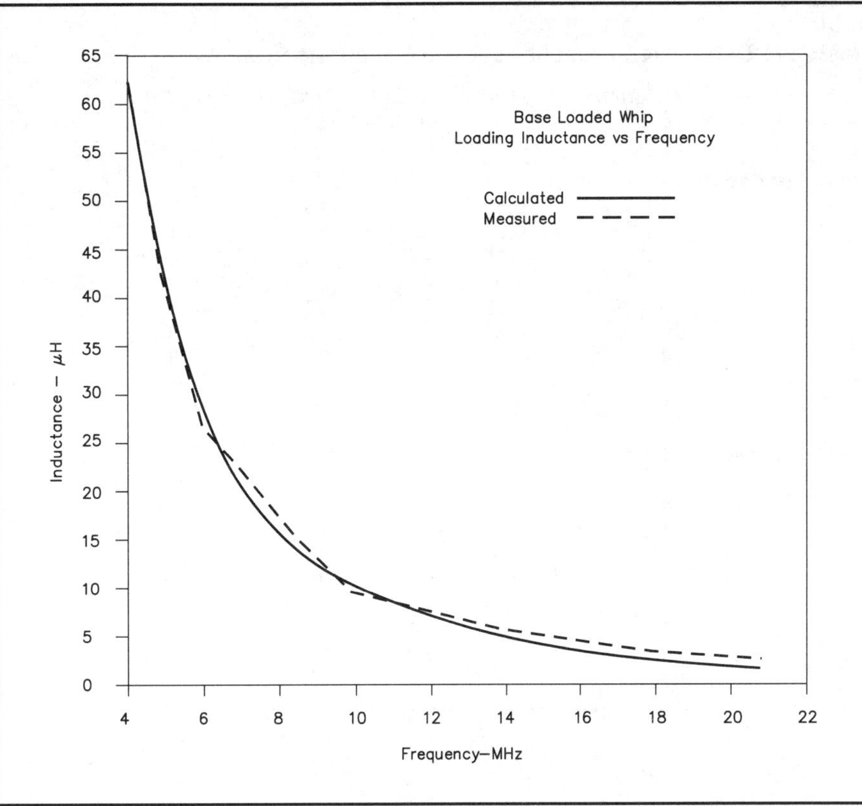

Fig 3—A plot of inductance required for resonance of the 9-foot test whip antenna as a base-loaded unit. The calculated and measured values were in close agreement.

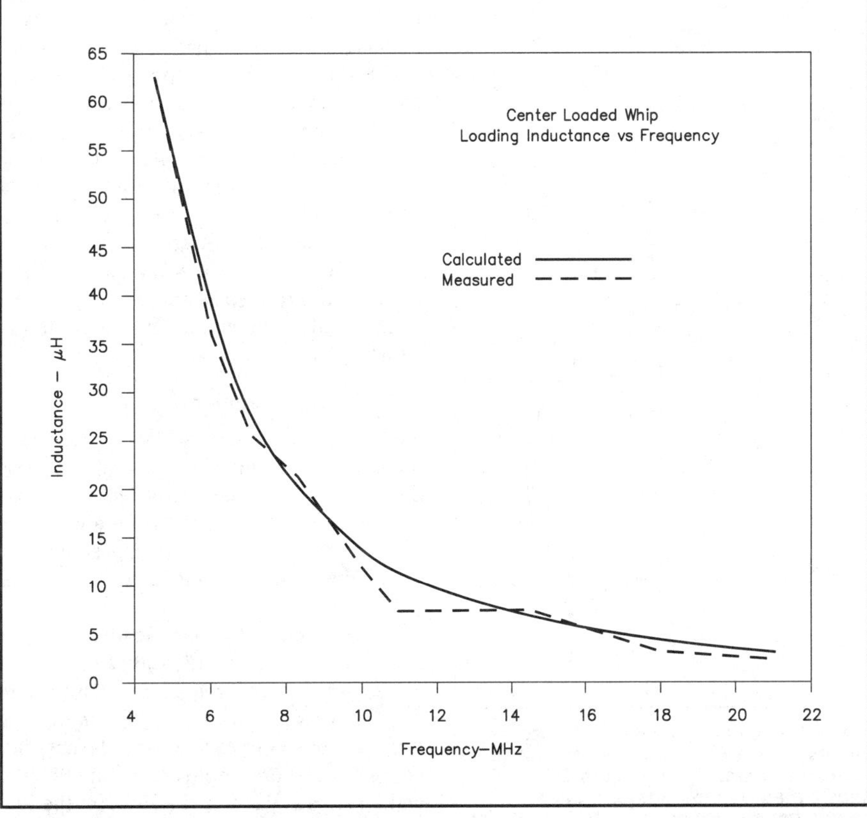

Fig 4—Slightly higher values of inductance were required for a center-loaded whip as compared to the values plotted in Fig 3.

Table 1
Resistive Data Taken From Impedance Measurements

	Frequency MHz	R_r Ω	R Coil Ω	R Meas Ω	% Eff
Base Loaded Antenna					
	4.1	0.55	7.11	15	3.7
	5.0	0.82	5.8	16	5.1
	6.0	1.19	4.79	15	7.9
	7.0	1.61	4.0	17.3	9.3
	8.5	2.38	3.19	17.5	13.6
	10	3.29	2.61	25.7	12.8
	14	6.45	1.6	36	17.9
	18	10.7	1.0	41	26
	21	14.5	0.6	52	28
Center Loaded Antenna (See text for comments on radiation resistance and efficiency)					
	4.7		11.8	29	
	6.0		9.18	34	
	7.0		7.18	38	
	8.0		6.8	34	
	10		5.4	43	
	14		3.8	63	
	18		3.8	57	
	21		2.3	93	

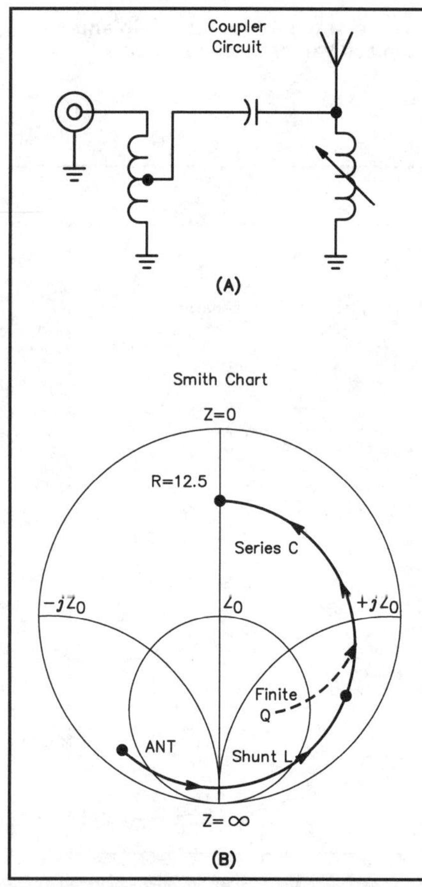

Fig 5—The coupler used in the tests consisted of a shunt inductance and series capacitance. The Smith Chart shows their use to transform the whip impedance to a value close to 12 + *j*0. The 4:1 autotransformer then converted this value to 50 + *j*0.

culating the radiation resistance of the center-loaded whip is on more tenuous grounds. I have therefore not presented a radiation resistance value or an estimated efficiency. The column showing % EFF for the base-loaded whip is obtained by dividing R_r by the R_{MEAS}. For our comparison let us rely on A/B signal strength tests.

One thing is striking about the impedance measurements—for a center-loaded antenna the measured resistances run consistently high. This is true even when you allow for a higher radiation resistance and a nearly doubled loss resistance. For the base-loaded antenna the difference between the measured resistance and the sum of R_r + R_{COIL} is interpreted as ground loss. For the center-loaded antenna the interpretation is a bit less straightforward.

The Antenna Coupler

Neither the center loading nor the base loading actually matches the impedance of the antenna. In general not less than two elements are required to match a complex impedance. In the previous measurements the loading coil simply is used to resonate the antenna.

Table 1 shows the base-loaded antenna has an impedance of 15 + *j*0 Ω or a 3.3:1 SWR on the lower frequencies. This SWR may either shut down your transmitter or cause the power to back down. The antenna coupler used here employs a shunt inductor and a series capacitor to transform the impedance to 12.5 + *j*0. This is then stepped up to 50 + *j*0 with a 4:1 transformer.

The circuit for this coupler is shown in **Fig 5** along with Smith Charts showing the technique for this matching scheme. The actual value of inductance used is smaller than the value required for resonance. The BASIC program ANTCOUP.BAS solves for the inductance and capacitance required for a 1:1 SWR for a given antenna accounting for ground losses and finite coil Q.

In many of the studies I have seen on mobile antennas, the values of coil Q discussed are unrealistically high for a coil fitted with weather protection and structural attachments. A huge "Texas Bugcatcher" with no weather cover and spaced windings might approach a Q of 300. The more typical long slender close-wound coil with metal mechanical ends and a shrink weather seal will have a much lower Q. I have measured many and generally the Q was closer to 50.

On-The-Air-Tests

Propagation tests or A/B comparison tests on HF links are always a bit difficult to make repeatable. It is generally necessary to go rapidly from one antenna configuration to the other to minimize fast changes in propagation. The process of moving the coil and insulator from the center to the base of the radiator and retuning takes several minutes. Therefore, the comparison test was made between a center-loaded whip fed directly with coax, and a whip without a loading coil but fed by a coupler.

Since a relatively complete spectrum was desired, the signal strength had to be kept very low in order to avoid creating interference outside of the ham bands. The Hewlett-Packard signal generator was set for 1 V output and the antenna was always fed through the fixed bridge used for tuning. A fixed bridge and a battery-operated voltmeter were used for sensing the tuned condition.

An SWR less than 2:1 was obtained at each frequency. The nominal power into a precise 50-Ω load through the bridge is 1.25 mW. With the 2:1 SWR the output could vary by about 1.2 dB. The signal generator was powered by an inverter and no power lines or other connections to the ground were used on the car.

At the receiving end the Tektronix spectrum analyzer was used as a receiver. This unit displays the frequency and the amplitude of the signal under the cursor, in printing, on the display. A 15-foot sectioned mast was used for the receiving antenna. A base-loading roller coil was used for this antenna. At 18 MHz one section of the mast was removed and at 21 MHz two sections were removed to prevent the mast from exceeding a quarter wavelength.

The car was driven about 1000 feet from the receiving location and parked by the curb. At each frequency, the center-loaded whip was resonated with the loading coil first, with an SWR less than 2:1. The center-loading coil was then shorted and the antenna base lead transferred to the coupler output insulator. The coupler was then tuned, with an SWR usually less than 1.1:1. When this operation was complete the receiving-end operator (W2TYO) was informed by a 2-meter hand held, and the signal strength read and recorded. The short was then removed from the loading coil and the base lead moved over to the BNC connector used with the center-loaded antenna. The receiving-end operator was again informed via 2 meters and the center-fed antenna reading taken. The time for the changeover was perhaps 15 seconds.

The coupler was expected to be at a slight disadvantage since it is a fully packaged and weathertight roadworthy unit, whereas the center-loading coil was without weather protection and structural protection. The center-loaded coil had a Q of about 150 and a stray capacitance of 2 pF. By comparison, the coupler inductor had a Q of 80 and a stray capacitance of 9.9 pF in the output insulator and the pigtails inside and outside of the case. The coupler Q had been in the 150 range and the stray capacitance around 2 pF before installation in the case.

The measurements were taken in a suburban residential neighborhood, and there were trees and other obstructions in or along the path between stations. **Fig 6** shows the relative signal strengths received from the two configurations in the first test. At 10 and 18 MHz the coupler is slightly better than the center-loaded configuration; however, at all other frequencies it is slightly worse. Since a single "S" unit is equivalent to 6 dB, the differences shown would probably not be distinguishable on an ordinary communications receiver.

The first test was run with the transmitter approximately 1000 feet to the west of the receiving site on relatively level ground. A second test was run with the transmitter 1000 feet to the east. Here the transmitting site was below a substantial hill, and the car could not be seen over the curve of the hill. The data in **Fig 7** shows the coupler is not stronger than the center-loaded antenna at any frequency but the difference is less than one "S" unit. There is some variation in the data. The first test at 21 MHz showed the coupler a little more than 4 dB lower and the second test only 0.3 dB lower. The measurements taken were repeatable.

The reason for the difference in the results is not known; however, there are some points to consider. Over this short range skywave components probably played no part. In the first test the antennas were actually within sight of one another and some line-of-sight propagation may have taken place in addition to the expected ground wave. In the second test the hump of the hill placed the mobile antenna well below the line-of-sight so propagation may have been primarily ground wave.

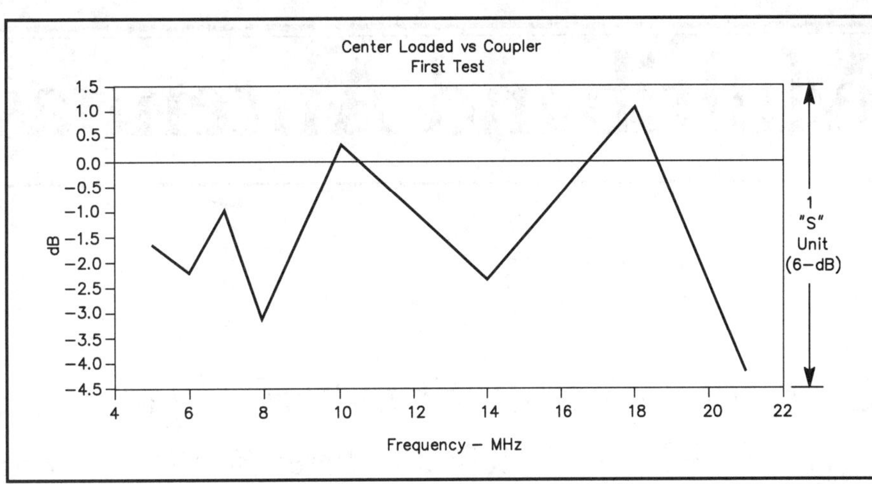

Fig 6—The vertical scale is the difference between the received signal strength using a center-loaded whip and a whip without a loading coil but with an antenna coupler. The maximum difference is under 6 dB or one "S" unit.

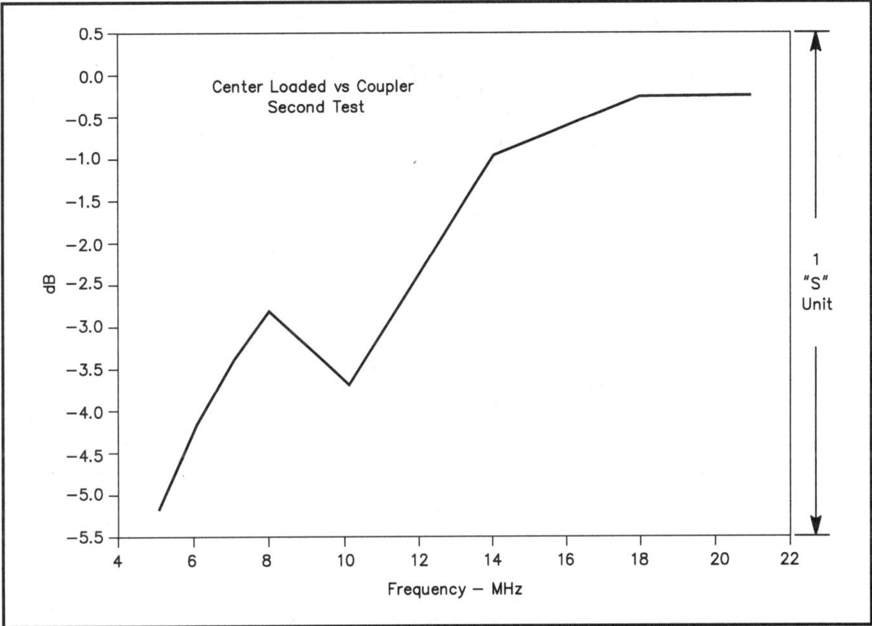

Fig 7—A second test provided a different set of variations when compared to the first test in Fig 6. Again the total difference was less than 6 dB.

Conclusions

The broadband tunable antenna coupler used in these tests is both road- and weather-worthy. It provides performance less than one "S" unit worse than a near optimum center-loaded whip design having no road or weather protection. With careful design to minimize the stray capacitance and to maximize the inductor Q it is possible to build an optimized base-matching antenna coupler. The coupler would be fully competitive with and perhaps even superior to a road- and weather-worthy center-loaded design. The advantages of broadband remote or automatic tunability and roadworthiness need not be sacrificed for performance.

Notes

[1] J. D. Kraus, *Antennas*, 2nd Edition (New York: McGraw-Hill Book Company, 1988), p 712.

[2] *The ARRL Antenna Book*, 13th Ed., p 288.

[3] J. A. Kuecken, *Exploring Antennas And Transmission Lines By Personal Computer* (New York: Van Nostrand Reinhold, 1986), p 281.

Multiband Antennas

On Center-Fed Multiband HF Dipoles

By John S. Belrose, VE2CV and Peter Bouliane, VE3KLO
17 Tadoussac Drive, Aylmer, QC, J9J 1G1, Canada
41 Leeming Drive, Nepean, ON, K2H 5P6, Canada

VE2CV and VE3KLO perform a thorough analysis on this popular multiband antenna, pointing out traps and pitfalls along the way.

Introduction

A multiband antenna is considered by most amateurs in radio to be one that has low SWRs on several bands of interest. This article reports on a detailed study of center-fed multiband dipoles. We present new data generated by numerical modeling on the radiation patterns for both horizontal and drooping (inverted-V) configurations. Previously published articles on the G5RV[1] dipole, lauding its multiband performance, have been concerned with achieving a low SWR on several bands, without consideration to radiation pattern. Yet the radiation pattern is a fundamental characteristic needed to assess the operational performance of any antenna. In this article we demonstrate that radiation pattern and dipole mounting orientation (horizontal or drooping) are more important considerations than merely SWR.

Since the opening of the 30, 17, and 12-meters bands, the center-fed dipole (CFD) with tuned feeders has seen a revival of interest and usage. The G5RV, a multiband center-fed dipole with a special feed line arrangement, has been popular for more than three decades. The 31.1-meter length is somewhat shorter than a quarter wavelength on 80 meters, making it an attractive antenna for some, since it will fit on a normal-sized city lot.

Many amateurs regard this antenna as some sort of panacea, particularly when it is erected as a drooping dipole (inverted-V). However, there is nothing magical about its particular length, or the method of feed.[2] Its performance is simply that of a center-fed dipole with a particular feed-line arrangement. Although an inverted-V dipole can be effective on a single band, the performance of a multiband drooping dipole is in general inferior to a horizontal dipole mounted at the same apex height. We now describe some of the history of the center-fed dipole.

1930-1940 Versions

The center-fed dipole with tuned feeders was a simple and widely used multiband antenna in the 1930 and 1940s[3] and various versions of it are still in use today. Because each half of the flat top is the same length, the feeder currents will be balanced at all frequencies. The length of the antenna is not particularly critical, and neither is the length of the feed line. However, some combinations are considered to be more attractive than others for impedance matching to the transmitter over a wide frequency range. The radiation patterns for this antenna are rarely shown explicitly in the amateur literature.

The dipole was generally fed with an open-wire transmission line. Antenna lengths of 41 meters and 20 meters were typical, with feed line lengths ranging from 13 to 24 meters. Series-parallel types of antenna couplers were used to resonate the antenna system, and link coupling was used between the coupler and the transmitter. The transmitter power amplifier was usually a pair of tubes operating in push-pull, so the entire transmitting and radiating system was balanced. Nowadays, however, since transceivers and Antenna System Tuning Units (ASTUs) are unbalanced devices, a balun must be used to maintain balance. This can lead to difficulties, since the balun may have to work into a very reactive mismatch load at some frequencies.

G5RV Version

In 1946 Louis Varney, G5RV, anxious to get on the air after the war, designed and erected a multiband antenna that would fit his average-sized backyard. The antenna consisted of a 31.1 meter flat-top, split in the center and fed by tuned feeders. Two versions were tested: one using full-length open-wire tuned feeders, Fig **1A**, and the other using a 10.36 meter open-wire stub fed at its base by either 72-Ω twin lead (transmitter grade) or 72-Ω coax, Fig 1B.

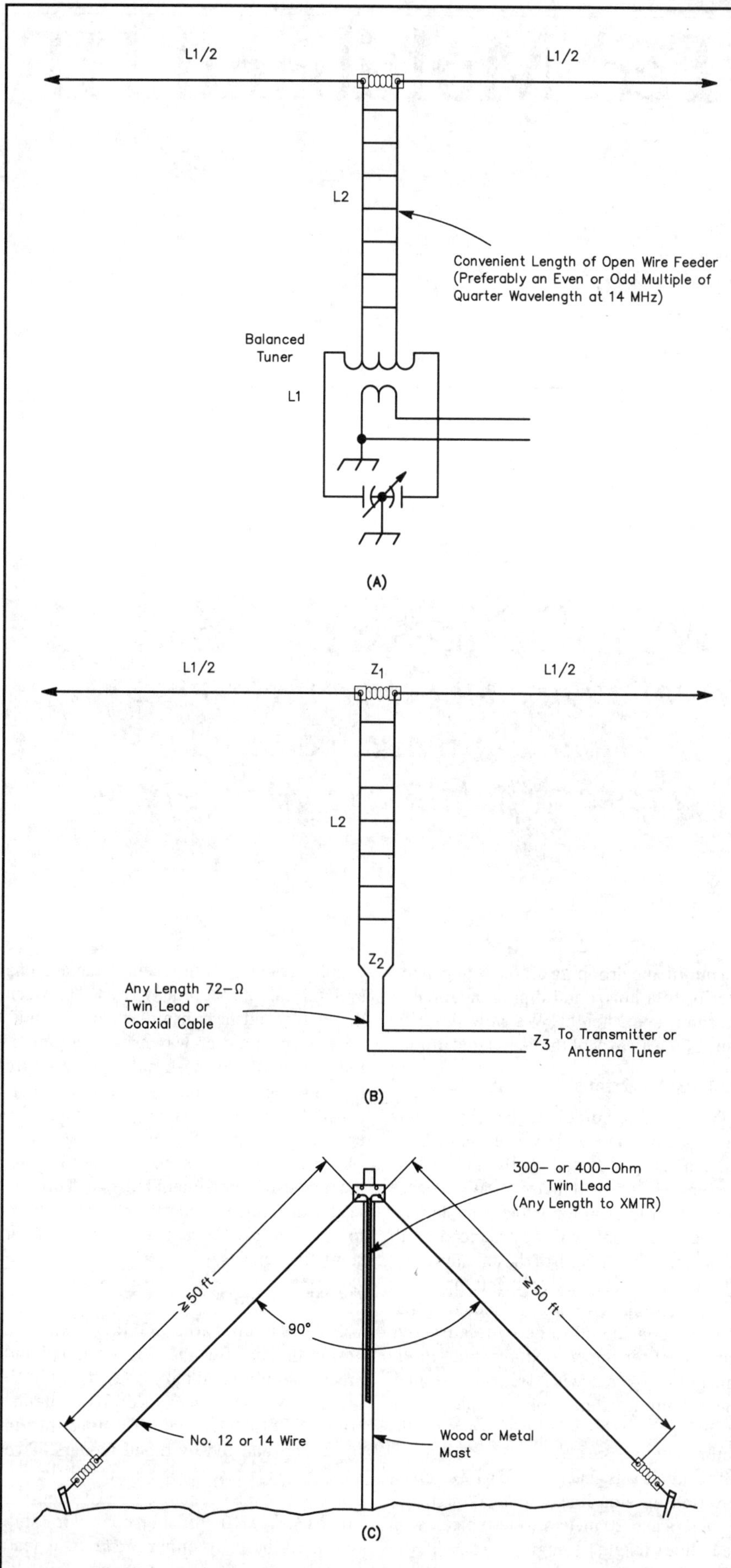

Fig 1—Multiband dipole antennas: At A, horizontal with full-length tuned feeders and balanced tuner in shack; at B, horizontal with an open-wire stub fed at its base by either 72-Ω twin lead or coax (the G5RV antenna), and at C, multiband dipole with drooping arms (inverted-V configuration).

The length of the stub was designed to be a half-wavelength at 14 MHz (the reason for this is discussed below). An alternative to the open-wire stub was also tested, using 300-Ω ribbon. In this case the stub was shortened by the velocity factor of the ribbon. The open-wire stub version, Fig 1B has become known as the *G5RV Antenna.* It was first described in a note by him in the *RSGB Bulletin* in July 1958.

With a suitable ASTU, this antenna can be used on all HF bands from 3.5 to 30 MHz. In his 1966 article, and in follow-on articles,[4] Varney referred to this stub as a *matching section.* This term has lead to considerable over-the-air discussion and confusion over the years. In retrospect, Varney concluded [private communications, 1990], that it was probably a mistake on his part to refer to this feeder arrangement as a matching section. Only on the 20-meter band is the dipole impedance (approximately 100 to 150 Ω at the center of the 3/2-λ flat-top) transformed by the λ/2 open-wire line to an acceptable match for 72-Ω twin lead or coax feeder going to the transmitter.

This matching section could be described more accurately as a *series-section impedance transformer.* Varney recommended using open-wire tuned feeders to reduce losses due to the large amplitude of standing waves on this section of transmission line. While he did not specify the impedance, the constructional details given correspond to a transmission line impedance Z_0 equal to 523 Ω. Note that on bands other than 20 meter, the entire feeder is also a series-section impedance transformer, a factor that is often ignored.

Clearly this feed line arrangement is special, since standard center-fed dipoles use a single feed line of constant characteristic impedance. In effect, the G5RV uses a combination feed line—the open-wire line connecting the antenna to the low-impedance feeder, which in turn connects to the transmitter. This complicates discussion about this antenna and results in having to define impedance at three places (see Fig 1B) on the antenna system: the dipole feedpoint impedance (Z_1); the impedance at the junction between the open-wire line and the coaxial feeder (Z_2); and the impedance at the transmitter end of the coaxial feed (Z_3). The confusion over Varney's use of the words *special, matching section* and *feeder* persists today.

A view held by some is that the G5RV can

be used without an ASTU on all traditional amateur bands. It is curious how radio amateurs came up with this notion, because Louis Varney never said that it could be used without an ASTU!

The antenna is usually not a perfect match on any amateur band, and since the SWR can be high on some bands, G5RV recommended that no balun be used. This is another topic of some controversy. A balun should be used to feed a balanced antenna by an unbalanced transmitter. However, the use of a balun can, as noted above, lead to difficulties when the SWR is too high. Without a balun there will be some feed line radiation, which can be particularly severe at some frequencies.

Modified G5RVs

Taft Nicholson, W5ANB,[5] describes a similar antenna, which produced for him an SWR less than 3:1 on the 7, 14 and 28 MHz bands. Brian Austin, ZS6BKW (currently GØGSF), a prolific writer on modified G5RVs, described a computer-aided study to optimize dimensions.[6] His ZS6BKW version exhibited SWRs less than 2:1 on the 7, 14, 18, 24 and 29 MHz bands. He found that the original version (using a 300-Ω impedance transformer section) had SWRs < 2:1 on only two bands, 14 and 24 MHz. Our own G5RV antenna, fabricated for this study with a 50-Ω coaxial feed line, had SWRs less than 2:1 on 3.5, 7.5 and 19.5 MHz, but not on 14 MHz, the design band for the dipole (see below).

While Austin's computer-aided design was on the right track, the dipole impedance values he used were those for a thin-wire antenna in free space (theoretical published values). His computer program used a table of these values and solved the transmission line equation to calculate the impedance Z_2, and so deduce SWR on the low-impedance feeder. Real antennas are suspended at a practical height over a building or over finitely conducting ground. Notwithstanding, his calculated SWRs were shown to be in reasonable agreement with experiment.

Andrew Griffith, W4ULD,[7] has recently revisited Nicolson's antenna, using a modern antenna analysis code, *ELNEC*. His SWR vs frequency plots show low SWRs on portions of the 40, 20, 17, 15, 12 and 10-meter bands. However, he had to fabricate a special 380-Ω ladder-line feeder of a particular length, for a dipole of a particular length, and he had to add an extra length of 450-Ω ladder-line, switch-selected to provide an impedance transformer for the 15-meter band.

Is all this effort to modify the G5RV center-fed dipole really worthwhile? We discuss this below.

G5RV Measured Impedance Characteristics

Before presenting results obtained by measurement and by numerical modeling, let us consider what to expect. To a first approximation, a dipole antenna with its open-wire feeder is resonant when the electrical length of one arm of the antenna together with the electrical length of the open-wire feeder is approximately an odd multiple of λ/4. Thus, in the frequency range from 3 to 30 MHz we would expect the G5RV dipole and its open-wire line to resonate at five frequencies—and indeed we shall see it is resonant at five frequencies.

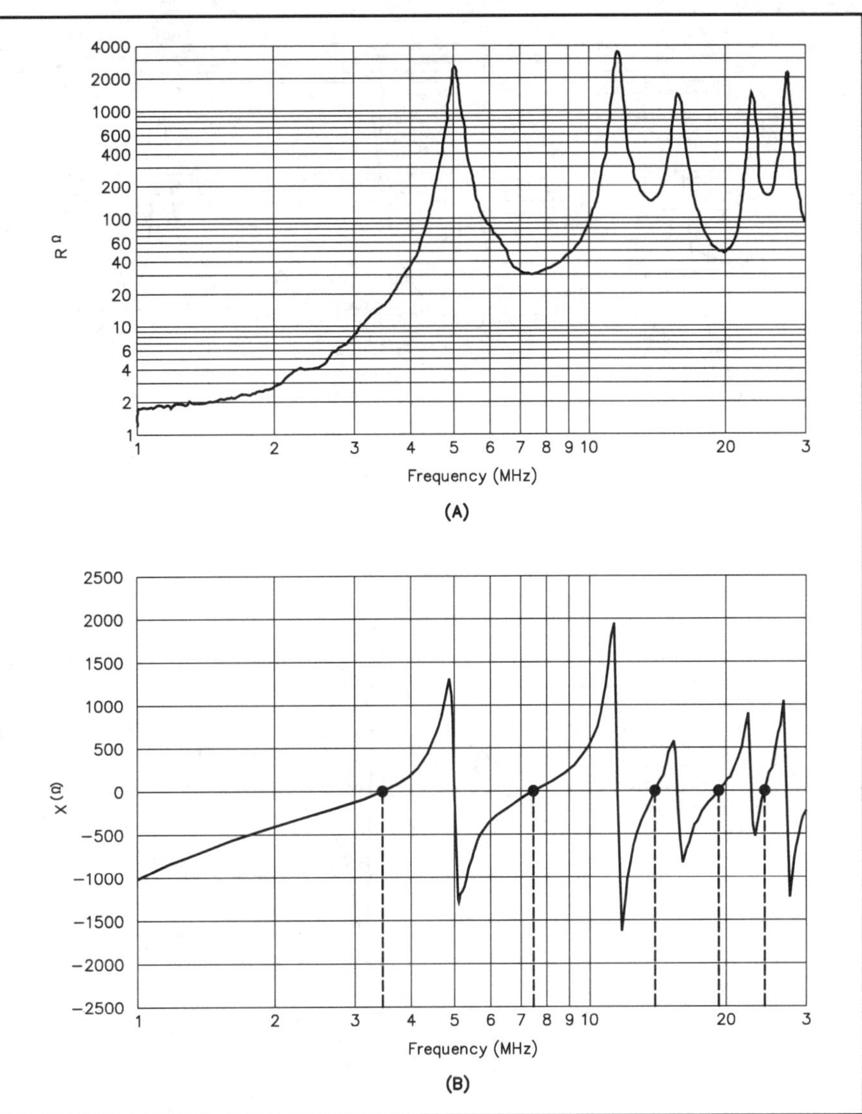

Fig 2—Balanced input impedance (Z_2) of the authors' G5RV in a drooping dipole configuration (Λ = 127°, approximately, apex height about 9 meters). At A, resistive portion of input impedance; at B, reactive portion of impedance. For dimensions see text or Table 1.

The authors' G5RV was installed as a drooping dipole, but with a shallow droop. The included angle (Λ) between the arms was about 127°, for an apex height of 9 meters. The dimensions are those given by Louis Varney: L_1 = 31.1 meters; L_2 = 10.36 × 0.95 = 9.85 meters of 450-Ω windowed transmission line. The velocity factor for the 450 Ω transmission line is 0.95, as shown in the expression for L_2 in **Table 1**.

The balanced impedance Z_2 (see Fig 1B) for this antenna was measured with a Hewlett-Packard Impedance/Gain/Phase Analyzer Model 4194A and an HP-supplied accessory balun transformer. Clearly the antenna is multiresonant; its reactance is zero for five frequencies in the 3 to 30 MHz band: 3.49, 7.52, 14.15, 19.5 and 24.6 MHz (see **Fig 2**). Note that the antenna was indeed resonant in the 20-meter band, at 14.15 MHz, as G5RV intended. The antenna's resistance at the resonant frequencies was 16, 31, 148, 48 and 162 Ω respectively. This variation in resistance illustrates in part the problem trying to achieve a low SWR at a number of harmonically resonant frequencies. For a 50-Ω feeder, the SWR at the corresponding resonant frequencies would be 3.1:1, 1.6:1,

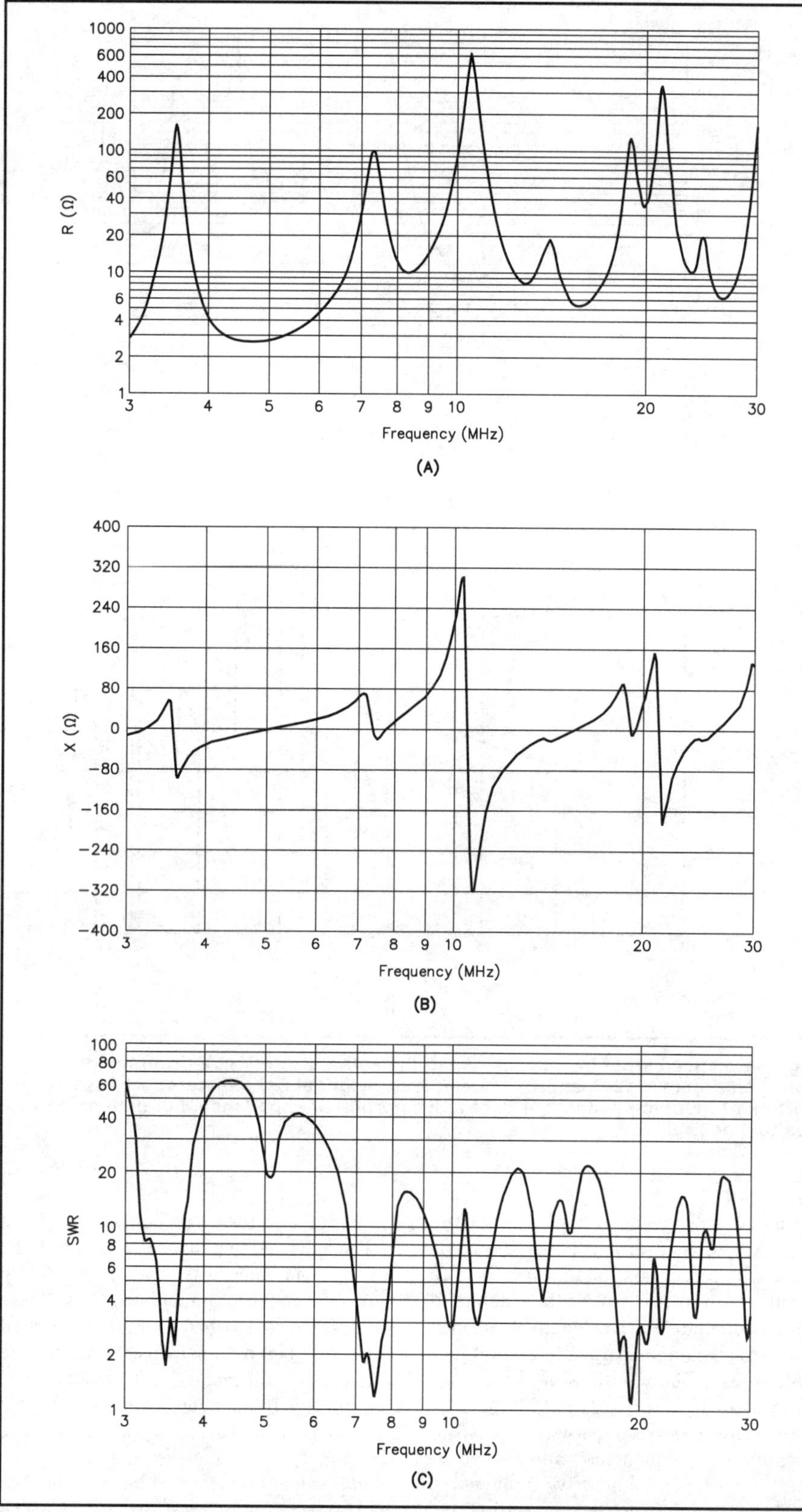

Fig 3—Unbalanced antenna system impedance (Z_3) and SWR for inverted-V antenna in Fig 2 when fed with 9.49 meters of 50-Ω coax (length includes the length of the 300-bead balun inserted between the coax and the 450-Ω ribbon). At A, resistive portion of input impedance; at B, reactive portion of impedance, and at C, SWR on 50 Ω transmission line.

3.0:1, 1.04:1, and 3.24:1. The other part of the problem is to dimension the antenna system (dipole length and feeder length) so that its multiple resonant frequencies fall within the desired amateur bands.

Author Belrose numerically modeled the antenna and its open-wire feeder, which was modeled as a part of the antenna system.[8] The calculated resonant frequencies (reactive component of the impedance Z_2 equal to zero), according to *MININEC* (*ELNEC* version[9]), are 3.56, 8.05, 14.67, 20.59, and 25.42 MHz. The antenna's impedance (resistance) at these resonant frequencies compared reasonably well with the experimentally measured values at the corresponding resonant frequencies. We would not expect an exact agreement, because the model calculation assumes for the impedance calculation that the ground beneath the antenna is perfectly conducting, whereas the real antenna was suspended over finitely conducting ground and over the roof of a field hut.

Our antenna was fed by an additional length of 50-Ω coax transmission line. In retrospect, following Varney's lead, we should have used 72-Ω coax. We used a 9.49 meter length of mini foam coax (total length including the length of a 300-bead current balun—to be discussed).

The antenna system impedance is changed by this feeder. In **Fig 3** the measured unbalanced input impedance (Z_3) versus frequency is given with this coaxial feeder and balun. Note in particular the anti-resonant responses in the 80-meter band, and just above the 40-meter band. The dipole with its 450-Ω feeder was resonant near these frequencies.

The SWR versus frequency is shown in Fig 3C for a 50-Ω line. While the SWR was low (less than 2:1) for three frequencies in the 3 to 30 MHz range, unfortunately it is high for most amateur bands. We should note that since a balun was used, there will be insignificant current on the outside surface of the feed-line coax, and therefore SWR should be independent of the length of the coax.

The G5RV type of antenna is clearly a multiband antenna. It is harmonically resonant on a number of frequencies in the 3 to 30 MHz band. For the radio amateur, the challenge seems to be to dimension the antenna so that the SWR is low for a number of amateur bands of principal interest. There are three parameters which can be changed: the dipole arm length, $L_1/2$, and the length (L_2) and impedance (Z_0) of the series section impedance transformer.

The difficulty in doing this and in duplicating the antenna can be judged by comparing some measured results given in Table 1. Here we compare the measured SWRs for two sets of almost identical antennas measured at different places and times:

- two G5RVs, the authors' and one fabricated and measured by ZS6BKW;[10] and

Table 1

Measured Bandwidth (SWR < 2:1) for Various Versions of the G5RV Type of Antenna

Physical Dimensions *Antenna Type*	L_1 *(m)*	L_2 *(m)*	Z_O *(Ω)*	*Velocity Factor*
G5RV*	31.1	8.7	300	0.85
Authors' G5RV	31.1	9.85	450	0.95**
ZS6BKW(2)	27.51	12.2	400	0.9
GØGSF	27.5	12.75	400	0.95

* Reference Note 6, SWR was measured with respect to a 75-Ω feeder.
** Windowed 450-Ω ladder line.

Antenna Type	*Frequencies in MHz for which SWR < 2:1*				
G5RV*	14.0-14.18	24.89-24.99			
Authors' G5RV	3.49-3.5	7.14-7.62	19.14-19.74		
ZS6BKW(2)	7.00-7.1	14.05-14.29	18.07-18.17	24.89-24.99	28.60-29.2
GØGSF	7.27-7.5	14.50-14.79	18.75-18.74	25.83-25.91	29.87-30.13

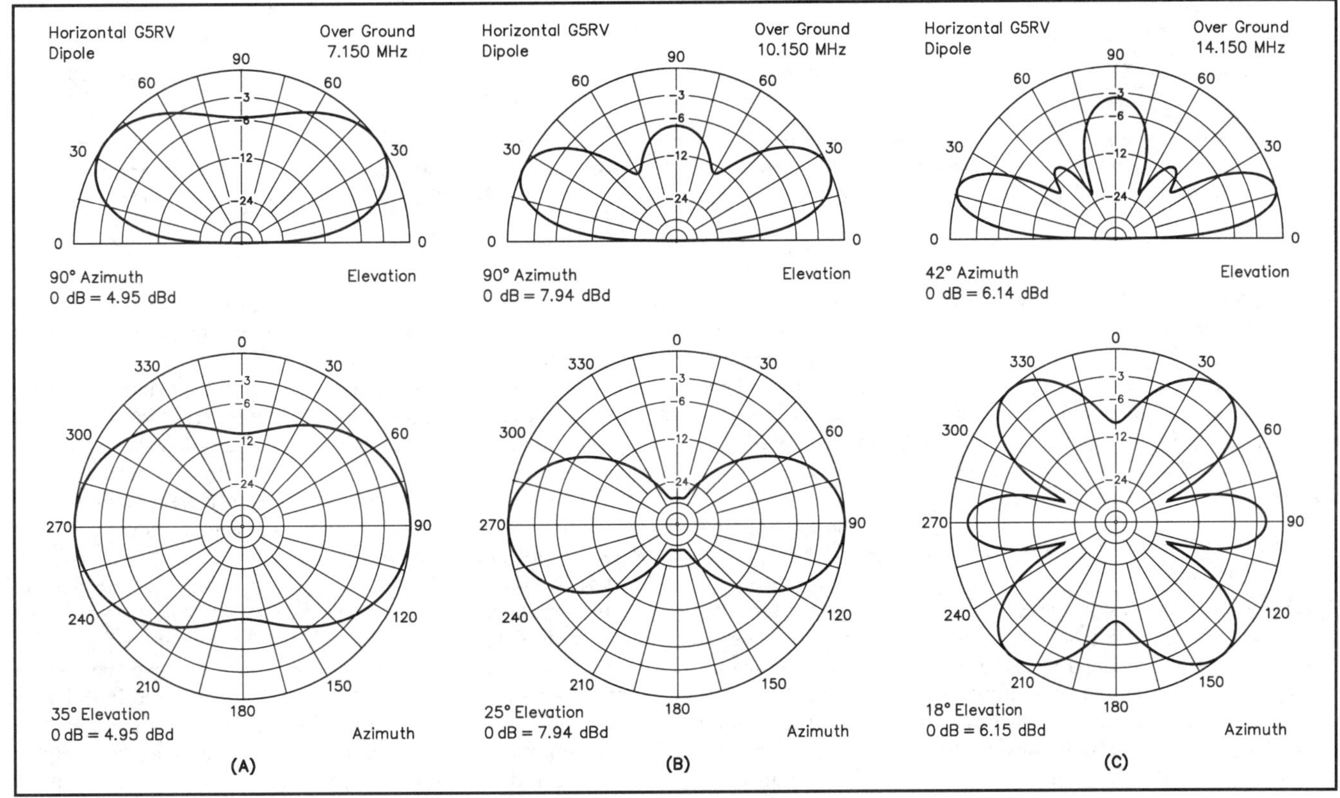

Fig 4—Principle plane radiation patterns (the azimuthal pattern at the elevation angle of maximum gain, and elevation angle at azimuth angle of maximum gain) for a horizontal (flat-top) 31.1-meter long dipole. Shown at A is pattern at 7.15 MHz; at B, pattern at 10.125 MHz, and at C, pattern at 14.15 MHz. The dipole height is 16.7 meters over average ground, and its orientation is along the 0° to 180° axis in the azimuth diagrams. Note how the pattern develops multiple lobes as the frequency is raised to 14.15 MHz.

- two so-called optimized versions, the ZS6BKW[2] antenna, and the GØGSF antenna,[11] which are almost identically dimensioned, but fabricated and measured at a different location and time.

For the two G5RV dipoles the lengths are identical, and L_2 is scaled in accord with the velocity factor for the transmission line used. Unfortunately, these antenna systems (dipole and feeder) are not quite identical, since ZS6BKW used 300-Ω twin-lead, and a 75-Ω coaxial feeder, whereas the authors used 450-Ω windowed ladder-line and 50-Ω coax. The authors also used a special balun (see below). Usually a balun is not used. However, the factor to notice is that the SWRs are low in different bands.

The ZS6BKW antenna exhibited low SWRs in five amateur bands. It was optimized in the field test to achieve this result.

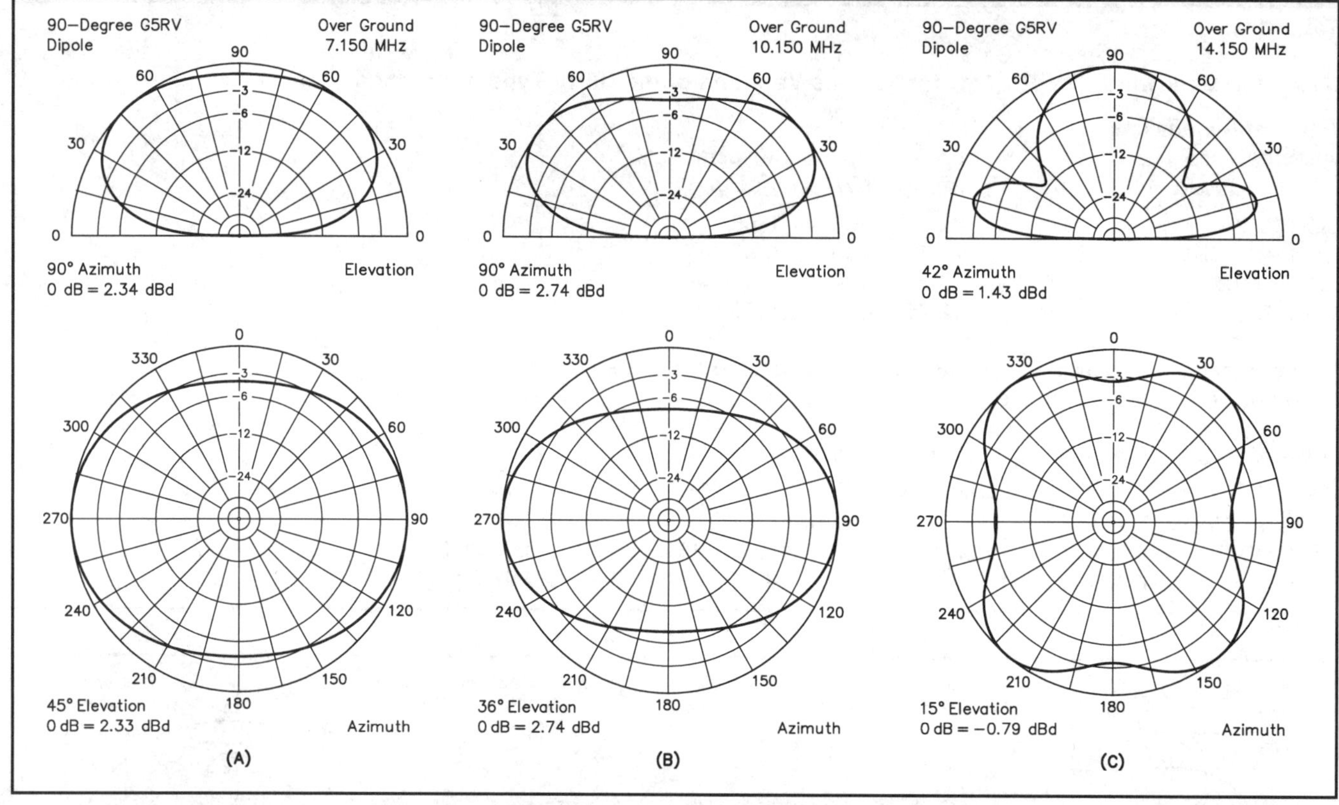

Fig 5—Radiation patterns for inverted-V antenna shown in Fig 1C: dipole length 31.1 meters, apex height 16.7 meters over average ground (conductivity of 3 mS/m and dielectric constant of 13). The included angle between legs of dipole is 90°.

The GØGSF antenna exhibited low SWRs in related bands, all just outside amateur bands.

The differences are easily explained. No two antennas are exactly identical due to differences in the characteristics of the feeder, the height above and the conductivity of the ground beneath the dipole.

Radiation Patterns

Dipole Orientation

Dipoles are often not installed in a horizontal straight line. Half wave dipoles are generally tolerant of bending, sloping or drooping to fit the antenna site. But this is not the case, as we shall see below, for multiband dipoles. We will consider here horizontal and drooping dipoles (often called inverted-V dipoles). We will not consider bent dipoles.

Louis Varney has always illustrated his multiband dipole mounted as a classical horizontal dipole, see Fig 1A and B. But radio amateurs frequently use multiband dipoles in a drooping dipole configuration, since only one support mast is needed. In fact this configuration is illustrated in two places in *The ARRL Handbook*, including the 1993 edition.[12] The sketch in *The ARRL Handbook* shows a multiband dipole with arm length ≥ 15.24 meters, fed with an open-wire transmission line. The caption under the sketch giving constructional detail states that "the included angle (Λ) between the arms of the dipole for best efficiency should be between 90° to 110°."

The authors have not seen substantiating data to support this recommendation.

Calculated Radiation Patterns

The azimuthal radiation pattern of the G5RV is similar to that of a λ/2-dipole, but its broadside directivity starts to break down into secondary lobes when the dipole arm length is greater than ⁵/₈ λ (dipole length > 1.25 λ). For our 31.1-meter dipole, this electrical length corresponds to frequencies > 10 MHz. While the antenna can be used on higher frequencies, the azimuthal pattern will be multilobed, which in the authors' view is not desirable.

Let us look at the radiation patterns. In **Fig 4** we plot the elevation and principle plane azimuth patterns for 7.15, 10.15 and 14.15 MHz of a horizontal 31.1 meter dipole at 16.7 meters over average ground (conductivity, σ = 3 mS/m, dielectric constant, ε = 13). The 3.75 MHz pattern (not shown) is dipole-like, with a maximum gain of 6 dBi according to *NEC-2* (*NEC/Wires Version 1.5*[13]), at a launch angle ψ = 90°. As frequency is increased, the directive gain increases, reaching a maximum just above the 30-meter band, where the dipole is a wavelength long. The azimuthal pattern is multilobed at 14.15 MHz, where the dipole is 1.5 λ long.

The elevation angle for maximum gain decreases continuously from 90° at 3.75 MHz to 7.5° at 29 MHz. When the azimuthal pattern becomes multilobed for amateur bands ≥ 14 MHz, the launch angles for all lobes at a particular frequency are the same.

A drooping dipole configuration is popular among amateurs who wish to use a single antenna for several bands of operation, and is considered by some to be even better than a horizontal dipole when used in this configuration. On **Fig 5** we plot the elevation and azimuth patterns for a drooping dipole at an apex height of 16.7 meters and included angle Λ = 90°. The radiation patterns are significantly different from those for the dipole in a horizontal orientation, and the gain is less at all frequencies, see **Table 2**.

The reduction in gain is particularly a problem at 14.15 MHz. (The detailed study looked at radiation patterns for all amateur bands.) The radiation pattern is complicated, and not well depicted by viewing patterns in two planes. The principle-plane pattern is the azimuth pattern at the elevation angle of maximum gain, and the elevation pattern at the azimuth of maximum gain. The azimuth pattern is very elevation-angle sensitive, since the vertical plane pattern changes with azimuth.

For example, compare the pattern for the horizontally mounted dipole at 90° azimuth (one of the two secondary azimuthal lobes) in **Fig 6A** with the pattern in Fig 4C for the same dipole, but at one of the four main lobes at 42° azimuth. In the horizontal configuration, the peak elevation angle is the same for major and secondary azimuthal lobes, at 18° elevation.

Now look at the pattern for the drooping dipole orientation at the same azimuth angles. In Figs 6B at 90° azimuth and in Fig 5C at 42° azimuth the elevation patterns are very different. In Fig 6B in the plane broadside to the dipole there is no low-angle lobe. At 42° azimuth in Fig 5C, the low angle lobe is a secondary lobe, since power is radiated upward (at $\psi = 90°$)—which is power wasted for useful communications. Fig 6C overlays for direct comparison the azimuth and elevation patterns for both dipole configurations at a 42° azimuth angle.

Let us try to interpret this result. The standing wave of current on a $\lambda/2$ dipole is a maximum at the center of the dipole and zero at the ends of the dipole, hence the current is a maximum at the apex height. Drooping the arms of the dipole has only a small effect on gain and pattern. When the length of the dipole is greater than $\lambda/2$, the place on the arms of the dipole where the current is a maximum will be displaced from the center of the dipole. These current maxima occur at a lower height when the arms of the dipole are drooping. We might therefore expect that the effective height of the dipole would be decreased, and hence the launch angle increased. But the pattern and gain changes are more complicated than this. When the dipole is horizontal, the take-off angle decreases continuously with increase in frequency, since the electrical height of the dipole increases with increase in frequency. When the pattern becomes multi-lobed, ψ_{max} has the same value for all of the lobes in the azimuthal plane. But this is not the case for the drooping dipole, and the differences become greater the greater the droop.

In the detailed study, radiation patterns for $\Lambda = 100°$ and 127° were examined (not shown here). As one would expect, the patterns for $\Lambda = 127°$ are more like those for the

Table 2

Gain vs Frequency for a 31.1 Meter Dipole Over Average Ground, for Two Dipole Orientations: Horizontal and Inverted-V, with Included Angle between Arms ($\Lambda = 90°$).

Frequency (MHz)	*Max Gain dBi Horizontal*	*Max Gain dBi Inverted-V*
3.75	6	4.3
7.15	7.1	4.5
10.15	10.1	4.9
14.15	6.4	1.4

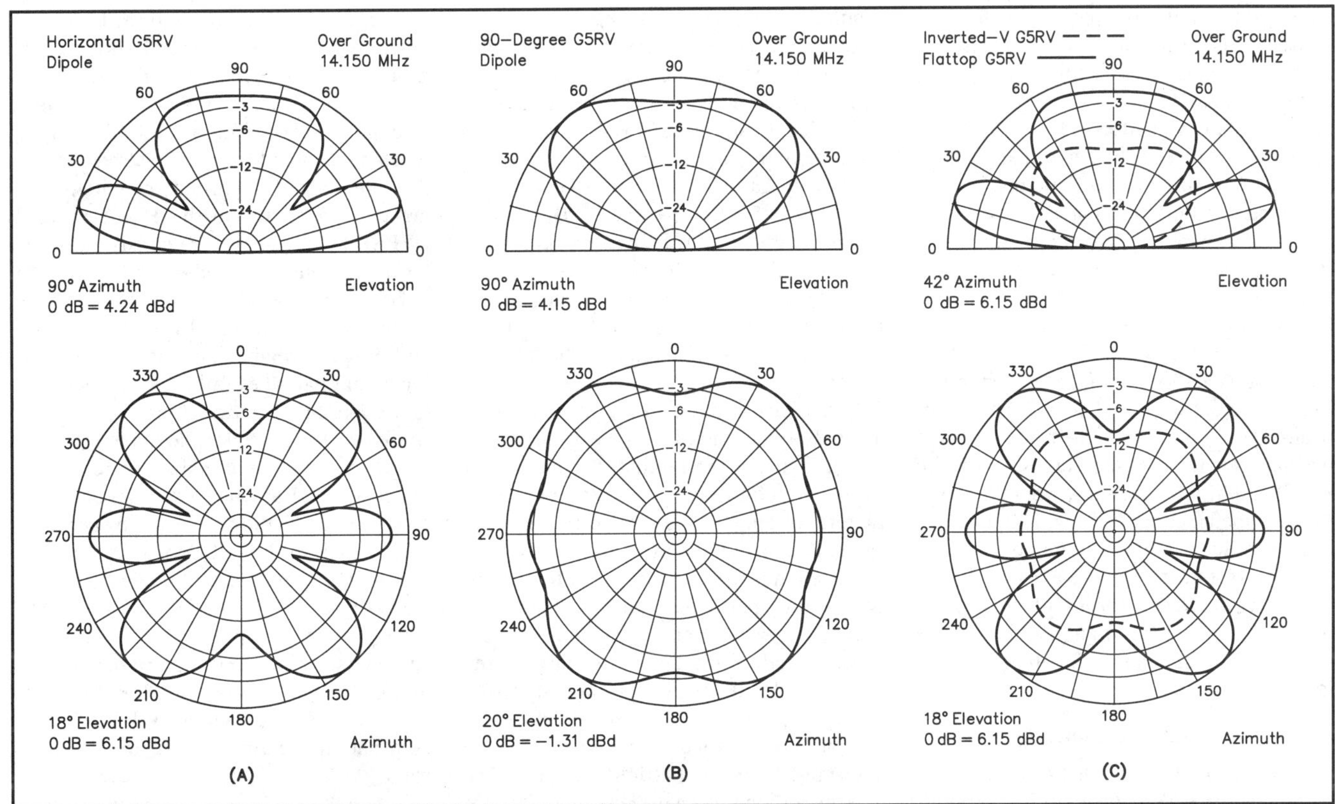

Fig 6—Radiation patterns for a horizontal dipole 16.7 meters over average ground, compared with a drooping-dipole (inverted-V) orientation ($\Lambda = 90°$, apex height 16.7 meters) at frequency of 14.15 MHz. The elevation patterns are for an azimuth angle of 90°, which corresponds to a secondary lobe clearly evident for the flat-top orientation shown in A. At B, the patterns for a G5RV dipole with 90° included angle between the legs is shown. The elevation pattern is for a 42° azimuth angle. At C, the patterns for the flattop and inverted-V are overlaid for direct comparison. When operated at a harmonic of its fundamental resonance, the pattern for the inverted-V configuration deteriorates considerably compared to the flat-top mounting configuration.

horizontal dipole and the patterns for Λ = 100° resemble those for Λ = 90°.

Clearly, a multiband dipole if used at frequencies where its electrical length is appreciably longer than $\lambda/2$ should be installed as a horizontal dipole, or as horizontally as possible.

Feeding a Multiband Dipole

In the authors' view, the correct length of the feed line for a multiband dipole is the length required to go from the output terminals of the antenna system tuning unit to the antenna terminals! Regardless of the length of the feed line, both the antenna and the feed line are made resonant by the conjugate match action of the tuner. Our recommendation, based on personal experience, is to use open-wire line for the *total length* of the required feeder—this results in low losses. This view is also held by Walter Maxwell, W2DU. An additional advantage in using a full-length open-wire feed line is that the balun can be placed inside the shack, so that its performance can be evaluated. Alternatively, a balanced ASTU can be used without a balun.

Note that the 150 Ω impedance of the 1.5 λ, 31.1 meter dipole on the 20-meter band could have been matched by employing two lengths of coaxial cable, to form a balanced transmission line, a type of transmission line often used by the authors. Varney designed his antenna to be resonant on the 20-meter band so this feeder arrangement would have suited him.

In their studies of the performance of dipoles and baluns, the authors have used a balanced transmission line as an extension to the open-wire line. Such a transmission line is easy to fabricate and feed into the shack, and losses are not too high unless a long feed line is used.

The balun we have used is a 300-bead 1:1 balun of type 73 mix material on mini-Teflon-dielectric coax. This W2DU ferrite-bead-over-coax balun is the best balun that we have so far devised for use with HF multiband dipole antennas, where the mismatch impedance can be high. The commercial 50-bead baluns are perfectly satisfactory for resonant dipoles, even for a 75/80-meter dipole, but for multiband dipoles where the mismatch is high (SWR > 3:1) the longitudinal impedance of a 50 bead balun is too low.

Conclusions and Recommendations

When reading about what has been written about the G5RV, or modified versions based on the G5RV principle, we must not think of the transmission line segment L_2 as a matching section. More generally, it is a series-section impedance transformer.

This view is particularly important when considering the combined effect of the open wire line and the coaxial feeder. Clearly the coaxial feeder of length L_3 is an additional series-section impedance transformer, and also includes the impedance transfer characteristics of the balun. The analysis approach (previously published) has considered only the impedance-transfer characteristics of the open-wire-line part, but has ignored the effect of the lead-in coaxial cable. This complication was avoided by considering only SWR rather than impedance.

While there is nothing fundamentally wrong with this approach, in our view this confuses the understanding of the characteristics of the antenna system. Besides this, since a balun is not often used, one can hear over the air discussion about the G5RV, where some amateurs have found it necessary to shorten or lengthen the coax to achieve an acceptable SWR on a particular band—because for their antenna the coax feeder is a part of the antenna system!

In spite of the studies by those who have tried to optimize dimensions for the G5RV type of multiband dipole with its combination feeder arrangement, there is in our opinion little to be gained by this endeavor; since after cut-and-try and cut-and-try, if one decides at a later date to improve performance by significantly raising the height of the dipole, everything will change.

Louis Varney in correspondence with author Belrose [private communications, 1990] expressed the view: "While I think that Brian Austin's ZS6BKW-designs of a multiband antenna are interesting, in view of the fact that whatever bands he manages to achieve a SWR < 2:1, there will be other bands (or portions of bands) for which the VSWR > 2:1, and therefore the use of a suitable ASTU is essential. I would prefer more simply to use an ASTU." This comment is in accord with the view held by the authors.

Considering that a dipole's impedance depends on its height above ground, the conductivity of the ground, the method of support (metal towers or trees), and other factors, attempts to optimize the antenna system to provide low SWR on several bands are not worthwhile, excepting if operation on a particular band (or bands) for Field Day use without an ASTU is desired.

The length of the dipole depends on the frequency bands of interest, and since the radiation pattern changes with frequency, the optimum length for the dipole should be determined by a particular user. The 31.1 meter length is a dimension that is approximately $^3/_8$ λ long at the lowest frequency of operation, 3.5 MHz. This is a reasonable length. A dipole with shorter arm lengths quickly becomes impractical.

The multilobed pattern for frequencies higher than 10 MHz is in the authors' view undesirable. If propagation in a particular azimuthal direction is desired, it is important that the azimuthal pattern not have a null in the direction of interest. Dipole lengths > 1.25 λ do not in general meet this requirement. For operation on the higher bands a scaled-down multiband dipole could be used, but other quite different antenna types could be used instead.

Concerning dipole mounting orientation, it is rather unimportant whether the dipole is horizontal or drooping, if the dipole length is not appreciably > $\lambda/2$ at the frequency in use. However, when the dipole's length becomes > $\lambda/2$, it should be erected as a horizontal flat top, or as horizontal as possible ($\Lambda \geq 120°$).

The authors of this article have tried to downplay the view that the length of the dipole and how it is fed are critical parameters affecting antenna performance. Certainly one dipole should not be superior compared with another whose length is not very different. For example, W4ULD[14] found that his dipole (28 meters long), at 7 MHz and above, is more efficient than his G5RV (a dipole 31.1 meters long). Such an observation is inconsistent with expectation. It is, however, clear that transmission-line loss is an important parameter. One cannot expect that a coaxial-fed half-wave dipole, dimensioned for a lower band, will perform well as a multiband antenna. See *New Ham Companion* article by Steven Ford, WB8IMY.[15] This author reported "stunning" success for his 20.1 meter dipole when he changed from coaxial feed to ladder line, judged in particular by its performance in a pileup on the 15-meter band. [WB8IMY's results were consistently better on *all* bands he tried.—*Ed.*]

G5RV designed his antenna for the 20-meter band. He has always been pleased with its performance on that band, providing azimuthal lobes fall in directions of interest. His dipole is 1.5 λ long on 20 meters. In his series of articles, he made reference to half-size and double-size versions of his antenna, which are of interest, depending on bands for which the dipole will be used, and size of back yard. WB8IMY's dipole can in accord with this line of thought be considered as a scaled down G5RV, since his half wave 40 meter dipole is 1.5 λ long on 15 meters. It should therefore perform for him on 15 meters like the G5RV does on 20 meters. Besides which, WB8IMY's dipole should be resonant in the 15 meter band.

In conclusion, it is curious that so much effort by radio amateurs has been devoted to devising a multiband antenna that is resonant in bands of interest. What is more important is that the radiation pattern has directivity in directions of interest and at a suitable launch angle for the desired path lengths. A good antenna is not necessarily one that has low SWR in bands of interest. Antenna systems can be tuned by an ASTU, but you should pay attention to keeping feed-line losses low, use a good quality ASTU, and give due consideration to the type of balun used.

A simple all-band antenna that avoids the difficulty of pattern change with frequency, and is particularly convenient for frequencies > 10 MHz, is a compact horizontal loop.[16] Another antenna type, which the authors favor, is to use several dipoles in parallel (a so-called *stagger-tuned dipole*). This antenna system can be resonant on several bands, and its dipole-like broadside directivity on several bands is a desirable feature.

The wire antenna is obviously a field for experiment in "do-it-yourself" antenna construction.

Acknowledgments

The authors work for the Communications Research Centre, Ottawa, and they therefore have access to instruments and computing facilities not available to the average radio amateur.

Notes and References

[1]Pat Hawker, "More on the G5RV/ZS6BKW Antennas," Technical Topics, *Radio Communication*, Jan 1993 pp 43-45; Feb 1993, p 34; and Apr 1993, pp 53-54.

[2]W. Maxwell, *Reflections—Transmission Lines and Antennas*, ARRL Publication, 1990, pp 20-13 to 20-16.

[3]*The ARRL Antenna Book*, Eleventh Edition, 1968, pp 186-188.

[4]L. Varney, "An Effective Multi-Band Aerial of Simple Construction," *RSGB Bulletin*, Jul 1958; "The G5RV Aerial—Some Notes on Theory and Operation," *RSGB Bulletin*, Nov 1966, pp 705-748; and "G5RV Multi-band Antenna—Up-to-Date," *Radio Communication*, Jul 1984, pp 572-575.

[5]T. Nicholson, "Compact Multi-Band antenna with No Traps," *QST*, Nov 1981, pp 26-27.

[6]B. Austin, "Computer-aided design of multi-band dipole based on the G5RV principle," *Radio Communication*, Aug 1985, pp 614-617, 624.

[7]A. S. Griffith, "The 1/3-Wavelength Multi-band Dipole," *QST*, Sep 1993, pp 33-35.

[8]J. S. Belrose, "Modeling HF Antennas with *MININEC:* Guidelines and Tips from a Code Users Notebook," *ARRL Antenna Compendium, Vol 3*, 1993, pp 156-164.

[9]*ELNEC* antenna analysis software, Roy Lewallen, W7EL, PO Box 6658, Beaverton, OR 97007.

[10]See Note 6.

[11]See Note 1, *Radio Communication*, Jan 1993, p 44, Fig 9.

[12]*The ARRL 1994 Handbook*, ARRL, p 33-9, Fig 15.

[13]*NEC/Wires* antenna analysis software, Brian Beezley, K6STI, 3523 Linda Vista Dr, San Marcos, CA 92069.

[14]See Note 7.

[15]S. Ford, "The Lure of the Ladder Line," *New Ham Companion* section, *QST*, Dec 1993, pp 70-71.

[16]J. S. Belrose, "An Update on Compact Transmitting Loops," *QST,* Nov 1993, pp 37-40.

[17]R. Holmes, "Hamtenna with the Lot," *Radio Communication*, Oct 1993, pp 69-70.

The Improved Telerana, with Bonus 30/40-Meter Coverage

By Markus Hansen, VE7CA
674 St. Ives Cres.
North Vancouver, BC
V7N 2X3, Canada

VE7CA revisits the Telerana, using computer modeling to improve the F/B on 20 and 15 meters. He also added 30/40-meter coverage in a clever fashion.

In the July 1981 issue of *QST*, YV5DLT, Ansyl Eckils, described an innovative method to construct a log-periodic dipole array (LPDA). Using four fiberglass poles emanating from a center hub and stringing rope around the perimeter of the ends of the fiberglass poles, he produced a light but strong framework to support an LPDA made of wire elements. This antenna was inexpensive to construct relative to purchasing a new triband Yagi antenna. Also, it was easily duplicated by anyone who has only limited experience with hand tools.

The antenna worked equally well on all amateur bands from 20 to 10 meters, including the WARC bands. Furthermore, only one length of coax was required to feed the antenna. Sounds too good to be true, doesn't it? Ansyl called it the *Telerana*, which means spider web in Spanish. The Telerana is described in the last three editions of *The ARRL Antenna Book*.

I constructed my first Telerana in 1987. The new antenna performed as expected with one exception—I was disappointed with the front-to-back ratio, particularly on the lower bands. This was confirmed by listening to local hams as I turned my antenna and made note of the S-meter readings. The front-to-back ratio on 20 meters was, at best, only 2 S units, increasing to approximately 3 S units on 10 meters. That equates to 12 dB on 20 meters.

Measuring the Telerana

I contemplated various possibilities to improve on an already-good antenna. Before attempting to modify the design of the Telerana, however, I felt it was necessary to be able to measure the radiation pattern before and after modifications were made. Otherwise, there would be no way of determining if the modifications were an improvement or not.

Wayne Overbeck, N6NB, described a method he used to determine the gain of VHF antennas when he was developing the Quagi antenna.[1] Wayne's method appeared simple and easy to duplicate. I will describe his method briefly, since the reference is no longer in print. The method requires:

1. A receiver with the ability to turn the AGC off.
2. A VU meter and a transformer to match the impedance of the VU meter to the output impedance of the receiver.
3. A stable signal source for the frequency range of the antenna to be tested.

To measure the radiation pattern of an antenna, connect the VU meter to the receive audio outlet and turn off the receiver AGC. Place a signal source several wavelengths away from the antenna to be tested. Turn the antenna so that the antenna's front lobe is pointed toward the signal source. Adjust the RF and audio gain controls so that the receiver is not saturated and the VU meter reads zero. It helps to have an accurate attenuator, as most VU meters have a scale that is only usable over a 10 dB range. I constructed the attenuator featured in both *The ARRL Handbook* and *The ARRL Antenna Book*. Turn the antenna and at every 10° or 15° record the reading from the VU meter onto an ARRL PATTERN WORKSHEET. These worksheets are available from ARRL order no. 1360 (100 sheets, 8.5 × 10 inches)[2] and should be part of any antenna experimenter's list of supplies.

The pattern produced by this method will show you a lot about the antenna you are testing. You can easily determine the front-to-side and front-to-back ratio, and you will quickly be aware of any sidelobes that are present. If two antennas are close to each other, and you suspect there is interaction between the two, you can run a pattern check first with both antennas in place and then again after removing one of the antennas.

The purist will say, and rightly so, that the radiation pattern depends upon the angle of arrival of the received signal. Further, because of surrounding antennas, power and telephone lines, there may be reflections affecting the shape of the pattern. Therefore this method does not always produce a totally accurate picture of the radiation pattern. However, the purpose of using this method is that it produces meaningful *relative measurements*. When changes are made to an antenna design, the experimenter knows with some certainty that the changes are either positive or negative.

The antenna being tested must always be in the same location and at the same height, and the signal source must always be in the same location when taking antenna field measurements. I cannot overemphasize this point. My QTH is located on the slope of a mountain, in a city environment. Electrical, telephone and cable TV lines are all overhead. You can just imagine the reflections occurring here.

For a signal source I constructed a 7 MHz VFO, followed by a low-power class C amplifier, rich in harmonics. Thus I could use the same VFO on 20, 15 and 10 meters. I coupled the output amplifier to two short horizontal lengths of wire using a few turns wound around the output transformer. Since the signal produced by the VFO is quite weak, I found that I have had to record my antenna pattern measurements at night when the bands are quieter.

The effect of the local environment became very evident when I moved my signal source to different locations, such as up the hill one block or down the hill one block. Though there was little change in the front lobe of the antenna pattern, the side and back lobes all showed different peaks and nulls. Make sure that you maintain your signal source in one particular location throughout the course of an antenna experiment!

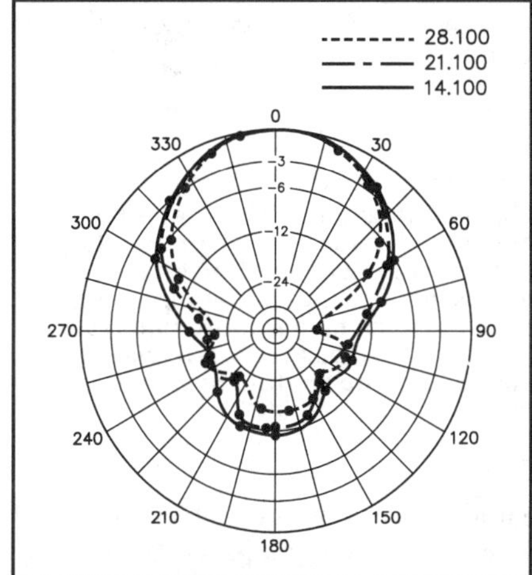

Fig 1—Measured azimuth patterns of original Telerana design, 60 feet high on tower, on three bands. The front-to-back ratio varies with frequency in a band, but is never very impressive.

Fig 1 represents the measured patterns of the original Telerana on 20, 15, and 10 meters at my QTH. Notice that the front-to-back ratio decreases with frequency. Peter Rhodes, K4EWG, mentioned this in his article, "The Log-Periodic Dipole Array."[3] He recorded front-to-back ratios similar to those I measured on the Telerana.

Improving the Telerana

Since I wanted to improve the front-to-back ratio on 20 meters in particular, I modified the Telerana by removing the two longest elements (which are resonant below 20 meters) and added a 20-meter parasitic reflector. I used this approach because I wanted to use the existing framework supporting the Telerana. I estimated the length of the 20-meter reflector by adding 5% to the length of a standard 20-meter wire-element dipole, resonant at 14.2 MHz.

Much to my disappointment, when I put the array back on the tower, the SWR was very high at the bottom of 20 meters—only at 14.350 MHz did the SWR start to decrease below 3.1:1 Furthermore, the front-to-back ratio was now only about one S unit! It was obvious that the longest remaining element of the modified Telerana was resonant above the 20-meter band and the reflector was not tuned to the correct frequency.

I decided to recalculate the lengths of the LPDA so that the longest element was resonant near the bottom of the 20-meter band. I also had to find a way to tune the 20-meter reflector where I wanted it to be.

My friend Darrell Wick, VE7FCR, had taken a keen interest in the Telerana design and my experimenting. He offered to write a computer program so that we could easily try different design parameters for the LPDA. He used the design procedure outlined in the 15th edition of *The ARRL Antenna Book*. Darrell had also purchased an antenna software program called *MN*, a derivative of a powerful antenna-modeling program called *MININEC* and was anxious to try something other than the sample files that came with the program.

Darrell's program also creates antenna files that can be used in other programs, such as *MININEC3*, *MN*, *ELNEC* and *NEC*. This became very useful when we began altering different parameters of the Telerana design, because of the complex geometry of such a wire antenna. (Darrell's program is called *LPDA* and is available from him directly. See Notes.) If you have an old computer and no math coprocessor, be prepared to wait for a long time for the computer to print out a

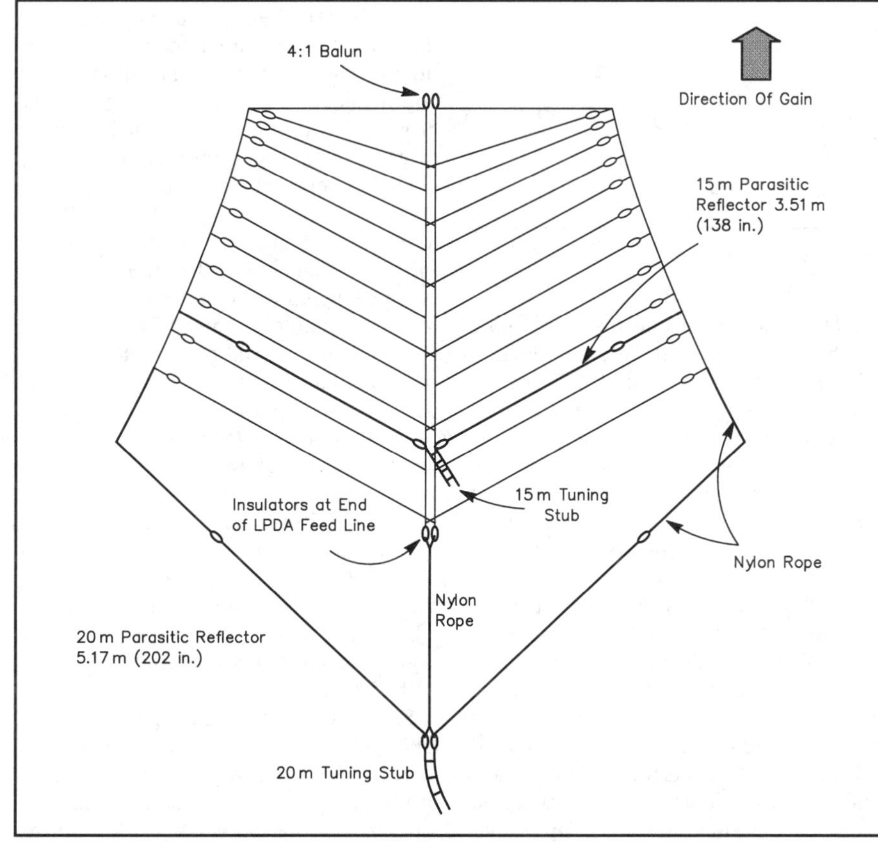

Fig 2—Physical layout of modified Telerana with 20 and 15-meter reflectors added. Note the tuning stubs for the added reflectors.

Table 1
Element Lengths and Spacings in Inches

Element Number	*1/2 Element Length (inches)*	*Spacing (inches)*	*Total Distance (inches)*
R1	202.0	102.0	0.0
L1	210.1	38.7	102.0
L2	191.2	17.6	140.7
R2	138.0	17.6	158.3
L3	174.0	32.0	175.9
L4	158.3	29.1	207.9
L5	144.1	24.5	237.0
L6	131.1	24.1	261.5
L7	119.3	22.0	285.6
L8	108.6	20.0	307.6
L9	98.8	18.2	327.6
L10	89.9	——	345.8

Note: The reflector element lengths do not include the lengths of the stubs.

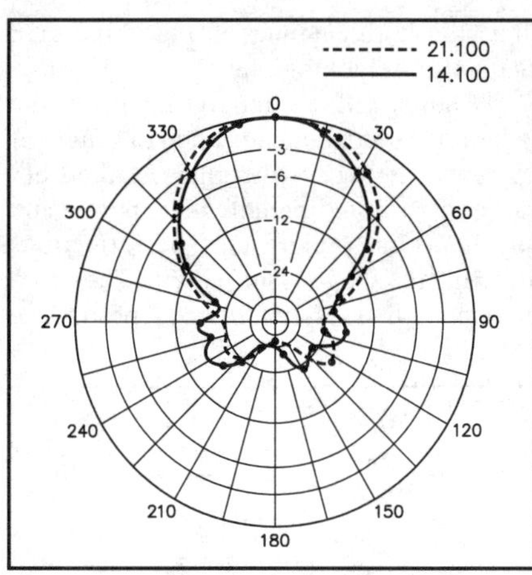

Fig 3—Measured azimuth patterns of modified Telerana after incorporating reflectors for 20 and 15 meters. These measurements are for the frequency showing maximum front-to-back ratio. Note that the peak rearward lobes are about 20 dB down from peak of forward lobe.

Table 2
Modified Telerana Design Parameters

Lower frequency	f1 = 14.05 MHz.
Upper frequency	fn = 30.00 MHz.
Design constant	tau = 0.91
Relative spacing constant	sigma = 0.046
Apex angle	2a = 52.1°
Cotangent of angle	Cot(a) = 2.0444
Design bandwidth (fn/f1)	B = 2.14
Active region bandwidth	Bar = 1.2275
Structure bandwidth	Bs = 2.621
Diameter of elements	= 14 AWG
Feed line impedance	R_0 = 208 Ω
Antenna feeder impedance	Z_0 = 433.81 Ω
Diameter of feeder wire	= 14 AWG
Feeder spacing	S = 1.9757 inches

pattern for a multi-element array such as the LPDA. The original Telerana design took *MN* over 8 hours to produce the pattern on an old IBM compatible XT 8088 computer! With a math coprocessor and a new 386-based computer it takes under 2 minutes.

Using Darrell's program and *MN*, we tried many different designs for the LPDA, with and without parasitic reflectors. The right parasitic reflector lengths gave a considerable improvement in the front-to-back ratio. Even on 15 meters, where the 15-meter reflector is positioned between elements of the LPDA, the front-to-back ratio increased by over 15 dB.

Fig 2 shows the location of the parasitic reflectors relative to the elements of the modified Telerana design. Refer to *The ARRL Antenna Book* for construction details of the support structure and feed line. The element lengths and spacings are shown in **Table 1**. These were chosen so that we could use the fiberglass pole lengths of the original Telerana design. This design has a spacing of 2.59 meters (8.5 feet) between the 20-meter reflector and the longest element of the LPDA, as measured along the transmission line connecting the elements. According to *MN*, this appears to be close to the minimum distance that both works and yet is not overly critical for dimensional tolerances. **Table 2** shows the design parameters for the new array.

Darrell spoke to the author of *MN*, Brian Beezley, K6STI, about whether the program can accurately model an array like the LPDA, with multiple driven elements fed with a transposed transmission line at each element. Brian indicated that trying to model transmission lines in *MN* by assuming multiple sources with a 180° phase shift at each dipole feedpoint probably is not entirely correct. However, there was good agreement between the pattern produced by *MN* and the measured results, despite not knowing the actual phase shift.

A new array was constructed using the element lengths shown in Table 1. This time, before mounting the antenna on top of the tower, I decided to tune the reflector by using a tuning stub in the middle of the reflector element, in the same way that Quad reflector elements are often tuned. The array was placed so that I could reach it when standing on the ground. Using a receiver connected to the array and my portable signal source placed several wavelengths away, I was able to adjust the length of the stub for minimum signal strength. I knew that the resonant frequency of the reflector would change when it was mounted on top of the tower, but I didn't know by how much. So I tuned it for 14.100 MHz. When the array was placed on top of the tower, I ran pattern checks on several frequencies. I found that the frequency where the front-to-back ratio was the highest decreased approximately 175 kHz when the array was raised 60 feet above ground.

Lowering the array once again, I added a 15-meter reflector and also readjusted the 20-meter reflector stub length so that when the array was on top of the tower the highest front-to-back ratio would be at 14.025 MHz.

Fig 3 shows the radiation patterns for 20 and 15 meters after the reflectors were added. The front-to-back ratio on both bands now exceeds 30 dB at the design frequency. Note that the worst-case lobes in the rear hemisphere behind the main lobe were down about 20 dB, as best I could measure

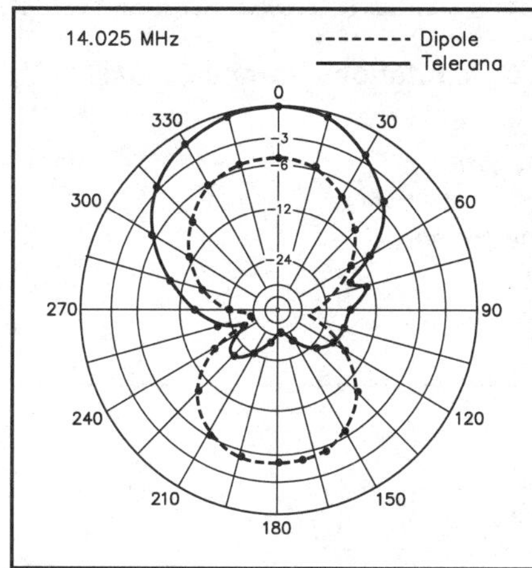

Fig 4—Measured 20-meter azimuth pattern for modified Telerana compared to reference trap dipole mounted 9 feet below Telerana. The height of antennas was made comparable for these tests by lowering tower to 48 feet for Telerana. The Telerana exhibits 5.25 dB more gain than reference dipole.

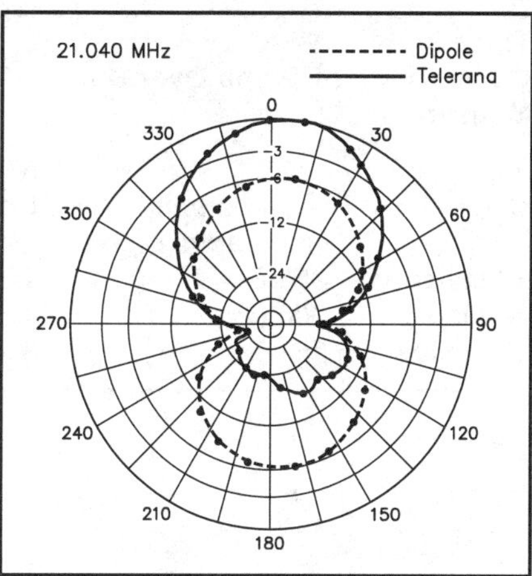

Fig 5—Measured 15-meter azimuth pattern for modified Telerana compared to reference trap dipole mounted 9 feet below Telerana. The height of antennas was made comparable for these tests by lowering tower to 48 feet for Telerana. The Telerana exhibits 6 dB more gain than reference dipole.

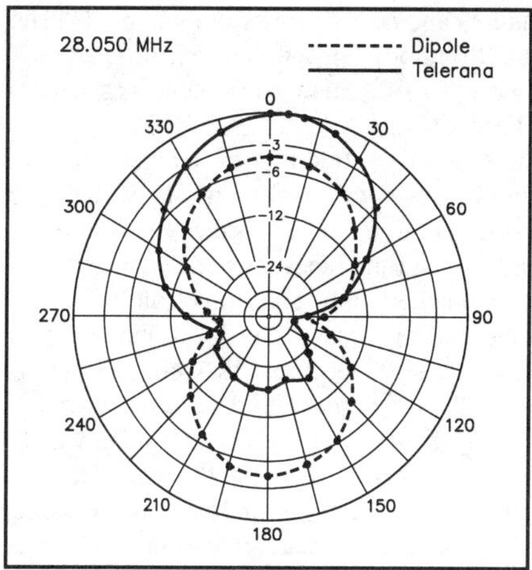

Fig 6—Measured 10-meter azimuth pattern for modified Telerana compared to reference trap dipole mounted 9 feet below Telerana. The height of antennas was made comparable for these tests by lowering tower to 48 feet for Telerana. The Telerana exhibits 4.25 dB more gain than reference dipole.

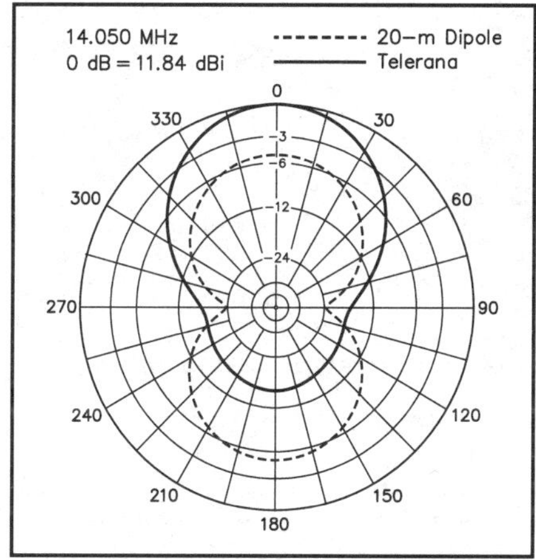

Fig 7—*NEC2* comparisons of Telerana's azimuth pattern on 20 meters, compared with reference dipole at the same height. Ground conductivity of 5 mS/m and dielectric constant of 13 were assumed for these computations. *NEC2* predicts that Telerana should have almost 5 dB of gain over reference dipole.

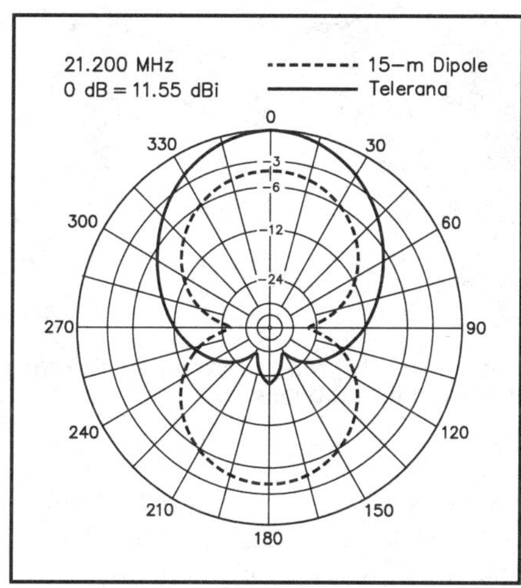

Fig 8—*NEC2* comparisons of Telerana's azimuth pattern on 15 meters, compared with reference dipole at the same height. Ground conductivity of 5 mS/m and dielectric constant of 13 were assumed for these computations. *NEC2* predicts that Telerana should have just over 4 dB of gain over reference dipole.

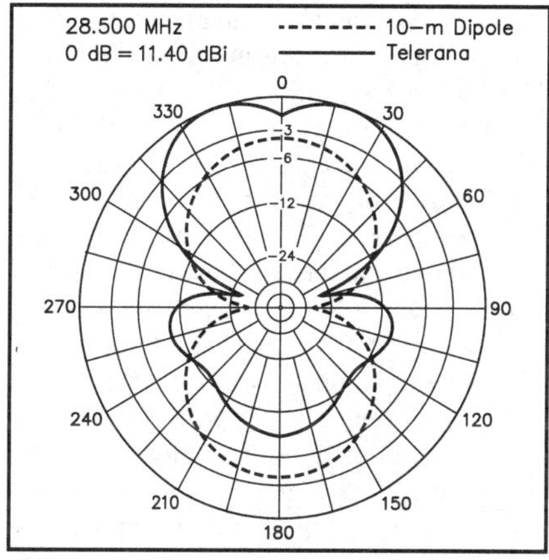

Fig 9—*NEC2* comparisons of Telerana's azimuth pattern on 10 meters, compared with reference dipole at the same height. Ground conductivity of 5 mS/m and dielectric constant of 13 were assumed for these computations. *NEC2* predicts that Telerana should have peak gain (on either side of small dip at nose of main lobe's response) of just under 3 dB of gain over reference dipole.

them. I tuned the 15-meter reflector when the array was on the ground. There was no change in the frequency of the 15-meter reflector when the array was mounted back on the tower. Presumably, this is because the 15-meter reflector has so much wire surrounding it that it is not affected much by proximity to the ground.

Antenna Gain

Have you ever wondered how much gain your new beam antenna really has over the dipole it replaced? Was it really worth the effort to put up a tower and build a new multi-element wonder? Until computer antenna-modeling programs came along, these kind of questions might have kept me awake at night. Modeling programs have done away with a lot of myth and wild exaggerations regarding antenna gain. Dean Straw, N6BV, Senior Assistant Technical Editor at HQ, offered to do some modeling of the modified Telerana using *NEC2*, the big brother to *MN*. As well, he suggested that I mount the driven element of my multiband beam and use it as a dipole underneath the modified Telerana. This way I could make a few hands-on comparative measurements.

I mounted the tribander's driven element 9 feet below and 90° to the front-back axis of the modified Telerana in order to minimize the interaction between the two antennas. My modified Telerana is mounted on a motorized tubular crankup tower, allowing me to easily change the tower height so that the antennas are at the same height when I make comparative measurements. **Fig 4**, **5** and **6** show measurements for 10, 15 and 20 meters, comparing the modified Telerana to the triband dipole. **Fig 7**, **8**, and **9** is what *NEC2* produced. There is a gain discrepancy between the field measurements and *NEC2*, but remember my QTH is anything but a proper antenna test range. A summary of gain measurements is shown in **Table 3**.

You will notice that on 10 meters *NEC2* predicts a notch in the middle of the front lobe. Though I never noticed this when taking actual field measurements, it may be that it is masked by interactions from surrounding house wires and the 30/40-meter dipole above the Telerana. As well, the *MN* or *NEC2* programs model antenna wires as being absolutely straight, with no bow in the center from the weight of the wire. The bow in the wire is only an inch or two, however this may account for some of the difference between the computer plots and the field plots.

The *NEC2* pattern plots in Fig 7, 8, and 9 are shown in dBi and are over real ground with the antenna at 60 feet. This adds another 6 dB of gain due to "ground reflection gain" at the peak elevation angle. **Fig 10** is a photograph of the completed Telerana with the reference dipole under it.

Fig 11 is a *NEC2* plot of a Cushcraft 20-3CD 3-element 20-meter monoband Yagi compared to the modified Telerana. The Yagi has roughly 2 dB more gain than the Telerana for roughly the same boom length. In my case, I am willing to accept a 2-dB reduction in gain in order to have 5-band capability in one antenna. As well, the difference in the cost between the modified Telerana and five monoband Yagis is significant, not to mention the cost of mounting them on separate towers!

I am generally pleased with the comparisons between the computer models and the actual field measurements, considering that my QTH does not qualify as an antenna test range. Here are a few comments from ham operators who were asked to observe the difference between the dipole and modified Telerana without telling them in advance what antenna they were listening to:

"Sounds like you turned your linear on."

"The second antenna is sure stronger than the first" (when the second antenna was the modified Telerana!)

When talking with long-distance DX stations such as VKs and ZLs, the difference was even more pronounced. I observed similar differences on receive as well. On receive, the most obvious benefit of a directional beam with a high front-to-back ratio is the increase in signal-to-noise ratio and reduction in QRM and QRN off the sides and back of the antenna. This became very evident when comparing the dipole to the modified Telerana.

However, I was actually surprised how effective the dipole was when mounted at 60 feet. Other than the fact that the dipole was typically 1 to 2 S units less than the Telerana, when conditions were good the difference was hardly noticeable other than when I looked at the S meter. Where I really appreciated the gain and front-to-back ratio of the Telerana was when conditions were poor, or when listening to a DX station in a pileup. The difference is like night and day.

Table 3
Comparison of Gains Over Dipole, *NEC* Computations Versus Actual Measurements

Band	*Actual Measurements Compared to Dipole*	*NEC Peak Gain Compared to Dipole at Same Height*
10 meters	4.25 dB	4.5 dB
15 meters	6.00 dB	4.5 dB
20 meters	5.25 dB	3.9 dB

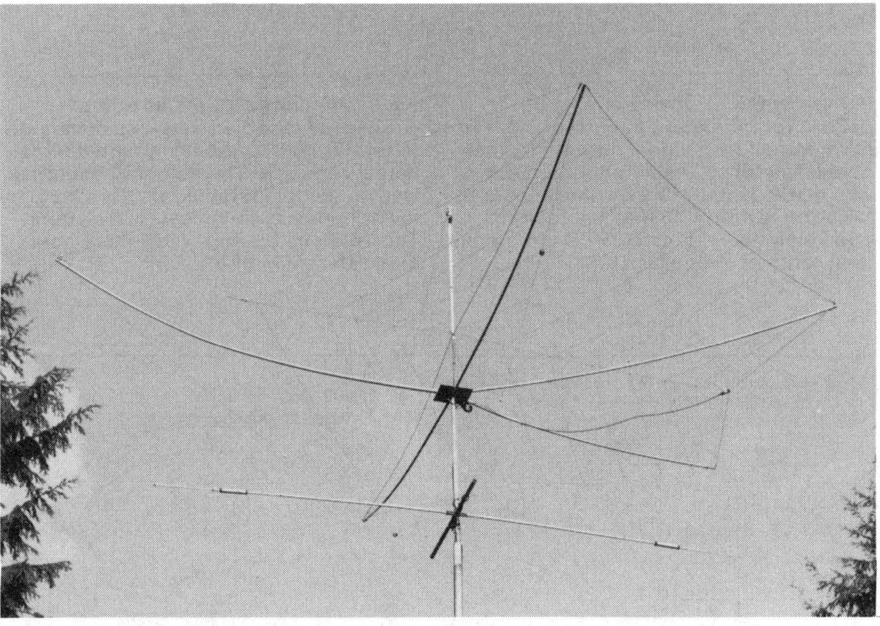

Fig 10—Photograph of modified Telerana in the air, with reference trap dipole mounted 9 feet below it.

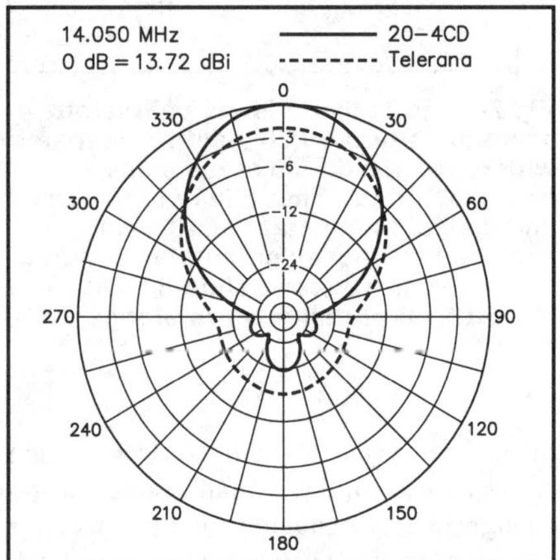

Fig 11—*NEC2* comparison of Cushcraft 20-4CD 4-element 20-meter Yagi at 60 feet with Telerana at same height. The boom lengths are comparable, but Yagi has about 2 dB more gain due to its narrowband response.

A Bonus—30 and 40 meters

During a storm, high winds whipped the fiberglass poles and the whole array inverted, just like an umbrella will sometimes do. To alleviate this problem I installed a vertical 3/4-inch plastic PVC pipe extending upward from the central hub where the fiberglass poles are attached. I strung 1/16-inch nylon cord from the top of the PVC pipe out to the ends of the fiberglass poles. Then I realized that I could replace two of the nylon cords with a 30-meter dipole. It turned out that the distance from the top of the PVC pipe to the end of the fiberglass poles was too short for a 30-meter dipole. I inserted an insulator in the wire nearest the end of the fiberglass poles and by extending the wire, brought it back to the antenna mast below the center hub of the Telerana. Later, I added 30-meter coaxial-cable traps to the ends of the 30-meter dipole and more wire to resonate the dipole on

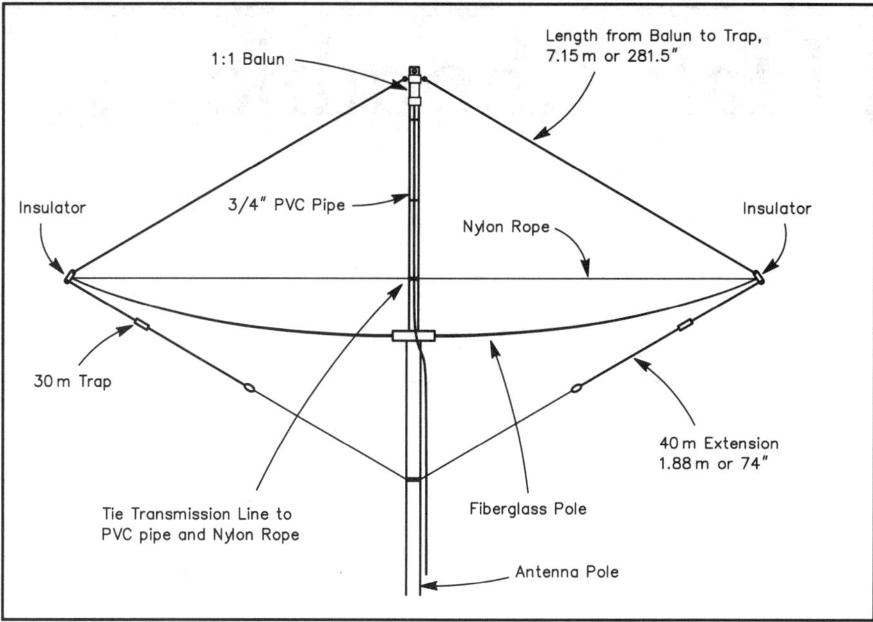

Fig 12—Side view of 30/40-meter addition to Telerana, using ¾-inch PVC pipe as vertical stabilizer and support for 30/40-meter trapped dipole.

40 meters. A separate transmission line is used to feed the 30/40-meter inverted V. See Fig 6 for dimensions and construction design.

Fig 12 shows the layout for the 30/40-meter inverted V dipole. When working Europe from this QTH, the null off the side is to the East Coast. This really helps decrease QRM. When listening to a weak signal, I can turn the dipole so that it points in the direction of the station, making a considerable difference. I checked the radiation pattern of the modified Telerana, by taking actual field measurements after installing the 30/40-meter dipole and I was pleased that the radiation pattern on 20 to 10 meters was not significantly affected by the addition of the 30/40-meter dipole. I did not model this addition on *MN*.

Conclusion

Adding parasitic reflectors to the Telerana for 20 and 15 meters has significantly improved the front-to-back ratio on these bands. When operating on 20 and 15 meters, I now have to be more conscious about where the antenna is aimed, otherwise I simply do not hear a weak signal if it is off the back of the array. When listening to DX stations arriving at very low angles, an S6 signal disappears into the noise when the array is turned 180°. Perhaps someone may want to add parasitic reflectors on 17, 12 and 10 meters as well.

The addition of the 30/40-inverted V with the apex at 70 feet has resulted in an increase in my DX totals on these bands, a good indication of the effectiveness of this addition.

Once again, computer antenna modeling has played an important role in optimizing an antenna design. Being able to make actual field antenna pattern measurements allows me to check whether my assumptions are true or not. A hands-on comparison with a dipole at the same height is the acid test. The modified Telerana really does have respectable gain and a high front-to-back ratio compared with a dipole. With the addition of 30 and 40-meter coverage, this is a very compact and effective antenna covering seven amateur bands!

Darrell also built the modified Telerana and he found that it was easy to duplicate the design. Radiation pattern checks of his array show similar results. Neither Darrell nor I would trade our modified Telerana for the narrowband triband Yagis they replaced! We both encourage anyone to build this excellent antenna.

I wish to thank Darrell for writing the LPDA design program, Joe Young, VE7BFK, for his helpful comments regarding this article, Dean Straw, N6BV, for modeling this array on *NEC2* and especially my wife and children for putting up with me and my constant raising and lowering of the antenna, wire and fiberglass poles and assorted clutter on the front lawn, during the last 6 years of intermittent antenna experimenting.

Notes

1. "Measuring Antenna Gain with Amateur Methods," *The ARRL Antenna Anthology*, (Newington: ARRL, 1978), p 145. (Out of print.)
2. Jerry Hall, "A New Look for *QST's* Antenna Patterns," *QST*, Jul 1980, p 26.
3. Peter D. Rhodes, K4EWG, "The Log-Periodic Dipole Array," *QST*, Nov 1973, p 16.
4. Darrell A. Wick, VE7FCR, 1491 Edgemont Rd, Victoria, BC V8N 4P7, Canada.

The Tilted Half-Rhombic Antenna

By Michael Orr, AA2PE
9 Jagger Avenue
Neptune, NJ 07753

AA2PE analyzes a unique multi-band wire antenna with switchable pattern.

This antenna is designed for general coverage use on the HF bands from 30 through 10 meters. It is broadband so that matching is possible without the need for tuned matching networks. For use at AA2PE, the antenna had to fit in a restricted area of about 100×100 feet. It can be built with materials readily available to amateurs. The most expensive part of the antenna are the two broadband matching transformers

Description

The tilted half-rhombic antenna resembles the more familiar "sloping V" antenna. The difference is that the tilted half rhombic is fed at an end, and a sloping V is fed at the center (apex). The feedpoint and terminating point of the antenna are placed on diagonally opposite corners of the given area. To keep the supporting structure from being an obstruction, it is located in one of the remaining corners. An illustration of the tilted half-rhombic antenna is shown in **Fig 1**.

The antenna, as installed at AA2PE, has 27.2 meters of wire in each leg. At 30 meters, this provides a leg length of nearly 1 λ. At 10 meters, the leg length is slightly over 2.5 λ, as calculated from the equation:

$$\text{Leg length} = 150\,(N - 0.05) / F_{MHz}$$

where N is the number of half wavelengths in each leg.

The height of the apex in the AA2PE antenna is approximately 9½ meters. With 27.2 meters of wire in each leg, the included angle between the wires is approaching a more acceptable value for radiation lobe alignment from each leg. More wire may be used, and is encouraged, especially to make each leg longer than 1 λ at the lowest desired operating frequency. If the increase in wire length is significant over that given here, the height of the apex must be increased to get a quality pattern. The approximate angle at the apex between the wires should be maintained at about 90° to 100°. More information on radiation tilt angles and lobe alignment in V and rhombic antennas may be found in Chapter 13 of the 15th Edition of *The ARRL Antenna Book*.[1] This antenna may be considered as two long wires end to end, and at an angle from each other. Radiation from a long wire is conical, with the wire as the axis.[2]

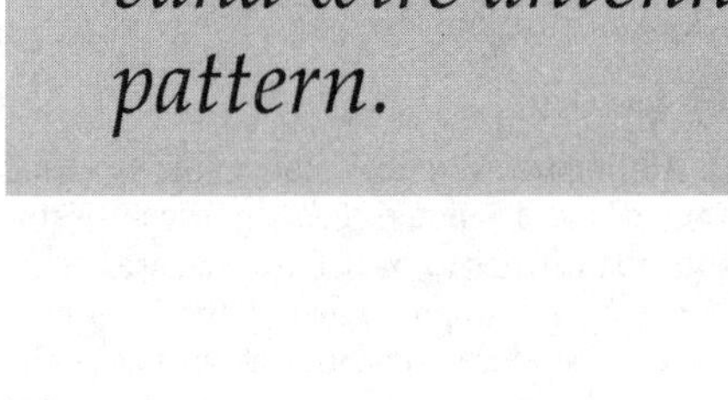

Fig 1—Layout of AA2PE tilted half-rhombic antenna.

Pattern

The antenna's major lobe radiation is along the axis drawn from the feedpoint to the terminating point, and in the direction of the terminating end (See **Fig 2**). A computer simulation plot of the antenna has been made using the *AO* antenna analysis and optimization program.[3] [Note that with wires close to the ground, *AO*, a variation of *MININEC*, will not give accurate results, typically inflating gain by as much as several dB. Running *NEC* for the same antenna verifies this.—*Ed.*]

Field-strength measurements were taken at a 2 km radius around the prototype antenna at 28 MHz. To get a comparison between model results and actual results, the data was entered on the plot created by *AO*. See **Fig 3.** The receiving antenna was a vertical whip mounted on an automobile and the measurements were taken with a calibrated field-strength meter. The field-strength meter readings in microvolts were normalized and converted to dB for plotting.

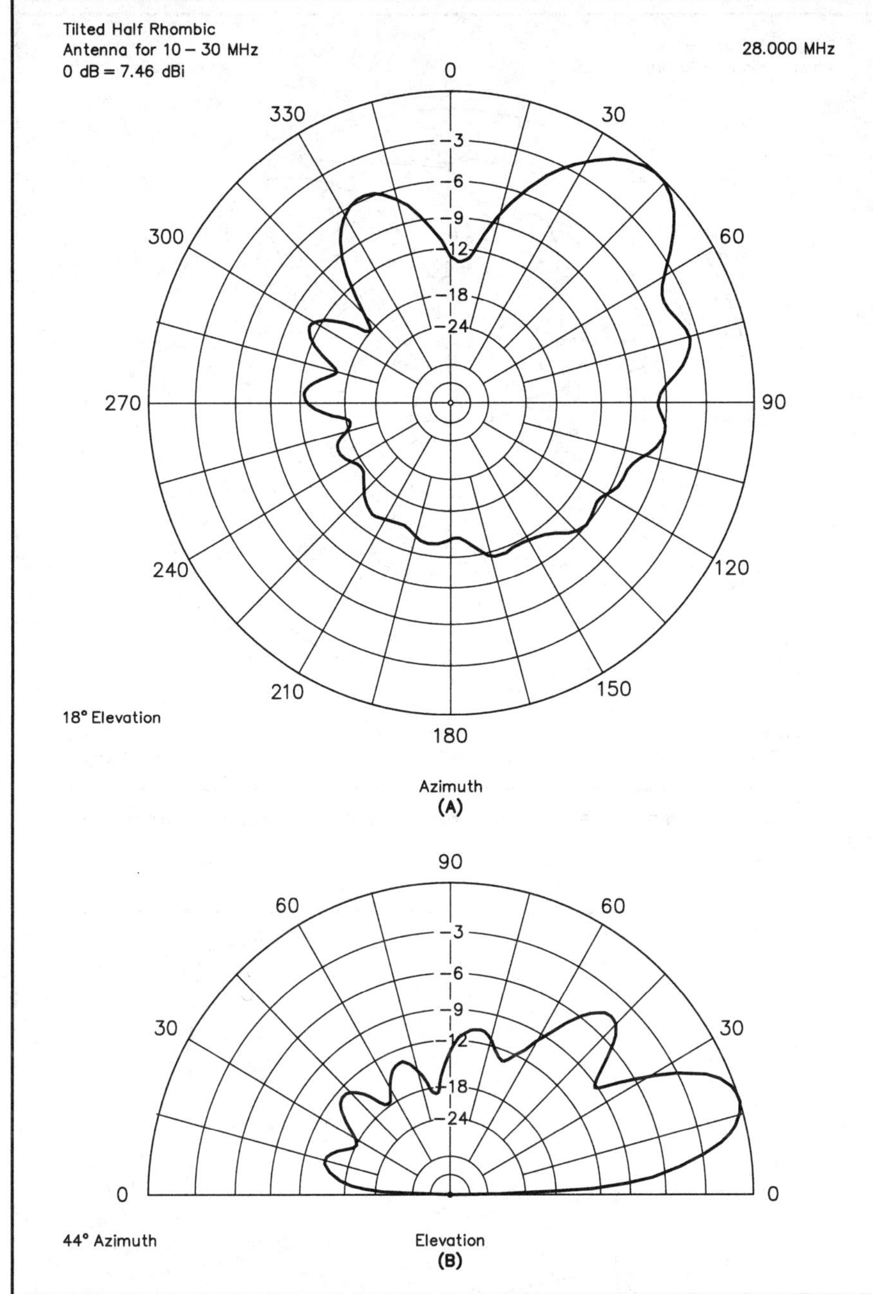

Fig 2—At A, azimuth response computed for AA2PE half-rhombic antenna at 28.0 MHz. Maximum radiation is along axis drawn from feedpoint to terminating end. At B, elevation pattern at azimuth of maximum radiation.

Table 1
Computed Feed-Point Impedance and Computed Versus Measured SWR of AA2PE Tilted Half-Rhombic Antenna

Freq (MHz)	*Series R*	*Series X*	*Calculated SWR*	*Measured SWR*
10.1	484	–132	1.34	1.10
14	477	–222	1.60	1.05
18.1	423	–96	1.26	*
21	396	–152	1.46	1.25
24.9	364	–115	1.72	*
28	268	–148	1.94	1.40

* Data unavailable

As Fig 3 shows, the actual radiation pattern is very well correlated to the model results. There are a few minor departures from predicted values. There is a spurious lobe appearing at 225°, exactly in line with the radiation axis. This represents radiation in the opposite direction due to a reflected wave on the antenna. The reflected wave is a result of imperfect match between the antenna characteristic impedance and the matching transformers and their loads. The reduced field strength off the "east" of the antenna is believed to be due to measurement error, caused by being in an area of higher vegetation where the measurement was taken.

There is a small, unaccounted for, shift in the main lobe's position. Power lines and metal rain gutters in the vicinity reradiate the energy from the antenna, causing a distorted pattern. As with other sources of measurement error, the actual antenna pattern will not always be identical to the model results. Hopefully, the differences are slight enough so that one can rely on the model results to predict actual antenna performance.

Impedance Matching

Broadband operation is accomplished by terminating the end of the antenna into its characteristic impedance. The tilted half-rhombic has an impedance about half that of a standard rhombic, around 400 Ω. To simplify the matching network design, a terminating impedance of 450 Ω has been chosen. The feed line is connected to the antenna through a 9:1 unbalanced to unbalanced transformer. The terminating resistance is provided by a standard 50-Ω dummy load, through a 50-Ω coaxial cable and a 9:1 transformer. Ground return is provided with a counterpoise wire along or just a few inches below ground. No other radials or additional ground points are required, beyond what should already be established for safety reasons. The broadband transmission line transformers used were of the Guanella design presented in *Transmission Line Transformers*.[4] This critical part of the antenna must be carefully built for optimum performance.

The input impedance was measured with an RX meter connected directly at the feedpoint, without the matching transformers. Since the feedpoint is on the ground, this is very simple to do. The measurements are given in **Table 1**. A noise bridge could be used to make impedance measurements, if one has a battery operated receiver. It would be more convenient to make the measurements through the matching transformer to determine the departure from a 50-Ω match. It may also be possible to determine matching network loss by observing the depth of null, compared to a pure 50-Ω load.

Depending on how much feed line you have, the measured SWR in the shack may be significantly lower. For instance, I measured an SWR of 1.4:1 on 10 meters, where the impedance measurement indicated that the SWR should have been 1.94:1. The difference is due to attenuation in the feed line of the reflected wave as it travels back to the transmitter.

Since the antenna is directive, it's desirable to have some control over the radiation direction. A coworker suggested bringing the 50-Ω coaxial feed line into the shack from both ends of the antenna, through the 9:1 transformers. If access is provided to the two feed lines, the connections can be exchanged between the dummy load and rig. This turns the antenna main lobe over by 180°. A diagram if this is shown in **Fig 4**.

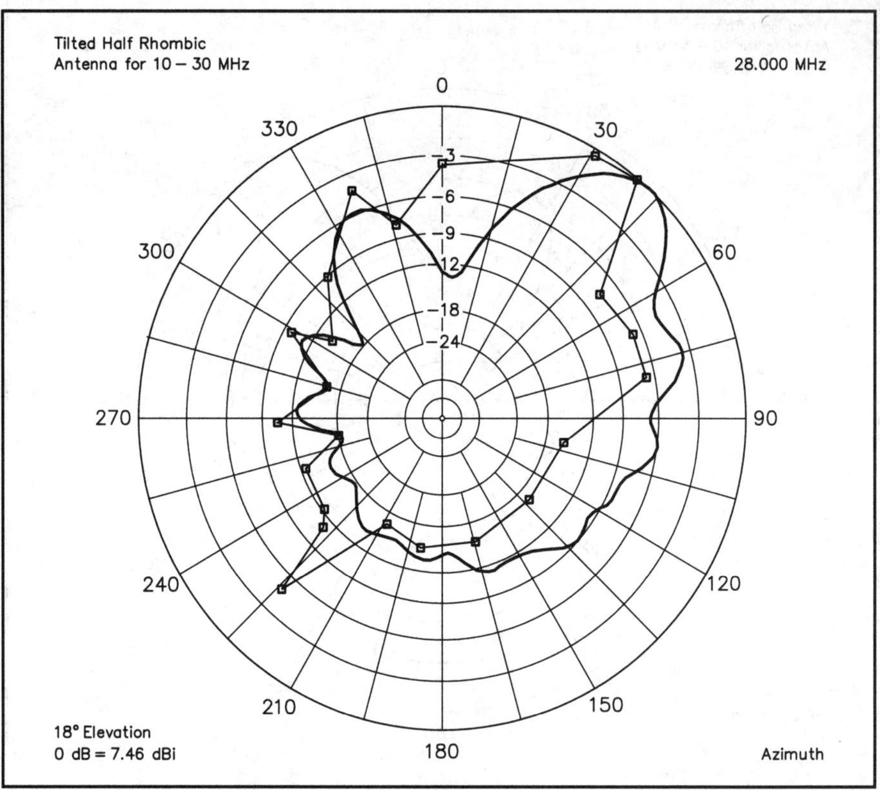

Fig 3—Computed and measured azimuth response of half-rhombic antenna at 28.0 MHz.

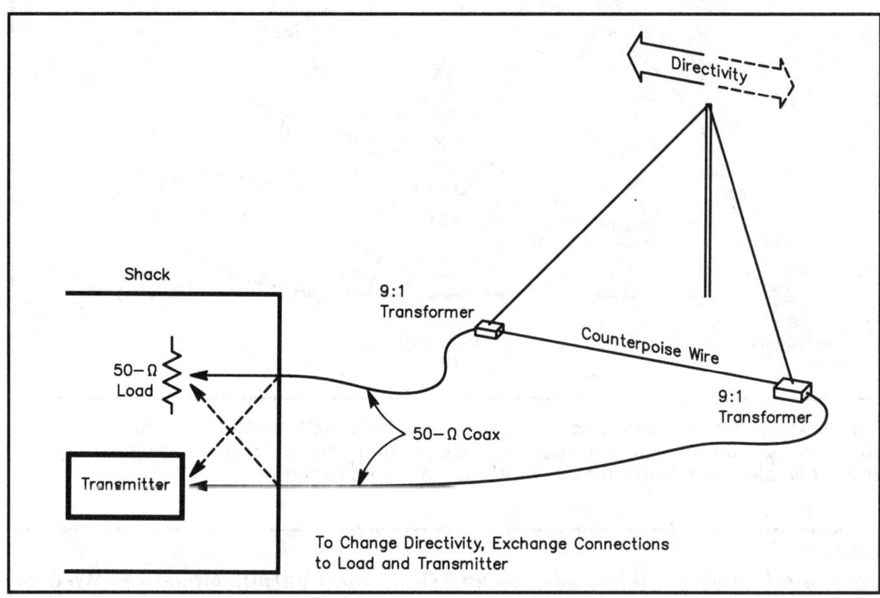

Fig 4—Changing directionality of AA2PE half-rhombic by exchanging 50-Ω coaxes in shack.

Performance

Does the tilted half-rhombic antenna work? Yes, it does! On the air, it has been used for DX and stateside QRP contacts on 10 through 30 meters. These contacts were made during a few weekend hours of operation in the Fall of 1993 and with an output power of only 5 W or less. With such low power, the received RST reports were not astounding, but the signal was copied successfully. Stations as far away as 15,000 km have been contacted as well.

I recognize that this antenna is not the perfect antenna. My goal was to provide a structure that could be used within the limitations of space and placement, and to get on the air with decent performance. Does the antenna have its operational limitations? Yes, it does! Attempts were made unsuccessfully several evenings to contact a station in Antarctica on 30 meters. On this band, though, the antenna has reached its limit of performance. As the frequency of operation is lowered, the performance of the tilted half-rhombic drops off. This is illustrated in a plot of the antenna at 30 meters. See **Fig 5**.

Comparing the radiation pattern of this antenna to a terminated sloping V demonstrates a significant gain and pattern quality advantage, resulting from the feedpoint being taken at the end rather than at the center (or apex). The antenna analysis program *AO* was again used to make this comparison. The same antenna structure is modeled, with the feedpoint at the center, and then with the termination at either end. The results are shown in **Fig 6**. It can be seen that the gain and pattern quality of the sloping V is considerably worse than the tilted half-rhombic.

Receiving tests were conducted on 13 meters by establishing a listening schedule of selected short-wave broadcast stations within the main radiation lobe of the pattern. A half-wave dipole was installed, and A/B comparison tests were conducted. A measurement of receiver AGC voltage with a digital voltmeter was made up to twenty times for each antenna, 10 readings for A, 10 for B, 10 for A, and so on. The data collected show measured gain of as little as 2 dB and as great as 11 dB, with the average about 5 dB. Many factors contribute to the variability in the data—mostly the arrival angle and polarization of the received wave. These factors are variable over time and station.

Pattern and Gain Optimization

The angle between the two legs of the antenna, called the *included angle*, plays an

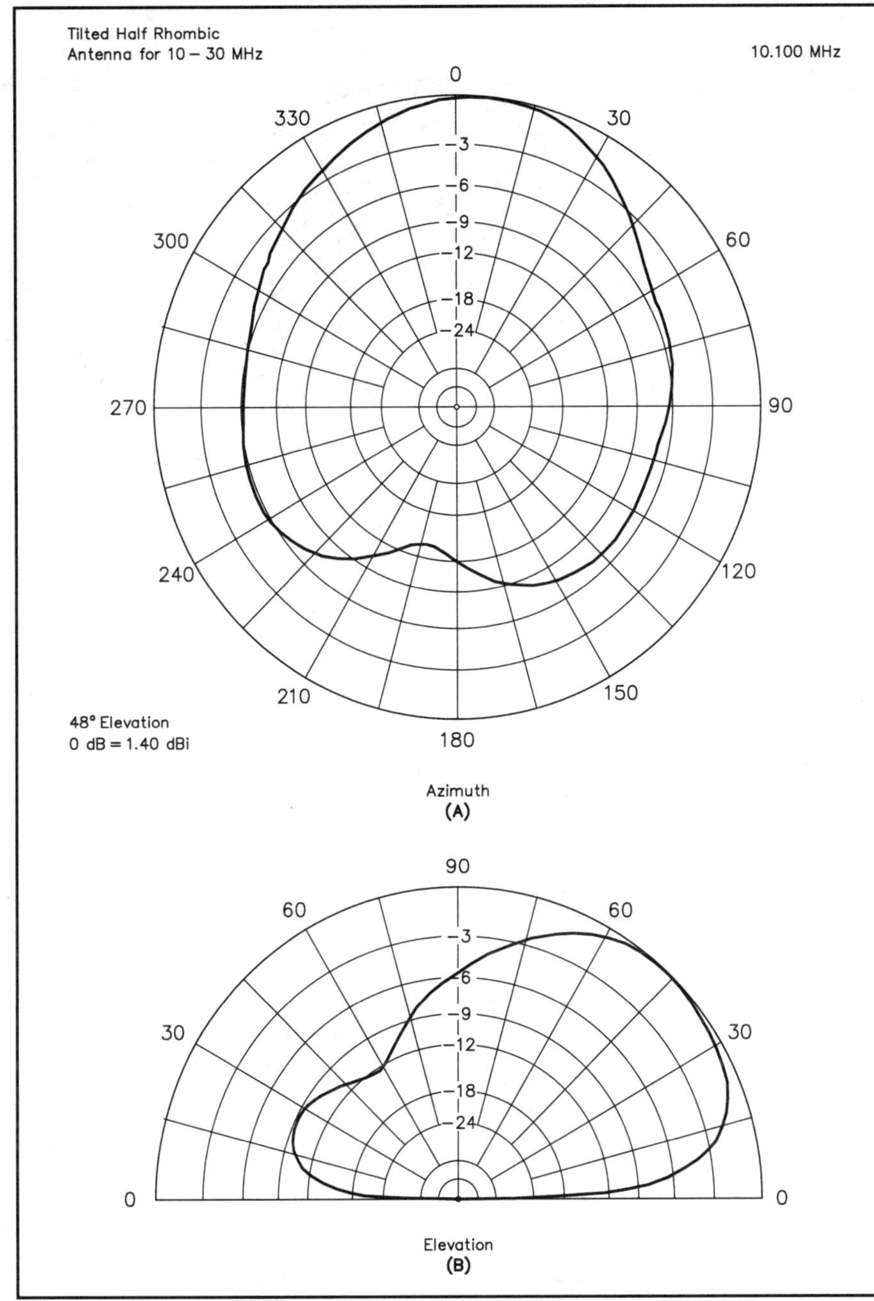

Fig 5—At A, azimuth response for AA2PE antenna at 10.1 MHz. At B, elevation response.

important role in the antenna's performance. Using this knowledge, some may have the flexibility to place the antenna in a more favorable position. As it was placed at AA2PE, the included angle is approximately 95°. One could predict, from the information in Reference 1, that the optimum included angle should be 110° for a leg length of 2.5 λ. This prediction was tested with the antenna modeling program *AO*, by invoking its antenna optimization feature. Left to its own problem solving, *AO* gave a result of 106°, with a gain increase of 0.6 dB on 10 meters. For each frequency, there will be a different optimum included angle. A comparison of the AA2PE antenna and the *AO* antenna optimized at each frequency is shown in **Table 2**.

Conclusion

By moving the feedpoint of a terminated sloping-V antenna from the apex to an end, a considerable improvement in antenna performance can be achieved. This antenna provides a good impedance match across the HF spectrum without the use of a Transmatch. The SWR is relatively low on the feed line. The tilted half-rhombic is a very good candidate for a Field Day antenna, since the counterpoise need not be buried, only one support is required, and the feedpoint is at ground level. No ground rods are necessary for the antenna to function properly. The termination may be removed, yielding a more omnidirectional pattern. This may be desirable for Field-Day use.[5] If more space is available and a higher apex can be provided, the antenna's operating range may be increased to cover the entire HF spectrum and to improve gain. Remember, the transmission line transformers have to be made sufficiently broadband as well.

Acknowledgments

The author wishes to give special recognition and thanks to Palemon "Dubie" Dubowicz for his suggestion to bring both ends of the antenna into the shack to allow changing the pattern direction. Dubie also provided much assistance in the area of broadband transformers, theoretical and practical, and supplied some units used for the initial prototype. The original vertical half-rhombic design on which this antenna is based was developed by Dubie for the US Army Signal Corps. The Army antenna kit (AS-303) can be erected within 20 minutes by two soldiers, and is truly a field antenna. I hope this effort will return to Dube a rekindled desire to obtain his Amateur Radio license, which he long ago let lapse.

Notes

[1]G. Hall, K1TD, Editor, *The ARRL Antenna Book* (Newington: ARRL, 1988).

[2]J. D. Kraus, *Antennas*, Second Edition (New York: McGraw-Hill, Inc, 1988), p 502.

[3]*AO 6.0 Antenna Optimizer* software, Brian Beezley, K6STI, 3532 Linda Vista Dr, San Marcos, CA 92069.

[4]Sevick, Jerry, W2FMI, *Transmission Line Transformers*, 2nd Edition (Newington: ARRL, 1990)

[5]It is recognized that the current that would have been dissipated by the terminating load is now reflected, and will raise the SWR. Experiments at AA2PE indicate that if the wire is sufficiently long (greater than 1 λ per leg) the SWR remains below 2:1. The caution is that this measurement depends on the length of the feed line and its loss, and whether the load is merely disconnected from the terminating feed line, or if the feed line itself is removed. The unterminated antenna should be tested for a specific installation before counting on it to provide an acceptable SWR.

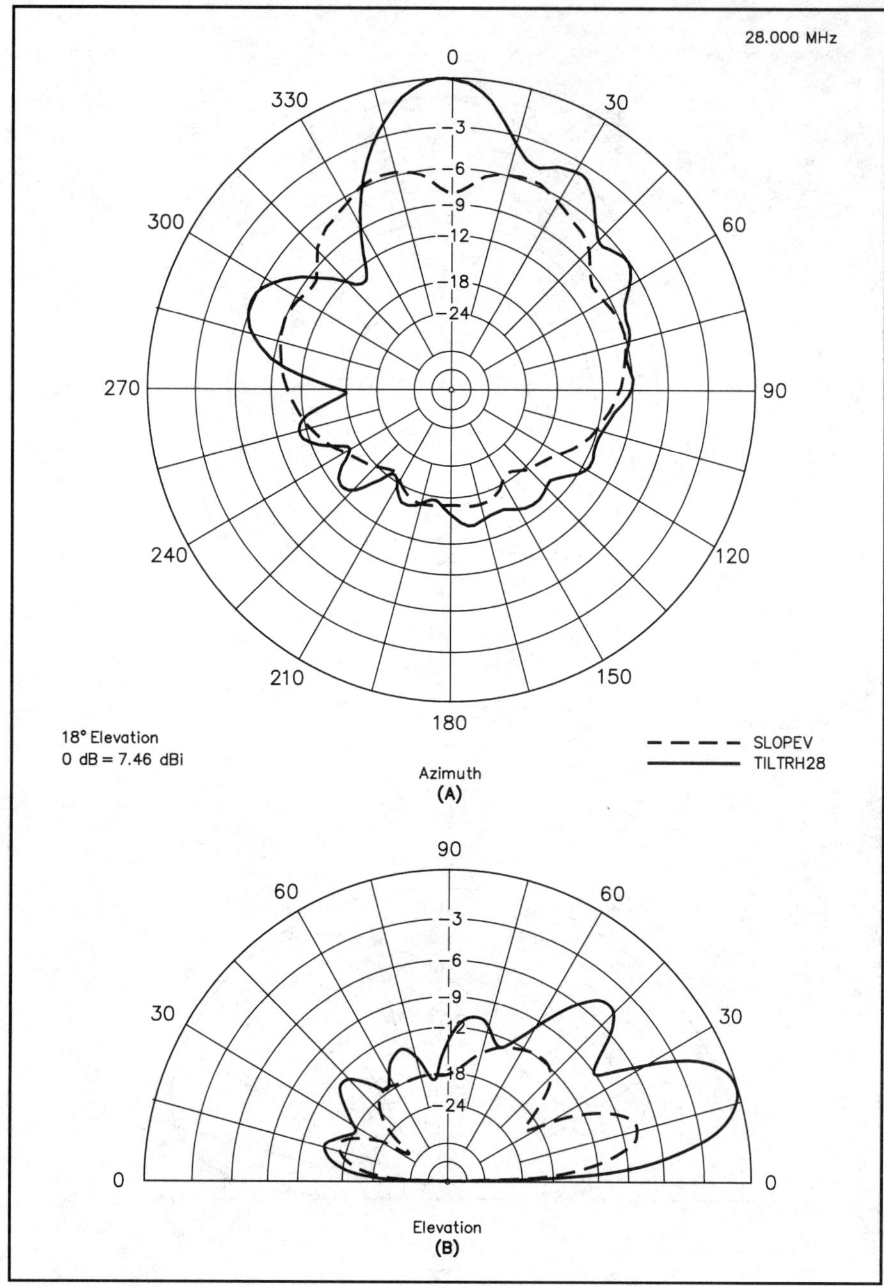

Fig 6—At A, comparison of half-rhombic (solid line) and terminated sloping-V antenna (dashed line) at 28.0 MHz. Note that pattern for half-rhombic has been rotated to line up at 0° azimuth. At B, comparison of elevation patterns for same two antennas.

Table 2
Gain as a Function of Included Angle for AA2PE Tilted Half-Rhombic

Freq (MHz)	*AA2PE Gain (dBi) (95° incl angle)*	*AO Incl Angle (Degrees)*	*AO Gain (dBi)*
28	7.5	106	8.1
24.9	7.1	102	7.5
21.0	7.0	97	7.0
18.1	6.0	88	6.0
14.0	4.0	81	4.7
10.1	1.7	59	2.5

Portable/Temporary Antennas

The Flying Field-Day Loop

By Ricky Bibby, AB5FW
3908 Wood St
Texarkana, TX 75503

A balloon-supported full-wave 40-meter loop for Field Day!

The Shreveport Amateur Radio Association is serious about the yearly Field Day outing. When they requested I take on the task of designing and erecting a 40-meter antenna, I knew it had to be more than just a dipole hung from a tree. The swing-tent coordinator Bob Matthews, KF5XV, confirmed that less than 500 to 600 contacts would be considered unsuccessful. Only nighttime contacts would be attempted, since 10 and 15 meters would be relied on during the day. This article documents the results of that 40-meter effort.

The Tower

I live in Texarkana, Texas, and the Field Day site is 75 miles to the south in Shreveport, Louisiana. This little geography problem meant that the entire structure, including the tower, would have to fit in my half-ton pickup truck for a two-hour trip. It also had to be light enough for me and my son, KB5VYD, to hand carry.

The tower would have to be lightweight and strong enough to support a wire antenna. A good solution was a simple wooden A-frame mast, as shown in older editions of *The ARRL Handbook*[1] and in *The ARRL Antenna Book*[2] for many years.

So we took a 20 foot long 2×6 inch board and sawed it lengthwise into three pieces. These would fit perfectly on the ladder rack on my truck. The three 20-foot long boards would be bolted together to form a structure about 37 feet tall with a 3-foot overlap where the boards came together. I intended to add a 13 foot long fiberglass pole (a cane fishing pole would work also) to the top, to bring the total height of the support structure to nearly 50 feet. See **Fig 1**, which shows the idea behind the support structure.

The Antenna

I consider the full-wave loop to be an exceptional antenna, and I've had considerable experience erecting them. Because the tower was only capable of supporting a wire antenna, not some sort of Yagi, a loop was chosen for the 40-meter antenna. There is a problem—even with a 50 foot support, the midpoint of the loop would only be 30 feet high, not even a quarter wave above ground. This would obviously not be an optimum situation.

We spent considerable time trying to find some way to get the antenna up higher so

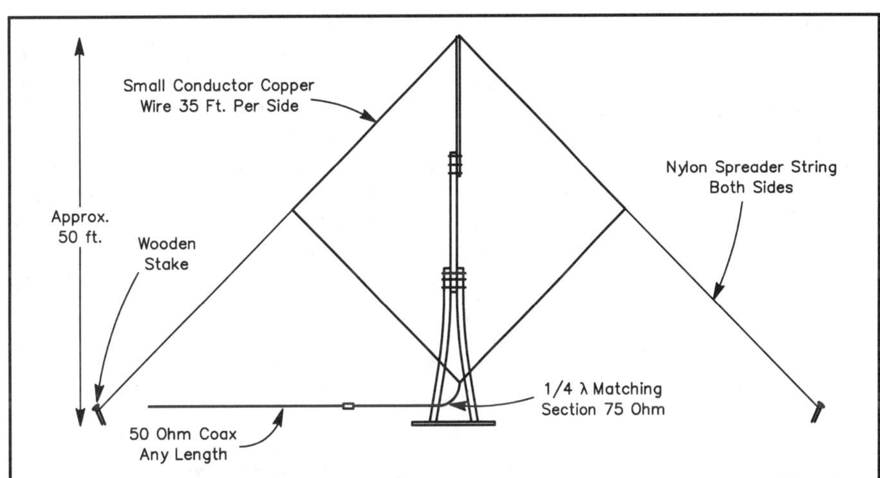

Fig 1—A-frame mast used to support the full-wave 40-meter loop. This was the first design concept for AB5FW's Field-Day antenna. The midpoint height, however, would only be about 30 feet off the ground. The A-frame mast mechanical design is shown in *The ARRL Antenna Book*.

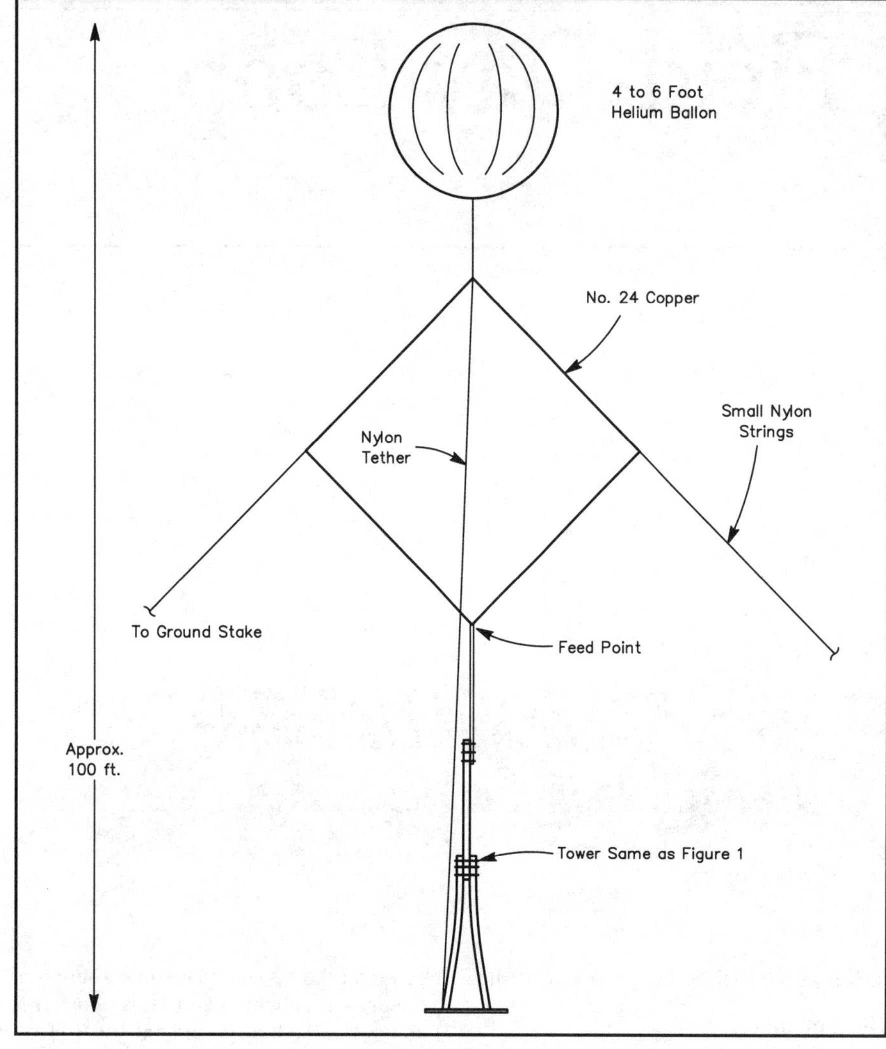

Fig 2—A balloon-supported extension of the tower design shown in Fig 1. Now the midpoint of the antenna is a half wave high on 40 meters. Make sure you use the nylon tether to the balloon, and stay well away from power lines and trees!

that the midpoint would be at least 70 feet high, a half wave on 40 meters. At first we considered a taller A-frame type of tower, but when we figured it would have to be 93 feet tall, we quickly abandoned that thought process! How could one get an antenna to hang over 90 feet tall in midair for Field Day?

Someone else had already figured it out. Stan Gibilisco, W1GV,[3] had been flying helium-filled balloons for some time. His antennas were simple verticals fed against ground. We wondered whether a balloon could be used to support our loop. The structure shown in **Fig 2** was what evolved.

Field Day, 1993

Materials were gathered, helium was purchased and the wooden tower was constructed to see if this idea would "fly." On June 26, 1993, we put it to the ultimate test. Here are some construction hints. We used #24 wire for the antenna element. This size of wire is the best compromise for weight and strength. In a pound of #24 wire there is about 900 feet of wire. I used a 75-Ω λ/4 matching section because the antenna was designed for a single band, but 450-Ω ladder line could be used for multiband use, with a Transmatch at the operating position.

The guy and tether lines should be made of nylon cord. I originally tried to use monofilament line, and one of the guys broke—during the middle of the night, as Murphy would have it. This unfortunate incident required taking the whole thing down and the loss of about an hour of operating time, not to mention a lot of aggravation.

At least six people are required to stand the tower up; three at the tower and three on the guy lines. This prevents the structure from getting twisted and broken. Make sure that you are familiar with all the safety rules for flying a balloon—stay away from power lines and trees; use a tether properly. Follow the cautions and guidelines outlined by W1GV.

So, how did we do? Despite the time lost to the monofilament guy-line fiasco, we made 541 QSOs on 40 meters. As anyone who has worked with antennas for a while knows, objective evaluation is a most difficult task, particularly during the chaos of Field Day. Unfortunately, I can't erect a structure this size for a permanent installation at my house, so I can't run a long-term evaluation. However, I think our Field Day results show that the concept works.

Our tent had a total of 1231 QSOs on 10, 15 and 40 meters. I'm pretty pleased with this, especially since we weren't on 20 meters—the "glory band." Of course, now the big problem is going to be: How are we going to top this for next Field Day?

My thanks to KB5VYD, who helped me at every stage of design and construction. He also told me in advance to use nylon guy strings, which proves the old proverb: "It's not so bad to be old as to be old and stupid."

Notes

[1] *The Radio Amateur's Handbook,* (Newington: ARRL, 1976), p 621.

[2] *The ARRL Antenna Book,* 16th Edition (Newington: ARRL, 1991), p 22-4.

[3] S. Gibilisco, "Balloons as Antenna Supports," *The ARRL Antenna Compendium, Vol 2* (Newington: ARRL, 1989), pp 142-144. A must-read article on the use of balloons to support wire antennas.

A Vertical Dipole with Tuned Feeders

By Stan Gibilisco, W1GV
2301 Collins Avenue, Apt A-632
Miami Beach, FL 33139

You can't get much more simple than this multiband wire antenna!

Several years ago I built a small vertical dipole antenna and fed it with TV ribbon. The idea was simple: stand a tuner-fed "random dipole" on end. It worked well, even though the ribbon line can become lossy when the SWR is high, and its characteristic impedance fluctuated maddeningly when it rained.

Last winter I visited my parents in Minnesota. They live on a hilltop. An FT-101EE, keyer and Johnson Matchbox sit in a basement bedroom, patiently awaiting my infrequent returns. It's fun to take a vacation from the subtropics to the Northland in the winter. (It's even more fun to get back to the subtropics afterwards.) The weather wasn't too bad this trip—none of that –25°F stuff!

I wanted to get on the air, so I built an open-wire fed version of the vertical dipole.

Open-Wire Feeders

When a horizontal dipole antenna is fed with open-wire (not TV ribbon, but air-dielectric open-wire line, such as commercial 450-Ω "window" line) and when a balanced-load Transmatch is used, one doesn't have to worry about matching the line at the antenna. The only real requirement is that the antenna be at least λ/4 long at the lowest operating frequency. Under these conditions, open-wire line has low loss, even if the SWR is high.

This antenna is balanced, no matter what the frequency, and regardless of the SWR, provided the feed line comes away from the antenna at a right angle for at least λ/2 on the lowest operating frequency. The radiation pattern resembles that of a λ/2 dipole over a wide range of frequencies.

The length L in feet of a λ/2 dipole antenna at a frequency f in MHz is given by the familiar formula: L=468/f. Maximum radiation is at right angles to the radiating element at all frequencies below 2.5 times the fundamental. At frequencies greater than the fundamental f and up to 2.5 f, there is a small amount of gain in the plane perpendicular to the wire. Above 2.5 f, the radiation pattern begins to change, with sidelobes developing.

Standing It On End

If the element is turned 90° so that it is vertical rather than horizontal, maximum radiation is toward the horizon in all directions. At frequencies greater than f and up to 2.5 f, there is a small omnidirectional gain at low angles. This is ideal for working DX. Above 2.5 f, the radiation angle increases. The antenna will still function at these higher frequencies, but might not work well for DX.

Theory predicts that an open-wire-fed vertical dipole should be an efficient low-angle radiator over a 5-to-1 frequency range, from 0.5 f to 2.5 f. A dipole cut for 14 MHz, for example, should cover 40 through 10 meters. This covers seven ham bands, and all frequencies in between.

In this antenna all current loops are well above ground level. This practically eliminates near-field ground loss, so often a problem with λ/4 verticals mounted on the ground. No radials are necessary. The lower portion of the dipole serves that purpose. The basic configuration for the open-wire vertical dipole is shown in **Fig 1**.

Although the effect of near-field ground losses are not a factor with this antenna, the far-field ground still affects signals, just as it does with any antenna, particularly vertically polarized antennas. If the ground has poor conductivity (such as sandy, rocky or dry soil), the far-field reflections will be affected greatly. If the ground conductivity is good (salt water is the best), ground re-

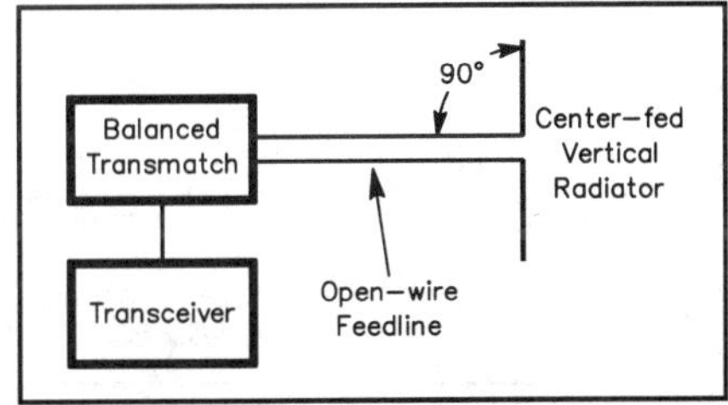

Fig 1—Basic configuration for center-fed vertical antenna, fed with open-wire feed line through balanced Transmatch.

flections at low elevation angles will be enhanced.

If the ground is highly conductive many wavelengths from the radiating element, and when the antenna is longer than $\lambda/2$ (but shorter than 1.25 λ), low-angle gain can exceed 3 dB compared to a dipole in free space. [For typical earth, the far-field losses negate the effect of low-angle gain achieved with a $\lambda/2$ or longer vertically polarized antenna. See "The Effects of the Earth" chapter in *The ARRL Antenna Book* for a complete discussion of ground losses for vertical antennas.—*Ed.*]

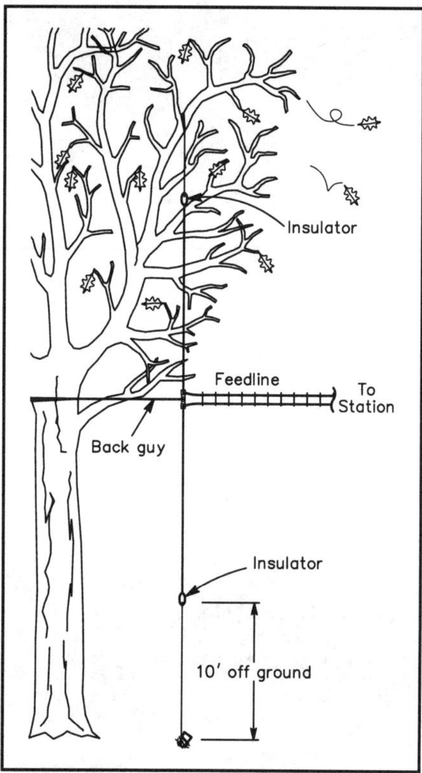

Fig 2—Drawing showing wire radiator suspended from tree branch. Keep bottom end high enough off ground so that people and animals cannot come in accidental contact with high RF voltages from antenna or feed line.

It Looks Slightly Bizarre

The feed line comes away horizontally from the vertical radiator. It looks rather peculiar; some might even call it ugly. I was more concerned with electrical performance than visual beauty. Structurally, the antenna must be guyed in the middle to balance the pull of the transmission line coming away from the element. Alternatively, it can be made of wire, suspended from a tree branch, and held taut by a stake in the ground.

A more real flaw is that the antenna is not really perfectly balanced. The lower half is closer to the ground than the upper half. Thus, the feed line currents are not exactly equal, and there is some line radiation. This increases the chances for problems such as "RF in the shack" and various forms of electromagnetic interference (EMI). In theory, a symmetrical, tuned-feeder antenna should not suffer from these problems any more than a well-engineered system fed with coaxial cable. The trick with this antenna is to equalize as nearly as possible the currents in the parallel-wire feed line, and to ensure that they are 180° out of phase with each other. That is, the system should be as balanced as is reasonably possible.

The balance can be improved by placing the base of the antenna a few feet above ground. The tree-branch scheme is good for this. See **Fig 2**. For a radiator made with aluminum tubing, a 10-foot 2×4 can be used. Placing the end above ground has an additional advantage—it puts the antenna high enough to be out of reach of inquisitive hands. This is important if you plan to run high power. The bottom of the antenna is at a high-voltage point. If anyone touches it while you are transmitting, an RF burn will result. The parallel-wire feed line should also be kept out of peoples' reach, because high voltages exist on this type of line.

In the middle of winter in Minnesota I did not have the time or means to elevate the base of my antenna. I just wanted a quick way to get on the air. I always use low to moderate power to test antennas anyway. (With 1500 W you can work DX with a dummy load—I've done it.) There were no trees in the vicinity with the appropriate height and branch structure.

Parts Needed

I wanted an antenna that would need a minimum of fuss, expense and installation time. Also, it was imperative that I not deface my parents' house or property in any way. I wanted the parts to be available in common hardware or electrical stores. It occurred to me that I should think of my visit as a sort of "wintertime Field Day," and set up the antenna with the idea that it might lend itself to the real Field Day. I visited the local hardware store and obtained the parts I needed, including aluminum tubing, hose clamps, plastic pipe coupler, nylon guy cord and cinder blocks.

A note concerning the feed line: Ideally, this would be prefabricated 450-Ω "window" line. This type of feed line is available from various sources who advertise in the ham magazines.

Since I didn't have any such line with me (and you can't get it in the hardware store either), I had to improvise. I used lamp cord, ripping it apart and then using the tension holding the two wires suspended horizontally to keep the spacing between the conductors fairly constant. The actual spacing and conductor size are not critical, as long as the two wires are close enough together so that they will work as a feed line rather than an antenna.

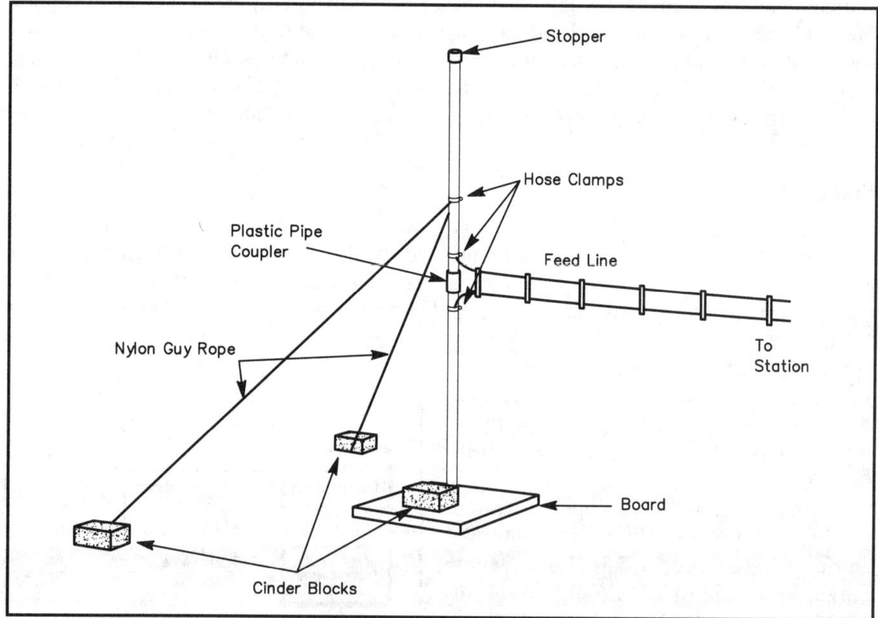

Fig 3—W1GV's temporary wintertime installation in Minnesota, using aluminum tubing and cinder blocks. Permanent installation using this construction technique would require a wooden fence around base of antenna to keep people and animals away from RF burns.

Putting It Up

My temporary antenna is shown in **Fig 3**. In my case the two sections of antenna were each 10 feet long, so the overall antenna height was 20 feet. This made the fundamental frequency of the antenna 23.4 MHz. The useful range was therefore 11.7 MHz through 58.5 MHz. Stretching the lower frequency limit a little, the antenna could be expected to function from 30 through 6 meters. My older radio covers only the "old" ham bands, so the antenna I built was workable only on 20, 15 and 10 meters.

On the Air

The antenna tuned well on 40, 20, 15 and 10 meters. I did not expect that it would work well on 40 meters, because 7 MHz is well below 0.5 f, but I saw no harm in trying it out on that band.

Conditions were fair-to-good on the higher bands. In a few minutes of operation I worked stations in Finland, Russia, Yugoslavia and Japan, using 20 through 10 meters. All contacts were on CW, and power output was 90 W. I got signal reports ranging from 559 to 599—always as good as, or better than, the report I gave out. (I concede that I don't give reports to flatter people. If they're S3, then I tell them they're S3.)

Conditions were poor on 40 meters, so I didn't even try to make any contacts there, except for calling CQ a couple of times without results.

Once I had seen that the antenna worked well on its design bands, I shut the radio off. I had other things to do in the short time that I was back home, among them visiting friends and eating my mother's home-cooked food.

Concluding Thoughts

To ensure that nobody gets burned by this antenna, the radiator and feed line must be kept out of human reach. Especially if you run high power, there will be very high RF voltages at the end of the antenna radiator (both top and bottom), and also along the feed line. Because of the random nature of this antenna and feed line, it is almost impossible to predict where a high-voltage spot might be. The safest course is to assume that the whole antenna and transmission line are hot. If the end of the antenna is close to ground, a wooden fence around it to keep people and animals away is needed.

This antenna could be used on camping trips, since a wire version can easily be hung from a tree. I did not make on-the-air comparisons of this antenna with others. It did seem to "get out" well, and true to theory appeared to be good for DX. It also exhibited that universal characteristic that plagues all vertical antennas—it was rather noisy for receiving.

Certainly my installation was a crude prototype, but the design can be improved.

- Use a height of 32 feet, for a fundamental frequency of 14 MHz
- Use prefabricated 450-Ω "window" transmission line
- Keep the lower end 10 feet or more off the ground
- Secure the base and guys with stakes
- Make a "kit" for next Field Day!

Propagation

The HF Illumination Profile and the Bright Leading Edge of Illumination

By William Alsup, N6XMW
1120 Ashmount Avenue
Oakland, CA 94610

N6XMW presents some very interesting, innovative ideas about how the ionosphere operates.

Have you ever tried to determine whether a QSO was one-hop or two-hop or both? Have you considered the best arrangement for a QRP QSO? Have you wondered about the relationship, if any, between the MUF and your antenna height?

This article investigates two important refraction characteristics of the ionosphere. *HF illumination profile* is a term I use to describe the illumination intensity pattern, as a function of range, of HF energy reflected to Earth from the F_2 layer. The *bright leading edge of HF illumination* describes a critical segment of the profile—the surprisingly high-intensity zone of illumination occurring just beyond the minimum skip distance. The bright leading edge of HF illumination, as we shall also see, is the *raison d'etre* of the *Maximum Usable Frequency* or MUF. These concepts help us visualize the RF energy patterns our stations cast over the landscape.

Ionospheric *focusing* and *defocusing* are also involved. Focusing refers to the disproportionate concentration of RF energy in a geographic region. Defocusing means the opposite. For example, if four times as much energy is reflected into one square mile as into another, the energy is more focused in the former and more defocused in the latter. Through a home-brew computerized tracing of signal paths, I have been investigating these patterns. Let's be clear first, however, on the difference between horizontal and vertical focusing.

Horizontal Focusing and Defocusing

In free space, the signal strength from an isotropic radiator varies according to the inverse square law. On Earth, however, the ground itself reflects upward one-half of the signal and the ionosphere reflects downward the other half (ignoring losses). The energy is thus trapped between two concentric spheres and spreads out horizontally, between the spheres, somewhat like a wave from a stone dropped into a pond. The original pulse of energy is stretched ever thinner around an ever-longer perimeter, until it reaches the *antipodal equator*, which is the circle around the Earth midway between the antenna and the antipodal point. The antipodal point is that point exactly on the opposite side of a sphere such as the Earth.

At the antipodal equator (one half way around the Earth), the signal strength is weakest, having been defocused on its travel one quarter way around the Earth. Thereafter, the ring shrinks, concentrating or "focusing" its energy. When the ring then reaches the antipodal point, the signal is fully reconstituted and fully refocused at its original power (except, of course, for unavoidable losses along the way). This is *horizontal* focusing and defocusing.

Vertical Focusing

In the vertical plane, signals are also focused and defocused, but in a different way. Two causes are: (1) the ever-increasing concentration inherent in the trigonometry of ionospheric reflections, and (2) the effect of Pedersen Rays.

Ever-Increasing Concentration of Ionospheric Reflections

For any given compass direction, frequency and ionospheric state, the usable vertical takeoff angles go from zero to the highest takeoff angle reflected back by the ionosphere. Although all possible takeoff angles are equally distributed about the point of radiation, the paths launched along the higher usable angles will return to Earth with more concentration than those launched at lower angles. Put differently, as the takeoff angle is gradually raised in equal steps, the hop distance, or range, is reduced by ever smaller steps and thus becomes increasingly concentrated. For example, as illustrated in **Fig 1**, the angles 0° to 20° fall to Earth over a span of 1500 miles (900 to 2400 miles), whereas the angles 20° to 40° fall to Earth over a span of 450 miles (450 to 900 miles). Although both span 20° of change, the latter sector is more

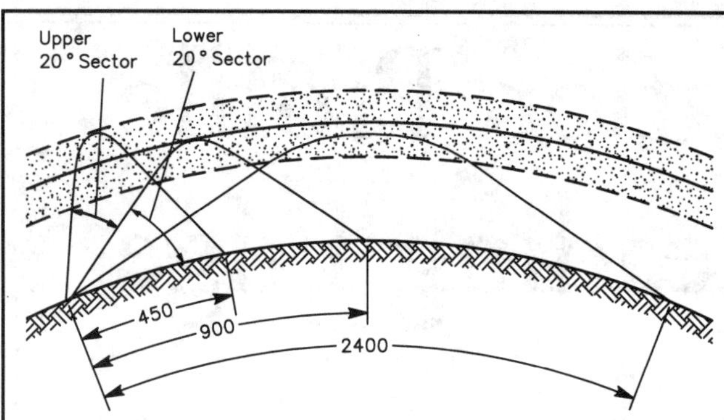

Fig 1—Signals launched in two different sectors towards F_2 layer: 5° to 20° and 20° to 40°. The signal launched in the higher-elevation sector is called the Pedersen ray. Signals launched in the lower sector return to Earth in a range from 900 to 2400 miles away, while signals launched in the higher sector return to Earth over smaller range—450 to 900 miles away. The higher-elevation signals are concentrated more, because they are spread out over a smaller number of miles.

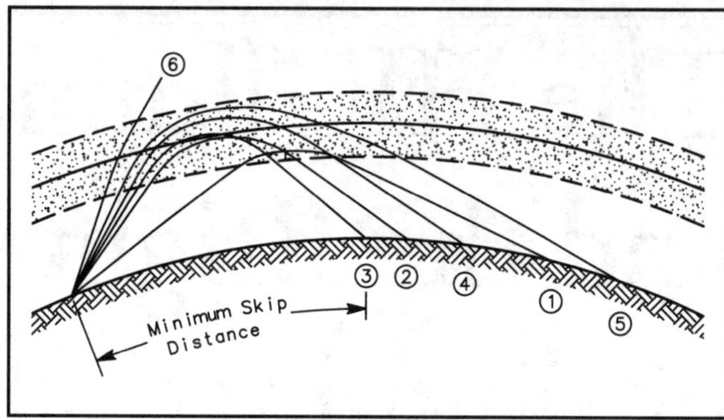

Fig 2—Signals launched at various elevation angles towards F_2 layer, which is shown with varying density either side of its center. The ray on Path 3 yields the minimum skip distance. Increasing the elevation angle higher results in the signal coming back to Earth at a greater distance, as in Path 4. Finally, the elevation angle is so high (Path 6) that it cannot be refracted back to Earth, and the signal goes out into space.

focused than the lower-angle sector. We are assuming that the relevant region of the F_2 layer is uniformly ionized.

The Effect of Pederson Rays

The second, although lesser, cause of vertical focusing is the reverse-range effect of high-angle rays, called *Pedersen Rays.* The F_2 layer is most ionized at its center line and decreasingly so below and above it (See **Fig 2**).[1] Signal paths launched at sufficiently low takeoff angles and sufficiently low frequencies are refracted to Earth before ever reaching the center line (Path 1). As the takeoff angle increases, the signal path approaches the center line and, with further increase, peaks above the center line and still returns to Earth (Path 2). At first, as the takeoff angle thereafter continues to rise, the range still shrinks as before. This much is well known.

Less known is the fact that as the takeoff angle continues upward and at an angle somewhat above that needed to peak above the center line, the range eventually reverses field. The angle at which this reversal occurs I call the *Reverse-Range Angle* (Path 3). At the Reverse-Range Angle, the range begins to *increase* and will do so at an ever faster rate as the angle rises further (Path 4). At some point, even a tiny increase in the takeoff angle produces a large increase in the range (Path 5). Finally, as the takeoff angle rises higher, the signal path escapes into space (Path 6).

This reversal in angle-range relationship occurs because once the signal path passes sufficiently into the upper reaches of the F_2 layer, it is only slightly refracted. Sometimes, in an extreme case, the refraction is just enough to match the curvature of the Earth. The signal thereby travels long distances. Pedersen Rays comprise only a small part of all vertical angles refracted back to Earth.

The Minimum Skip Distance and the Illumination Profile

It is now easy to see why there is always a minimum skip distance. Once the takeoff angle rises to the Reverse-Range Angle, it is mathematically impossible for the range to shrink further—indeed the range will increase even though the takeoff angle also increases! The minimum skip distance is not, as the author once thought, caused by finally reaching a takeoff angle that passes through the F_2 layer into space, but is caused, somewhat before that angle is reached, by the reverse-range effect of the Pedersen Ray.

It is also easy to see why there will normally be, just beyond the minimum skip distance, a great concentration of signal strength. First, that immediate area receives paths originating below and above the Reverse-Range Angle. Second, there is a disproportionate concentration or focusing in this zone. This zone is what I call the *bright leading edge of HF illumination.* For example, if the minimum skip distance is 400 miles, there will typically be strong signals in the 400 to 600 mile range (**Fig 3**).

Fig 3 illustrates the natural illumination profile of an isotropic antenna radiating equally at all vertical angles, with no ground reflections in the near vicinity of the antenna. The minimum skip distance in this illustration, about 400 miles, is totally dark. It is immediately followed by the brilliance of the bright leading edge.[2] Thereafter, illumination falls off sharply and then gradually fades to minimal levels. Intensity is shown in linear relative units. Takeoff angles 0° to 20° spread their energy across fully 1500 miles of the illuminated profile, while angles 21° to 45° concentrate their entire energy across only 600 miles. In this example, the Reverse-Range Angle is 45°; if it were to go lower, more and more of the illumination profile would dissolve, starting from the left. Fig 3 assumes a virtual height of 200 miles for all angles.

The actual height of the F_2 layer is 180 to 240 miles or more, with a dense center at 230 to 260 miles, on a summer day. On a winter day, it is 150 to 210 miles or more, with a dense center at 160 to 180 miles. At summer and winter nights, the F_2 height is 140 to 210 or more miles, with a dense center at 165 to 190 miles. The signal path is curved through the F_2 layer. The "virtual height" of the F_2 layer is the height of the triangle formed by extending upward the up-leg and the down-leg of the signal path until they meet at a "virtual apex" above the curved cap of the actual path. The up-leg and the down-leg are equal when the ionosphere is uniformly ionized in the refraction region, as is assumed throughout this paper.

Thus, for a virtual apex of 250 miles, the peak height of the actual path is lower. The calculations are based on this trigonometric model, taking into account the curvature of the Earth and using a circumference of 25,000 miles. The profiles shown take into account only vertical focusing and defocusing. If the horizontal defocusing were also included, the bright leading edge, being closer to the transmitter, would be even brighter compared to the "toe" (right-hand edge) of the profile.

The Reverse-Range Angle and the MUF

Why call it the Reverse-Range Angle? The MUF, of course, is the highest frequency usable between two given locations under given ionospheric conditions. Amateurs generally try to work at or just under the MUF for a given circuit. The bright leading edge explains why.

The trick is to pick the frequency, if possible, for which the bright leading edge of HF illumination falls right on the other station—and vice versa under the rule of reciprocity.[3] When we succeed, we are working at the MUF and the relevant takeoff angle is the Reverse-Range Angle. Any other usable (and therefore lower) band may illuminate the other station but not with the bright leading edge for that band, which will fall somewhere between the two stations. For any higher band, the bright leading edge will move beyond QTH_2, leaving QTH_2 in the dark.

You can usually get an idea of what the Reverse-Range Angle is for a band by tuning around the band, identifying the nearest and loudest stations, and then consulting **Fig 4**. The MUF and Reverse-Range Angle are controlled completely by the ionosphere and have nothing to do with your antenna height.

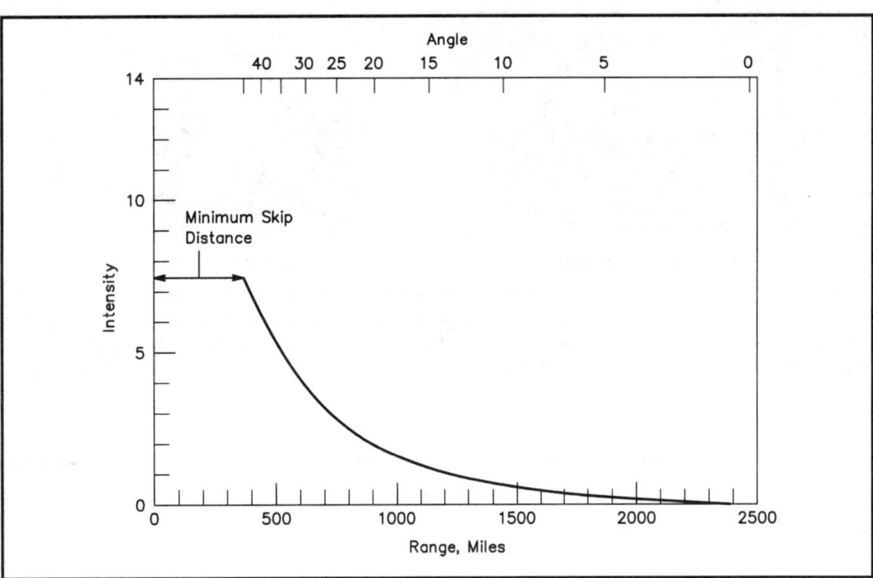

Fig 3—Single-hop ray-tracing analysis of signal intensity versus range in miles for an isotropic radiator. N6XMW calls the high-intensity region between 400 to 600 miles the *bright leading edge of illumination*. This analysis assumes that the virtual apex of the F_2 layer is 200 miles high

The Role of the Refraction Index and Frequency

The boundaries and, to some extent the intermediate shape, of the illumination profile depend on the refractive power of the F_2 layer and on the frequency. As the F_2 refraction index increases, the entire illumination profile shrinks toward QTH_1, and vice versa, for a fixed frequency. Put differently, the entire profile, including its bright leading edge, retracts and extends across the landscape, as the case may be, by as much as a few hundred miles as the refraction index goes up and down.

The profile, however, is not simply shifted left or right without changing its shape. As the refraction index increases, more and more vertical takeoff angles are bent back to Earth (rather than passing into space), adding more and more energy to the profile. The down-leg angles are also bent more, and thus become more concentrated, raising the intensity of the entire profile. When the index decreases, the profile loses strength, grows less bright, and moving away from QTH_1. When the lowest angles reach the maximum possible range (about 2500 miles), the toe can go no farther. As the index decreases yet more, the leading edge gradually approaches the toe, and the profile grows dimmer all the way, until finally there is no profile left.

There is, of course, a similar relationship when the refraction index is held fixed and frequency is varied. For any given ionosphere condition, decreasing the frequency will draw the entire profile closer to QTH_1; and so on as above. This illustrates a fundamental relationship

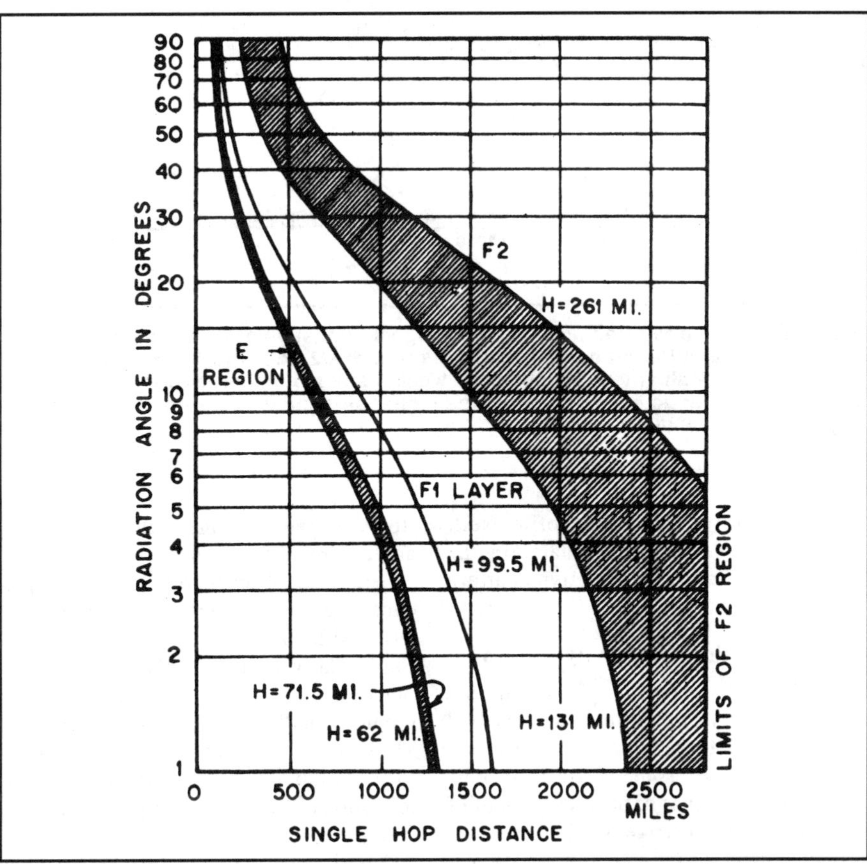

Fig 4—Distance plotted against wave angle (one-hop transmission) for the nominal range of heights for the E and F2 layers, and for the F1 layer. (From *The ARRL Antenna Book*.)

between the refractive index, the frequency and the illumination profile. If we could legally operate on any frequency, we could slowly change our frequency as the refractive index (and critical frequency) changed. We could thereby maintain the same illumination profile, subject only to one caveat—to avoid

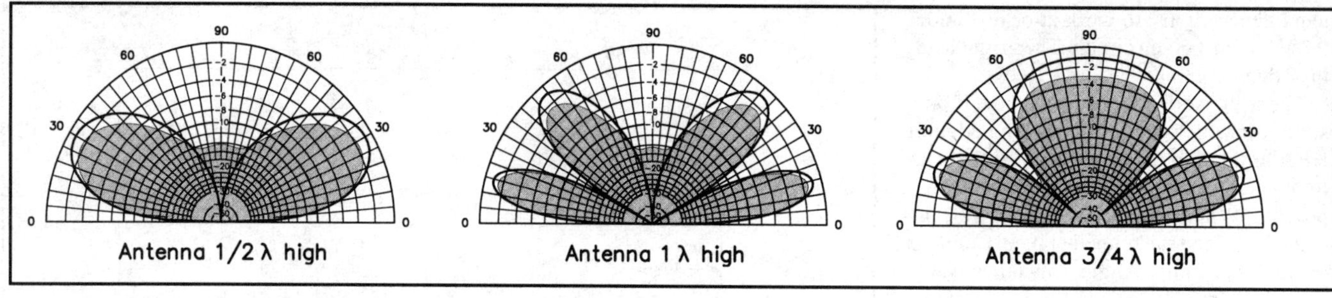

Fig 5—Vertical directivity patterns for various heights above ground. The solid-line curves are the perfect-earth patterns (broadside to the antenna wire); the shaded curves represent the effects of average earth (dielectric constant = 13, conductivity = 5 mS/m) at 14 MHz. Add 9.15 dB to values shown for gain in dBi. (From *The ARRL Antenna Book*.)

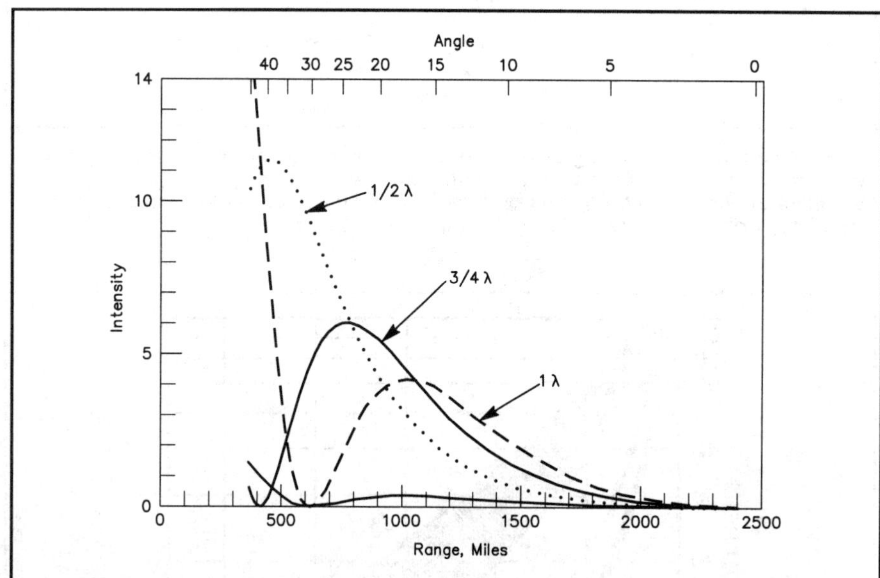

Fig 6—Single-hop ray-tracing analysis of signal intensity versus range in miles for three dipoles at different heights over flat ground: λ/2, ³/₄ λ and 1 λ. For close-in ranges the 1-λ high antenna produces a weaker signal compared to the lower dipoles, as would be expected, while the higher antenna is better for distances beyond about 900 miles.

brightness shifts within the profile itself, it would be necessary to maintain the same vertical radiation directivity pattern as the frequency changed.

Antenna Vertical Directivity Pattern

Unlike a free-space isotropic antenna that radiates equally in all directions, practical antennas on Earth, because of ground reflections, emphasize some vertical takeoff angles at the expense of others. One-half of the radiated power strikes the ground and is reflected upwards (at least for flat, perfect ground). The reflected waves either add constructively in phase with the remainder of the signal or destructively out of phase, to greater or lesser extent, in accordance with Huygen's Principle. At one or more takeoff angles, there is a complete phase coherence, which doubles the field strength at that takeoff angle. At certain other angles, the phases totally cancel and create a null. Between the extremes, the intervening angles result in intermediate phase coherences and net energy levels. This produces the familiar vertical directivity patterns with lobes and nulls illustrated in **Fig 5**. The vertical directivity pattern is a function of antenna height above effective ground, usually assumed to be flat ground.[4] The pattern is not a function of ionospheric focusing, obviously being fixed by ground reflections well before the lobes reach the F_2 layer.

A horizontal beam, by the way, has lobes and nulls at the same vertical angles as a horizontal dipole of the same height, but the beam radiates more energy into the lower lobes and less energy into the higher lobes than a dipole, thus brightening the far field illumination even more.[5] When the antenna's maximum power angle coincides with the Reverse-Range Angle, the bright leading edge will be even brighter, a fact of perhaps special interest in QRP operations. The two angles need not (and usually do not) coincide, since they are driven by totally independent considerations—one by antenna height and ground reflectivity, and the other by the physics of F_2 reflection. For example, ionospheric focusing might favor stations 1000 miles away, whereas a high antenna may favor stations 1700 miles away.

This clarifies a not-so-apparent contradiction between the proposition that we should work at the MUF and the proposition that we should favor low angles and high antennas. The Reverse-Range Angle for a station 1000 miles away may be 24°, whereas our high antenna may favor 12°. Indeed, it may even exhibit a null at 24°! A higher antenna will put more energy into the toe of the illumination curve and less into the steep angle near the MUF, as illustrated below. You may not necessarily want this effect. The ideal would be to work at the MUF (for QTH_2) and to raise (or lower) your antenna to match your strongest takeoff angle with the Reverse-Range Angle!

Significantly, a change in antenna height over flat ground will always modify the shape of the illumination profile, but it will not change the mathematical start and end points of the illumination profile. The impact of a change in antenna height is confined to varying the degree of illumination for the intermediate points. This is because a path launched along a given takeoff angle is bent by the ionosphere to the same degree regardless of how much energy it carries. Antenna height only dictates how much energy is directed at various takeoff angles, not how much that path will be bent when it reaches the ionosphere. One practical result of a higher antenna is to put more energy into the otherwise right-hand toe of the illumination profile. The effect is illustrated in **Fig 6**.

Fig 6 compares the range/signal intensity profiles of three horizontal dipoles: at ¹/₂ λ, ³/₄ λ and at 1 λ, assuming a constant ionospheric virtual apex of 200 miles. The intensity of these curves is modulated by the antenna vertical radiation patterns at various takeoff angles. Thus, at 1 λ high, the main lobe of radiation translates to an obvious swell of illumination at about 1000 miles in range for the 1 λ profile and at about 500 miles for the λ/2 case. At most, ground reflection will double the strength over that of an isotropic antenna. On the other hand, when the two phases are opposite, they cancel, forming a null that prevents illumination. For example, at a range of 600 miles for a 1 λ height there is a null. There is always a null at 0° for any horizontal antenna over flat ground.

As the takeoff angle rises, the virtual apex of the signal path also rises. In Fig 6, we let the virtual apex rise (increasingly so) as the takeoff angle ticks upward. The lowest virtual apex is 200 miles (at 0°) and the highest is 250 miles (at 45°), according to the formula:

$$\text{apex} = 200 + 200 \times (1 - \cos\theta)$$

where θ is the takeoff angle in degrees.

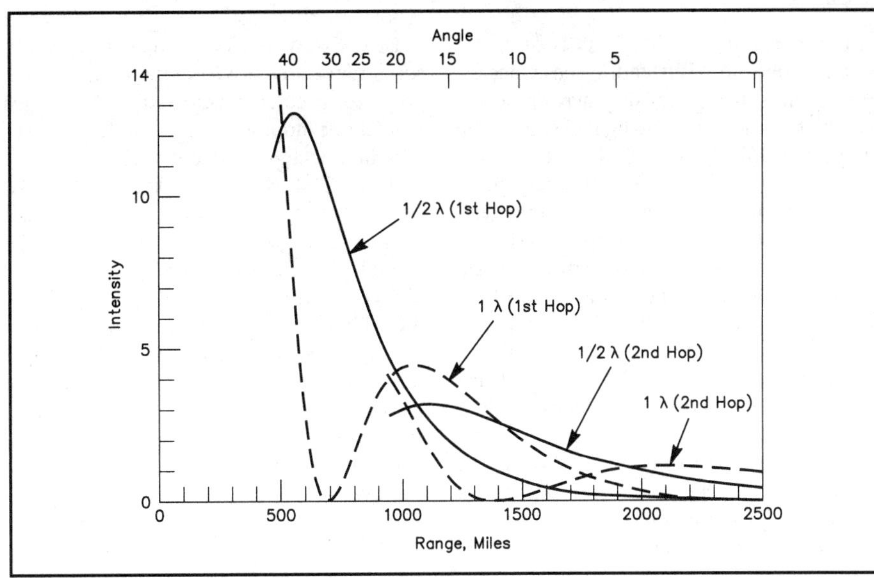

Fig 7—Ray-tracing analysis for both one-hop and two-hop signal intensity versus range for dipoles at λ/2 and 1 λ heights, assuming that two-hop signal is 6 dB down compared to one-hop signal.

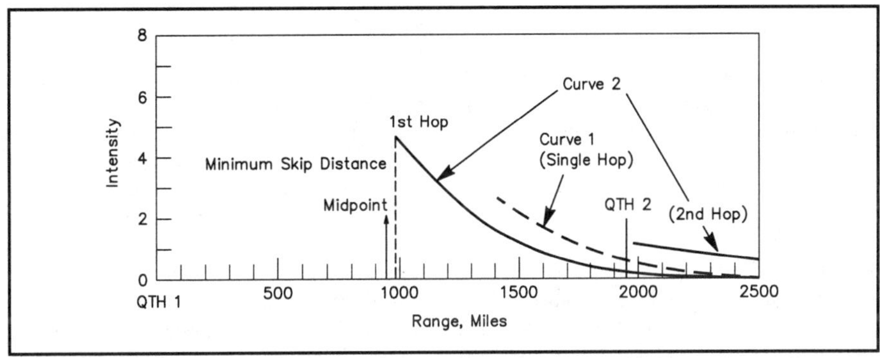

Fig 8—Ray-tracing analysis showing disruptive transition from one-hop to two-hop propagation for dipole antenna at height of ¾ λ over flat ground. Curve 1 is for single-hop case when refractive index is low, and Curve 2 is for two-hop case as the refractive index rises.

One-Hop Versus Two-Hop Propagation

The illumination profile of the second hop has two main characteristics. First, it covers twice as much ground as the first profile. For example, if the first profile covers a 1000-mile zone (range of 1500 to 2500 miles), then the second-hop profile covers a 2000-mile zone (range of 3000 to 5000 miles). This implies defocusing. Second, the intensity is also diminished by a second round of ionospheric absorption and reflection losses. Absorption losses are greater for lower frequencies and denser D-layer conditions. Ground reflection losses are greatest over dry desert and least over seawater. The lower the radiation angle, the lower the reflection loss, even over dry desert.

Fig 7 adds the illumination profile for the second hop for heights of λ/2 and 1 λ. It assumes that the second hop is one-quarter the field strength (–6 dB) of the first. For the 1-λ case, the Hop 2 signal strength in the 2000 to 2700+ mile range is adequate, with virtually no direct contribution from Hop 1. Where the Hop 1 and Hop 2 curves overlap, they will add or subtract depending on the phase differences. This will result in fading, which will be particularly severe where the two amplitudes are about equal (at about 1750 miles for the 1-λ height and 1100 miles for the ½-λ height).

Now consider the interesting case of the disruptive transition from one-hop to two-hop propagation. When the refraction index is lower, QTH_1 and QTH_2 maintain a single-hop QSO, as shown in Curve 1 in **Fig 8**. Each station falls on the weaker but readable end of the other's single-hop profile (reciprocal propagation). Two-hop propagation, however, is impossible because, for the distances involved, the necessary midpoint falls in the unlit minimum skip zone. As the refraction index rises, there's bad news: the entire profile shrinks toward QTH_1 and the single-hop profile is no longer readable at QTH_2; but the shift is not yet enough to illuminate the two-hop midpoint—so the circuit fails (Curve 2). As the refraction index rises yet more, however, the Hop 1 profile will finally illuminate the midpoint and two-hop propagation will be possible to QTH_2 (curve not shown). Indeed, the two-hop signal should come blasting in as the bright leading edge initially passes over the midpoint.

This progression typically occurs each day between certain locations. The F_2 refraction index rises progressively from sunrise until the sun is directly overhead (or even later in the summer afternoon). Thereafter, it remains about constant until late afternoon when it begins to decline. Thus, every morning, there are circuits that transition from one- to two-hop propagation, in some cases with, other cases without a gap, depending on the distances and frequencies used.

It is generally said that, between the same two stations, single-hop propagation will be stronger than two-hop propagation. For example, *The ARRL Antenna Book* (1988) states at page 23-14: "Assuming that both waves do reach point B...the low-angle wave [the single hop] will contain more energy at point B." The reasoning behind this is that absorption losses will be greater for 2-hop propagation, since it must pass through the D-layer an extra time (at least during daytime) and must also suffer ground reflection losses at the midpoint reflection.

However, the two-hop path is often the stronger one, notwithstanding these factors. First, ionospheric focusing may favor the two-hop path. As shown, the single-hop may well fall on the defocused toe of the illumination profile. The double-hop, however, may take advantage of the bright leading edge of illumination. Second, as shown, all practical antennas favor certain vertical angles more than others. For example, a horizontal dipole ³/₄ λ above flat ground will concentrate far more energy at a 20° takeoff angle than, say, at 5°. The latter ray will take off with 8 to 10 dB less power than the former. If the latter ray is single hop, and if the former is double hop to the same QTH, then the double-hop ray can afford to lose some energy to absorption along the way and still be stronger on arrival.

Example No. 1—Two Hop Or One?

Every Saturday morning, at 7:15 California time, N4CLT (Junior Feild), my oldest friend (since age five), and I maintain a twenty-meter SSB schedule. We both use horizontal antennas about 55 feet up with 100 W out. We are 2250 miles apart.

Even at the highest virtual apex (during the summer daytime), my one-hop signal arrives in Alabama on the weak toe of the illumination curve. The midpoint of my two-hop signal, however, falls close to the bright leading edge of HF illumination. This early in the day, the illumination profile may not have built up enough and the midpoint may fall in the minimum skip zone. In fact, we have had a few such occasions in which propagation was so poor that we could not maintain our QSO. Our schedule is near the transition from one-hop to two-hop propagation. The midpoint is thus just on the active side of the bright-line between the darkness of the skip zone and the brilliance of the near illumination field.

Empirical observations support this. While waiting for our schedule, I tune around and hear strong stations near the midpoint for the two-hop path, so I know the midpoint is open. Also, I don't hear hams much closer, so the bright leading edge is just over the midpoint.

How do I know we are *not* using one-hop propagation? Actually, a one-hop path may contribute a little, but not much. The one-hop path would require a vertical takeoff angle of approximately 5° or less in the summer to almost 0° in the winter (when the F_2 layer is lower) (Fig 4). These low angles imply significant vertical defocusing and great attenuation as they go through the lossy D layer. Also, our antenna heights favor about 18° to 20° takeoff angles, so our one-hop signals carry far less power than our two-hop signals. Finally, there are location-specific obstacles, namely the East Bay Hills, that block any really low angles for me in that direction.

Dean Straw, N6BV, compared these observations to *IONCAP* computations, which do indeed predict that the best mode changes from a one-hop, low-angle mode to a two-hop, higher-angle mode just about the time of the N4CLT-N6XMW schedule. *IONCAP* shows that in January and July, the transition occurs between 1400 and 1500 UTC for virtually all Smoothed Sunspot Number (SSN) conditions. We meet at 1515 UTC in the winter and 1415 UTC during Daylight Saving Time (in both cases 7:15 AM California time). The two-hop F_2 takeoff angle varies between about 12° and 17° in January and about 19° to 21° in July. The computer predictions are consistent with our observations.

Example 2—The Two-Hop Anomaly

Now, let's consider a different time of day and what could be called the "two-hop anomaly." Let's say I want to reach N4CLT in Alabama in the early evening at 0300 UTC on twenty meters. During the day, as stated, we regularly have solid QSOs. Why is it that at this time of day I can continue to hear Florida and Texas stations clearly but I can't hear N4CLT very well?

Usually during the day, we employ two-hop propagation. However, as the sun first sets on the easternmost hop, the F_2 layer will support only shallower angles and will no longer support angles steep enough for two F_2 hops from Alabama to California. The F_2 layer will, as stated, still support the two-hop path with its moderate angle between Florida and the Midwest, at least for a while longer. The Texas signals are *one*-hop. Thus, Alabama falls into a twilight zone or, more aptly, a *dark zone*. The Hop 1 profile is unaffected, but the Hop 2 profile has faded such that the two-hop path originally destined for N4CLT simply passes into space. (As before, one-hop propagation between us requires a takeoff angle sufficiently low to be unusable as a practical matter.)

Once again, *IONCAP* computations predict the two-hop anomaly just described. At 0300 UTC, Auburn, Alabama, can be reached only by a single-hop F_2 path at a very low takeoff angle—1° to 4°. This low an angle is a special problem for me, given the nearby East Bay Hills blocking my signals. Meanwhile, however, at 0300 UTC South Florida still booms into Oakland by two-hop F_2 propagation, and Dallas by one F_2 hop, all at higher angles that get over the hills.

Example 3—Antenna Height Control

Assume you are a QRP backpacker wanting to maximize the chance you'll be heard at all. You use a 20-meter QRP rig and a dipole, which you can usually string at heights up to 50 feet. What should you do to maximize your signal strength for short- and medium-distance QSOs?

You may need to resist your first instinct to string the dipole as high as possible. The way to maximize your signal is to overlay the main lobe with the Reverse-Range Angle, which varies with the solar flux, time of day, and season. In the summer in the afternoon, the Reverse-Range Angle for twenty meters can be as high as 60° or more, much lower in the morning and evening.

To match a high angle of 60°, our antenna height should be low, say about ¹/₄ λ (17 feet). To match a Reverse-Range Angle of 30°, we would raise the dipole to ¹/₂ λ (33 feet). To match a Reverse-Range Angle of 20°, we would shoot for 50 feet. Of course, we would want to consider whether the target range is populated sufficiently to have hams listening (as vast areas of the West's Great Basin are not).

Example 4—Frequency Control

Assume that the single-hop MUF to QTH_2 is 24 MHz but you can operate only on 14, 21 and 28 MHz. Do you try 28 MHz? No, that would leave QTH_2 in the dark, in the minimum skip zone. Do you drop to 21 MHz? Perhaps, but not necessarily. On 21 MHz, the single-hop path may be far from the bright leading edge. On the other hand, 14 MHz is close to the MUF for the two-hop midpoint. In this case, 14 MHz would be better, especially if your antenna's optimum takeoff angle is better for that band. By trying to visualize the illumination patterns your station casts across the earth, you can make these judgments more intuitively and accurately.

The Role of the E Layer

So far we have assumed that the F_2 layer is the only refracting layer. However, when the E layer is sufficiently ionized, it also can refract HF signals. The centerline of the E layer is about 75 miles above the Earth. Being lower, it catches the sun's radiation after the F_2 layer every morning and loses the sun's radiation earlier than the F_2 layer every evening. The E layer disappears at night. (Sporadic E persists into the evening but is a different phenomenon caused by wind shear, not solar radiation.)

At dawn, the F_2 layer is the only refracting layer. As the morning wears on, the E layer builds up sufficiently to intercept and to refract signals back to Earth. At first, this occurs only at the lower frequencies, and only for the shallowest angles. As E-layer ionization builds, higher frequencies and higher takeoff angles are intercepted and redirected to the Earth. Yet higher angles and frequencies still penetrate the E layer and enter the F layer. Let's call the breakpoint the *E-F Transition Angle*.

The critical frequency of vertical incidence for the E layer varies from 1 to 4 MHz, depending on the phase of the eleven-year sunspot cycle and the solar flux at the moment. At shallowest angles, however, the critical frequency is said to be about 3.5 or more times the critical frequency at vertical incidence (less at intermediate angles), according to *The ARRL Antenna Book*.[6] Thus, frequencies up to the 20-meter band are refracted by the E layer, at least at the shallowest angles. The E layer intercepts a considerable portion of the 40-meter and lower bands during the central daytime hours.

The main point is that when the E layer is active, the lower takeoff angles are caught and redirected by the E layer, resulting in an E layer illumination profile pretermitting what would otherwise be the far field of the F_2 illumination profile. The far field of the F_2 illumination profile will vanish so long as it lies in the shadow of the E layer. Given the trigonometry on both sides of the E-F Transition Angle, a corollary is that the E and F_2 illumination profiles will *always* partly overlap. If the E-F Transition Angle is sufficiently low, the F profile will overlay the entire E profile. In the overlap region, there will be both E and F paths to a station. This implies stronger than usual signals with fading.

So, to wrap this article up, let's now turn to a grand visualization.

Conclusion: A View From Space

One reason for this analysis is to help us visualize illumination profiles of our stations. Let's try it in three dimensions. Let's pretend we are high above the Earth and can actually see HF illumination. Let's pretend that it is noon below us, that the ionosphere beneath us is equally ionized in all directions, and that we are studying the pattern of a single antenna up one-half wavelength over flat ground. Let's again assume a Reverse-Range Angle of 45°.

Ignoring ground waves, we see that the antenna is in the middle of a dark circle (the minimum skip zone), which is surrounded by a brilliant perimeter of illumination (the bright leading edge of the HF illumination). Enclosing all of this is a wider circular and concentric belt of ever-reduced illumination at greater distances from the antenna.

Since all horizontal antennas have at least some horizontal directionality, the favored direction is more brightly illuminated. Such directionality, however, does not affect the radius of the belt of illumination, but only its brightness. To put it differently, the circular shape of the belt of illumination we see from space differs from the shape of the horizontal radiation patterns so familiar to amateurs. The circular belt of illumination is still circular but its brightness varies according to the horizontal radiation pattern.

If the antenna is raised (over flat ground), the start and end points of the circular belt will remain the same but the intermediate intensities change. The minimum skip zone remains the same; the bright leading edge remains in the same position but is dimmer, perhaps even obliterated if it coincides with a null in the vertical directivity pattern, and the toe of the profile is brighter, particularly where it coincides with a lobe. If a horizontal beam at the same height is substituted, it has two effects: (1) in the horizontal pattern, the beamwidth is more concentrated and the favored direction all the more illuminated, and (2) in the vertical pattern, the lower lobes of radiation are favored, which translates to a brighter toe and dimmer leading edge, both conditions favoring long-distance communication.

In all cases, there is also a second, more faint, circular belt of illumination at about twice the first radius, and a third, yet fainter circular belt of illumination at about three times the original radius, all representing multiple hops in the signal, and so on. In each case, the inside perimeter will be brightest. The belts overlap when the degree of ionization is high enough. As shown in the accompanying figures, they overlap at a Reverse-Range Angle of 45°.

Now, let's visualize the bands of illumination over the course of a day. In the early morning, if the F_2 layer does not yet support any propagation at all at the given frequency, there is no illumination. As the sun rises, the circle begins to build gradually, starting as an arc on the easternmost side. At this time, the minimum skip zone has a larger radius than later in the day. This is because, at first, the takeoff angles refracted are bent less sharply back to Earth, and the Reverse-Range Angle is low.

As the sun rises further, the belt of illumination gradually forms a full circle and the inside perimeter of the belt marches toward the antenna and becomes brighter as more and more takeoff angles are returned to Earth and more and more concentration occurs. A second-hop belt can also now be seen. As the refraction index increases, the Reverse-Range Angles rises, and the two belts eventually overlap. Before midday, the belts stabilize for several hours.

Near dusk, a reverse process occurs. The circular belts begin to stretch to the east. As the belts stretch eastward, communication can be maintained at greater distances to the east, even if for brief periods, as the profile and its bright leading edge, race across the land. Eventually, the easternmost arc of the belt fades and vanishes, forming an open oval, like a horseshoe. Once the circular belt vanishes in part, communication cannot be maintained in that direction.

Keep in mind that your reception profile is the same as your illumination profile. You can get a good idea of how your station illuminates the landscape by tuning around and noting the locations and strengths of stations heard and by noting the locations where you hear *nothing* (the minimum skip distance and nulls in your pattern). By listening on different bands, and comparing the signals of WWV (and W1AW) on various bands you can fill in even more data. Pretty soon, you can visualize the three-dimensional illumination patterns for various bands and decide how to make best use of them for your immediate objective and how to adjust your antenna to rearrange your profile.

Notes

1George Jacobs, Ted Cohen, *The Shortwave Propagation Handbook*, 2nd Edition (CQ Publishing, Inc., 1982), pp 6-7. Includes an explanation of why the F2 layer is most ionized at its center and for actual height data of the F2 layer.

2Kenneth Davies, *Ionospheric Radio Propagation*, (National Bureau of Standards Monograph 80, 1965), pp 167-169, pp 226-227. Experimental evidence indicates that the focusing factor is 6 to 9 dB.

3For a discussion of the rule of reciprocity, on which the aphorism "If you can hear them you can work them" is based, see p 2-11, 16th Edition, *The ARRL Antenna Book*.

4Les Moxon, *HF Antennas For All Locations*, (RSGB, 1991), Chapter 10.

5William Orr, *Radio Handbook*, 23rd Edition (Howard Sams & Co., Inc, 1988), p 22.3 and *ARRL Antenna Book*, 16th Edition, p 11-7.

6George Jacobs, Ted Cohen, *The Shortwave Propagation Handbook*, 2nd Edition (CQ Publishing, Inc., 1982), p 6-14. The 3.5 factor is not free of doubt—the highest frequency at any lower angle is governed by $F=F_c/\sin(\text{angle})$ where F_c is the critical frequency at vertical incidence.

7Harry Hooten, *The Amateur Radio Antenna Handbook* (Howard W. Sams: Indianapolis), p 29. He says the factor is two.

10-Meter Long Path During Solar Cycles 21 and 22

By Carl Luetzelschwab, K9LA
1227 Pion Road
Fort Wayne, IN 46845

K9LA puts together the whole story on 10-m long-path propagation. DXers take note: Many things happen at once along the path between you and the other station.

Introduction

Although I've been a ham since 1961, my first 10-m long-path QSO (with VS6DO) was in the 1986 CQ World Wide DX Phone Contest. VS6 was a new country for me and it started my interest in 10-m long path. But a move from the Dallas/Fort Worth area back north kept me relatively inactive until the winter of 1988. Then, spurred on by N6AV's "10-Meter Long Path" article in the January 1989 issue of *The DX Bulletin*, I began working Japanese amateurs early in the morning during the spring and summer of 1989 before going to work.

Soon I was interested in gaining a better understanding of this mode and I received help from a number of people. Gus, K2ARO, supplied me with his 10-m long-path log data going back to early 1979. Yuu, JH3DPB, maintained daily schedules with me to help better understand the duration of 10-m long-path openings to Japan. I had many conversations and correspondence with Bob, NM7M, who later wrote a booklet *Long-Path Propagation—A Study of Long-Path Propagation in Solar Cycle 22*.

My intent in this article is to present the guidelines and indicators of 10-m long-path propagation so others may enjoy this mode. Although I will focus on 10-m long path from the East Coast to Australia (VK6) and Japan (JA) due to the excellent amount of data from K2ARO, I will also comment on 10-m long path to and from other geographical locations. At the end of this article is a bibliography listing all of the reference and background material quoted and utilized in this article.

I strongly suggest reading NM7M's booklet cited above. *The Shortwave Propagation Handbook* by Jacobs (W3ASK) and Cohen (N4XX), *Radio Amateur's Guide to the Ionosphere* by McNamara, and *Ionospheric Radio Propagation* by Davies are also very helpful. The monthly periodicals carry many articles on HF propagation. One very pertinent to this article is "Propagation Broadcasts and Forecasts Demystified" by NJ2L in the November 1991 issue of *QST*.

What Is Long Path?

In general an HF radio wave follows a great-circle route. This is the shortest route between two points on the Earth's surface. **Fig 1** is an azimuthal map of the world centered on K2ARO. The path was computed using Oldfield's *DXAID* software. K2ARO would normally beam northwest over Canada to work JA (6843 mi) and VK6 (11378 mi) by short path.

Under certain conditions, though, K2ARO can beam on the reciprocal heading (southeast over South America and Antarctica) to get to VK6 (13467 mi) and JA (18002 mi). This path going southeast is designated the long path.

Some Observations About the Path

In **Fig 2** I have plotted the long path from K2ARO to VK6 and JA on a Mercator projection of theEarth. The path is broken into 1610 km steps. NM7M's *ULTMTLP* software was used for analysis. It goes through the southern auroral zone, and was affected to a larger degree than a non-auroral path by the activity of the Earth's magnetic field. The terminator (sunrise/sunset line or the *gray line*) at 1100 UTC is shown for the month of April in Fig 2 and the portion of the Earth in darkness is cross-hatched. A later discussion will cover the roles of the auroral zone and the terminator on 10-m long-path propagation.

Long-Term Look at the Path

Fig 3 is a plot of K2ARO's log data from 1979 to early 1993. The vertical axis is the number of days in a given month Gus made

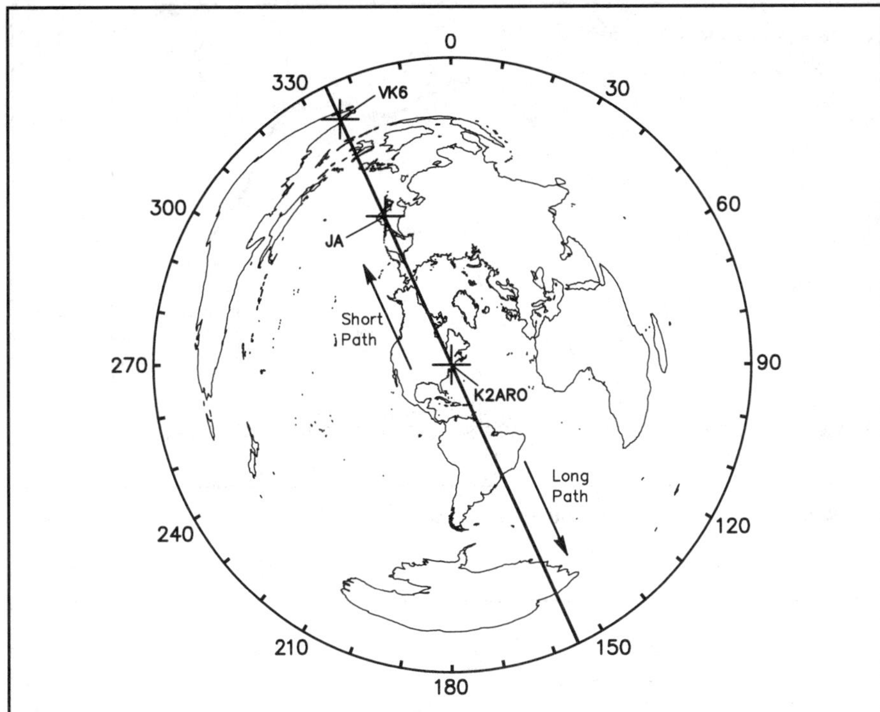

Fig 1—Plot of the long and short paths between K2ARO and VK6/JA locations.

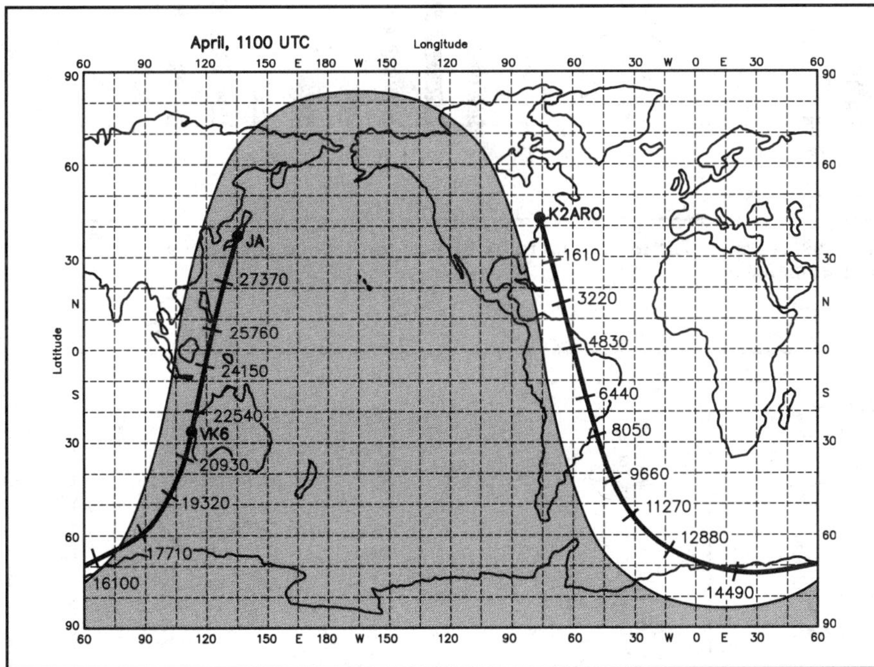

Fig 2—The same long path on a Mercator projection map. The path appears to follow the gray line.

a long-path QSO to either JA or VK6 on 10 m. No contacts were made in 1983 through 1987. The graph is compressed in the 1984 to 1986 time frame to save space.

The data clearly shows 10-m long-path was not a rare occurrence. It occurred regularly throughout certain portions of Sunspot Cycle 21 (June 1976 to September 1986) and Cycle 22 (September 1986 to late 1993). In order to provide a correlation with the sunspot cycle I plotted on the same figure the *smoothed sunspot number*. This is a 12-month running average of the monthly mean sunspot numbers and is called the *SSN* or R_{12}.

Daily *solar flux* numbers are broadcast by WWV and the ARRL station, W1AW. Use daily solar flux numbers with care, because they do not correlate well with *smoothed sunspot numbers (SSN)*. A full discussion of the calculation of the smoothed values is on page 23-13 of *The ARRL Antenna Book*, 17th Edition. Smoothed numbers can also be found in *QST* and W3ASK's Propagation column in *CQ*.

Fig 3 indicates 10-m long-path was a possibility when the smoothed sunspot number was above about 70. Unfortunately, this also shows 10-m long path is over for Cycle 22, as the smoothed sunspot number declined through 70 in early 1993.

A Seasonal Look at the Path

Note the data in **Fig 3** is plotted in terms of either a JA or a VK6 QSO. Even though the long-path heading to VK6 and JA from K2ARO is identical (as can be seen in Fig 1 and Fig 2), in reality the 10-m long-paths to VK6 and JA exhibits different characteristics when examined separately.

To demonstrate this, **Fig 4** is a plot of the number of days per month (averaged over all years) K2ARO made a JA QSO or a VK6 QSO. The path to JA opened in the spring, held up rather well throughout the summer, and gradually tapered off by the fall. In comparison the path to VK6 is more bi-modal in nature—it peaks around spring and fall, with an obvious reduction in availability in summer. The reason for this difference has to do with the path crossing the equator to get into JA and will be discussed later. The data for July is limited since it came from only one year, 1989, at the peak of Cycle 22.

A Daily Look at the Path

The long path to JA and VK6 is an early morning path for K2ARO. **Fig 5** is a plot of the earliest time and latest time (both averaged over all years) by month for a QSO into JA and VK6. It shows the path into VK6 consistently opening about a half hour earlier than the path into JA. Generally the openings to both JA and VK6 occur earlier in the summer than in either the spring or fall. The path into VK6 goes away about an hour before the JA path goes away. You can see the paths to both JA and VK6 close earlier in the summer than in either the spring or fall.

What Opens the Path

K2ARO's log data shows why the path to JA and VK6 opens early in the morning at his location. Multi-hop propagation (successive ionospheric refractions and Earth reflections) is assumed. Predictive data was taken from the F2 MUF maps in the 1971 publication *Ionospheric Predictions* from the Institute for Telecommunications Science. MUF is *maximum usable frequency*, and refers to the highest frequency propa-

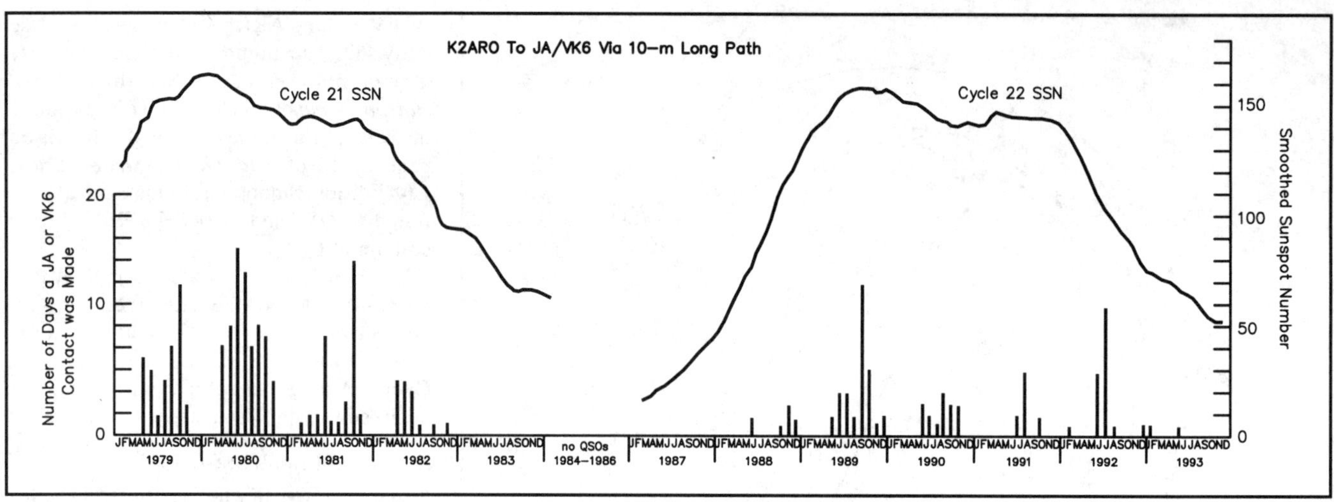

Fig 3—K2ARO's log data. Notice the correlation with the value of the smoothed sunspot number.

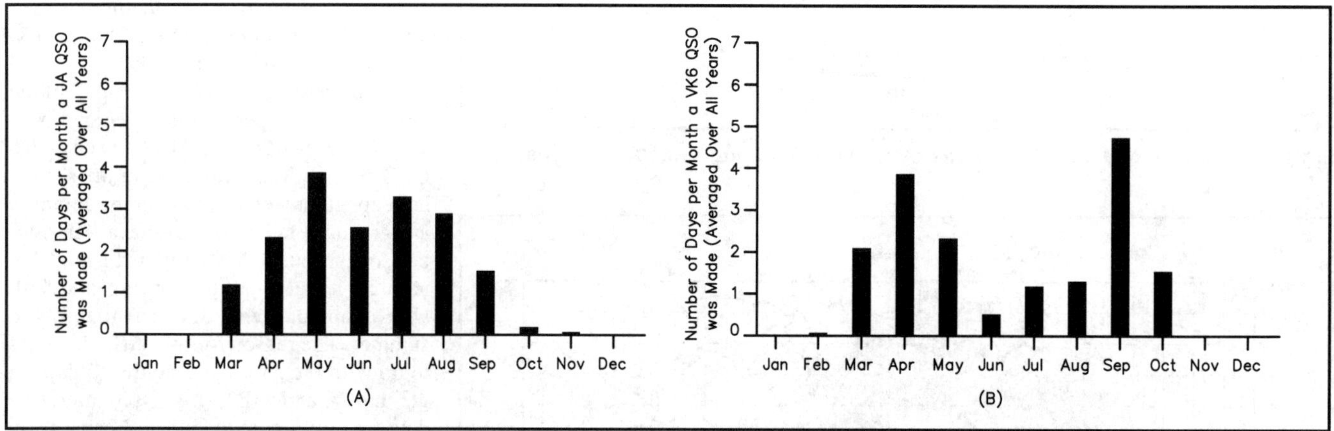

Fig 4—Seasonal variation in the number of successful QSOs. The histogram to JA shows continuous successes from spring through to fall while the histogram to VK6 has a summer null.

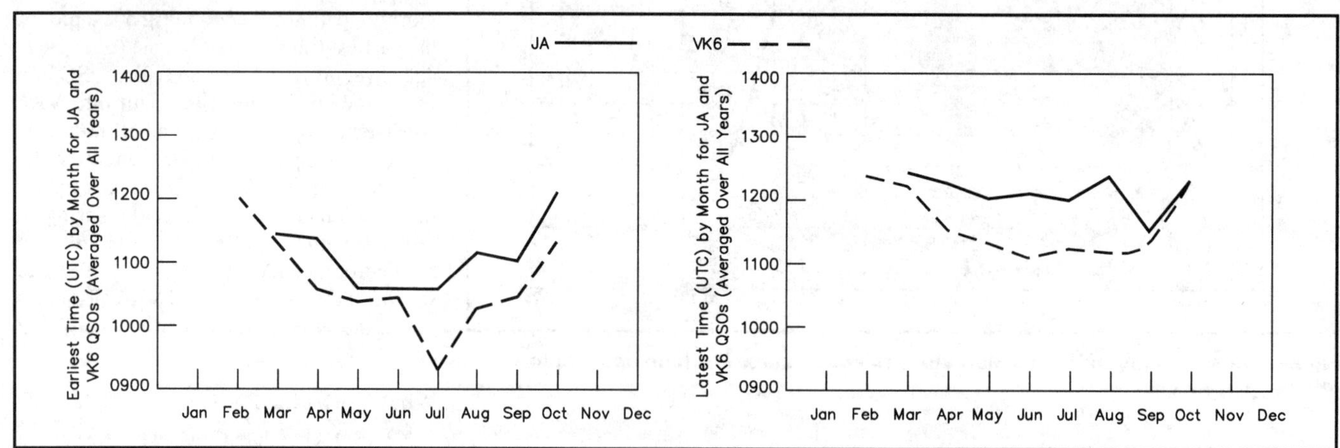

Fig 5—Early and late times for the QSOs. VK6 QSOs start and end first. Those to JA seem to lag by about a half-hour to an hour.

gated between two designated points at a given time.

This publication consists of four volumes. Volume 1 shows how to use the data, and Volumes 2, 3 and 4 consist of constant contour lines of median MUF(ZERO)F2, MUF(4000)F2 and MUF(2000)E for smoothed sunspot numbers of 10, 110 and 160, respectively. For a detailed explanation of these maps and the terminology used, refer to "High-Frequency Propagation Estimations for the Radio Amateur" by K1PLP (now

MUF (3220 km) F2 in MHz, K2ARO to VK6/JA, Apr 89, R_{12} = 160

Distance (km)	09	10	11	12	13	14	15	16
	Time, UTC							
0 (K2ARO)								
1610	20	21	27	33	33	35	35	36
3220	23	25	31	37	39	40	41	41
4830	23	27	33	39	39	39	39	39
6440	23	26	33	39	37	36	35	35
8050	23	26	34	42	43	45	44	43
9660	23	28	34	42	45	47	46	46
11270	22	28	33	38	41	44	43	42
12880	24	28	31	33	35	37	36	35
14490	26	28	29	30	31	30	28	25
16100	29	30	30	30	28	25	22	20
17710	31	30	28	27	23	20	19	17
19320	32	31	29	26	23	20	18	16
20930 (VK6)	34	33	30	27	23	20	19	17
22540	39	37	34	32	30	28	26	23
24150	42	39	40	41	42	43	42	40
25760	33	32	32	31	32	32	35	39
27370	44	42	43	44	45	47	47	45
28980 (JA)								

Fig 6—The crosshatched area shows MUF values greater than 28 MHz.

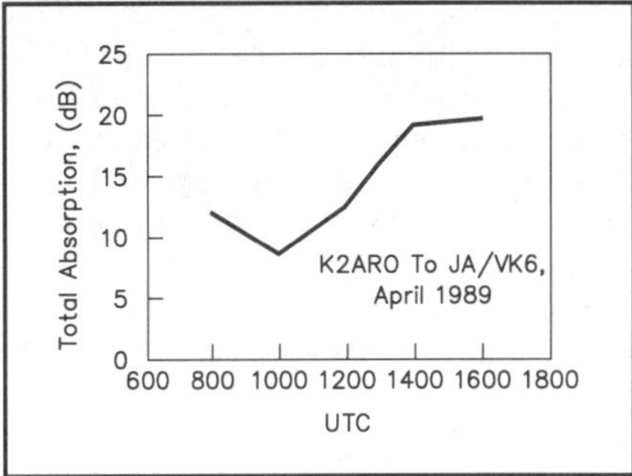

Fig 7—The path is also affected by absorption. This factor increases from 1000 until 1400 UTC.

Skeds with JH3DPB

The 2 hour, 42 minute opening Gus had to JA on 6/18/89 took me somewhat by surprise, as I had thought the 10-m long-path to JA was a rather short duration event. I was curious how long an opening the Midwest would have. To find the answer, JH3DPB and I maintained daily schedules during April 1992. April was selected only because it was a convenient month for both of us. We began calling each other at 1030 UTC and checked every 15 minutes thereafter. If contact was established, we would continue checking every 15 minutes until the path closed.

Although contact was established on many days, two days were very productive in terms of long-duration openings. On these two days contact was established at 1100 UTC (6 AM for me, 8 PM for Yuu) and continued until 1315 UTC (8:15 AM for me, 10:15 PM for Yuu). The Boulder A-index was less than 7 on both of these days. This 2 hour, 15 minute opening between the midwest and JA translates to about a 3 hour opening for K2ARO (because of his earlier sunrise). This agrees fairly well with his 6/18/89 opening.

The opening of the path was rather sudden in nature, indicating we had to wait (as expected) until the MUF on my end passed through 28 MHz. The closing was a gradual decrease in signal strength to the noise level, indicating the role of increasing absorption.

These skeds also enabled me to confirm K2ARO's data presented in Fig 5. The VK6s indeed came in earlier and dropped out earlier.

During these tests JH3DPB ran a TS-930 barefoot to a 6 element monobander at 110 feet, and I ran a TS-180 barefoot to a 4 element monobander at 72 feet.

K1TD) in the March 1972 issue of *QST*.

Generally 4000 km is used as the one-hop F2 limit. However, propagation beyond this distance can be predicted by evaluating the ionosphere at two control points, each 2000 km from the ends of the path. This technique, empirically based, is usually effective.

In Fig 2 the path is broken into 1610-km segments to present the MUF all along the path, not just at the common 2000-km control points. This will give us excellent insight into what is happening all along the path. The 3220-km MUF is used since it is the hop distance assumed in NM7M's *ULTMTLP* software. The path requires radiation at an elevation angle of 5°, an angle readily achievable on 10 m. Nine hops are needed to cover the distance. The 3220 km MUF is interpolated from the MUF(ZERO)F2 maps and the MUF(4000)F2 maps using Fig 5 of Volume 1 of *Ionospheric Predictions*.

Ionospheric Predictions

Fig 6 shows the results of this task. The choice of the MUFs for April 1989 allowed me to directly use Volume 4 (R_{12} = 160) of the *Ionospheric Predictions* maps (refer back to Fig 3 for R_{12} during April 1989). The figure shows how the 3220 km MUF varied all along the path from K2ARO to VK6 and JA beginning at 0900 UTC and going to 1600 UTC. F2 layer refraction, assuming multi-hop propagation, would occur at 1610, 4830, 8050, 11270, 14490, 17710, 20930, 24150, and 27370 km.

The cross-hatched MUFs in Fig 6 exceed 28 MHz and therefore allow 10-m propagation. Remember, all values given are median values. Therefore the actual MUF could be several MHz above or below the values in the table on any given day of April 1989.

The path is predicted to open up around 1100 UTC in April due to the increasing MUF on the western end of the path (K2ARO's end). This agrees very favorably with the earliest time data in Fig 5. The increase of the MUF with time (top horizontal axis) corresponds to K2ARO's sunrise and the subsequent increased ionization as the sun moves west (okay, the sun doesn't move west—the Earth rotates eastward).

What Closes the Path

The data in Fig 6 also shows the path is predicted to close after 1200 UTC due to an ever-widening area of decreased MUFs. This is shown by the expanding non-cross hatched areas between roughly the 17710 km distance point and the 20930 km distance point. The latest time data in Fig 5 agrees favorably with this prediction of closing. The area of decreasing MUFs corresponds to the part of the path between Antarctica and Australia.

The decreased MUFs may not be the only factor closing the path. I have estimated the total absorption over the entire path for the 0800 to 1600 UTC time period based on the solar zenith angles at the F-layer refraction points along the path. This data is presented in **Fig 7** and was calculated by using NM7M's *ULTMTLP* software and Figures 7.5 and 7.6 in *Ionospheric Radio Propagation*. Two items should be noted.

First, the absorption gradually increases starting at the time the path opens (about 1100 UTC). Thus the path probably begins to close as a result of both higher absorption as the western part becomes increasingly illuminated by the sun as well as the decreasing MUF in specific areas along the path as discussed earlier.

Second, the absorption is minimum at about 1000 UTC. This corresponds very closely to the time when both K2ARO and JA/VK6 are on the terminator. This condition of minimum absorption when both stations are on the terminator is not unique to K2ARO and JA/VK6—in fact this is the rule, and is in my opinion what causes the often-observed signal increase in gray-line DXing (especially on the lower-frequency bands).

The Duration of an Opening

Not only did K2ARO's log data provide the days during a month with 10-m long-path openings to JA and VK6, it also provided times (as we saw in Fig 5). Therefore, the length of an opening could be determined.

Gus made JA QSOs on 163 days and VK6 QSOs on 161 days throughout the 14-year period. I will venture an opinion that the path was open more often, but no one was available to take advantage of it.

The average duration of a JA opening was 23 minutes. The longest duration opening was 2 hours, 42 minutes (on June 18, 1989). There were a large number of short-duration openings of only several minutes.

The average duration of a VK6 opening was 17 minutes. The longest duration opening was 1 hour, 40 minutes (on April 17, 1992). Again there were plenty of short-duration openings of several minutes.

The Role of the Earth's Magnetic Field

Those of you who have read NM7M's work or those who have more than a passing interest in HF propagation realize the activ-

Ray Tracing

K2ARO and I undertook an effort to do a ray trace during a summer month to confirm a signal could overfly low MUF areas around VK6 but still proceed into JA. We picked July 1989. For this month the smoothed sunspot number was 160. We started with NM7M's *PRTYBIRD.BAS* program written to do ray tracing for satellite signals to the ground.

Gus (a retired software development manager) modified *PRTYBIRD* for our Earth-ionosphere-Earth condition, and I supplied the critical frequencies and layer heights in equation form. This information was obtained from curve fits from the MUF map data and from US Army Signal Radio Propagation Agency *Technical Report No. 9 Analysis and Prediction of Sky-Wave Intensities in the High Frequency Band*. From these equations and assumptions of electron density profiles, refractive indices versus height along the path in 2 km steps were available. Using Snell's Law in differential form, a ray was traced through the desired region.

To keep things relatively simple in this preliminary effort, we ray traced out of JA towards K2ARO for about 12,000 km. We were looking for a chordal hop to overfly VK6 and reach ground somewhere south of VK6. Our initial result at 1100 UTC is shown below. We had expected to get a single chordal hop (designated as a ^{2}F mode), but note our result is two chordal hops (designated as a ^{3}F mode) giving the first ground reflection at around 11,000 km. The first chordal hop was centered over the geomagnetic equator as expected, and the second was centered about over VK6 (see accompanying **Fig A**).

At first this double-chordal hop bothered us. But a review of Villard, Stein and Yeh's paper show they believe their backscatter-sounder records also showed ^{3}F modes. We also believe our initial result is accurate, but we're still testing the ray-trace program and we're still refining the model of the ionosphere. It's sure encouraging to see the possibility of ^{3}F modes reported elsewhere—this ^{3}F mode would certainly help explain the drop out of VK6 in the summer while JA is still maintained.

We are continuing with our efforts, focusing on starting at K2ARO and going all the way to JA. This should include both conventional multi-hops and chordal hops. K2ARO's log data will be used to confirm our results. Perhaps we'll have more to report at a later date.

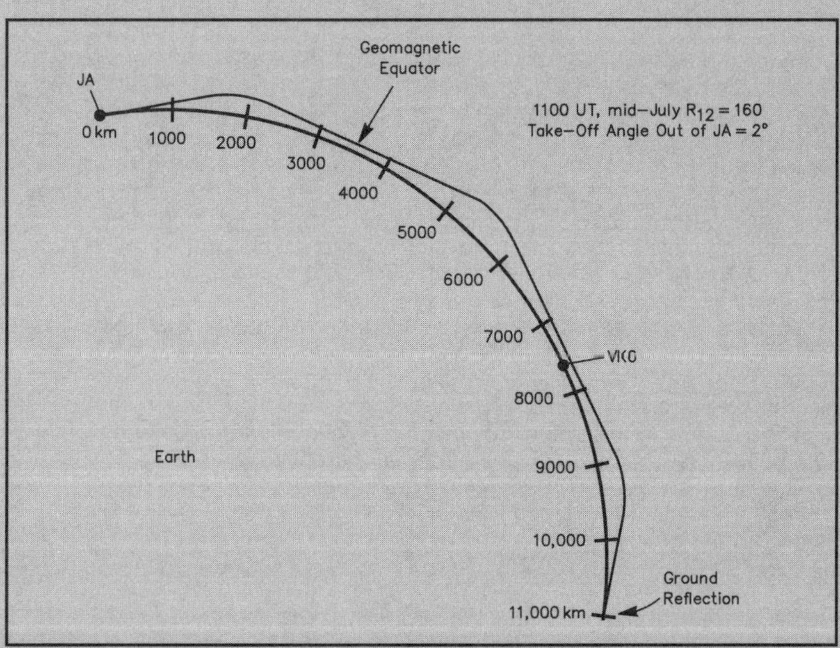

Fig A—Preliminary results showing double-chorded ray trace from JA to K2ARO.

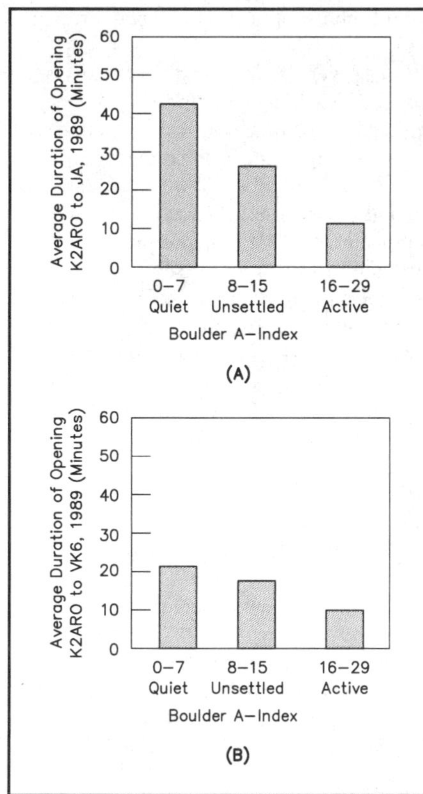

Fig 8—The plot shows the relation between the Boulder A-index and the length of openings. The effect is more pronounced to JA.

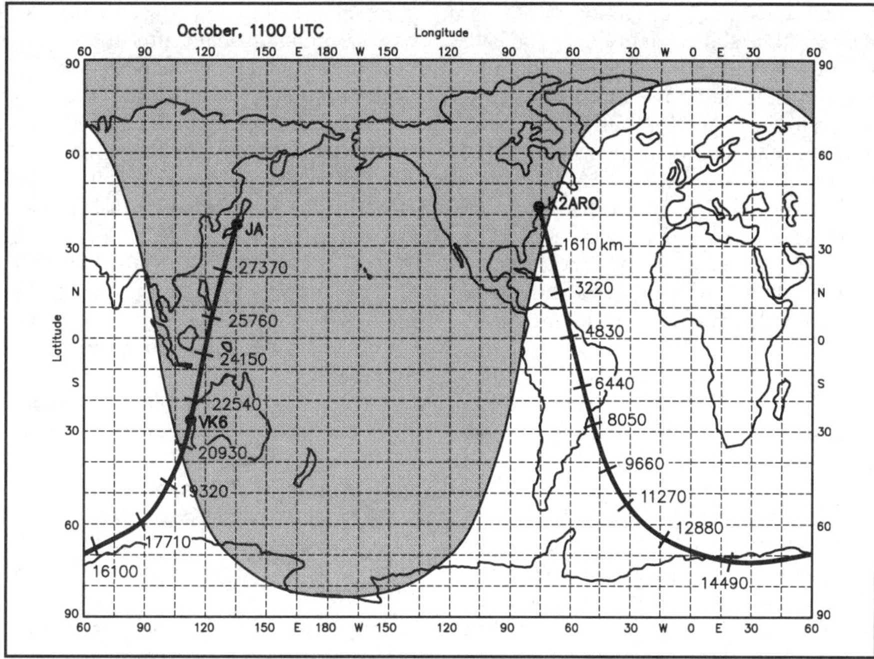

Fig 9—Long paths do not have to follow the gray line. Gray-line propagation helps but it is not a necessary factor in 10-m long path propagation.

ity of the Earth's magnetic field plays a large role in HF propagation. Taking the lead from W3EP's Table 2 in his "Predicting Transatlantic 50-MHz F-Layer Propagation" article in March 1993 *QST*, I have plotted in **Fig 8** the average duration of an opening to JA and VK6 during 1989 (based on groupings of the Boulder A-Index). I chose 1989 because it was the peak of Cycle 22 and I have back issues of QRZ DX with geomagnetic indices reported by KH6BZF. Fig 8 shows an obvious relationship of duration to the activity of the Earth's magnetic field—the quieter the magnetic field, the longer the opening.

The Role of the Terminator

The K2ARO to JA and VK6 appears to be a nearly gray-line path in Fig 2. You might wonder if this is a necessary condition for 10-m long-path. The answer lies in K2ARO's log data for QSOs prior to March 21 and after September 23. These dates are the spring equinox and fall equinox, respectively, when the terminator "flips over." Gus's path to JA and VK6 prior to March 21 and after September 23 did not hug the terminator as in Fig 2. Much of the path is in a highly sunlit area well away from the terminator, as shown in **Fig 9**.

Gus's log shows numerous QSO days occurring before March 21 and after September 23—in fact the numbers work out to about 10% of his JA total and 23% of his VK6 total. Thus the terminator is not a necessary condition for 10-m long-path. The bar graphs of Fig 4 do indicate most of the QSOs occurred when the path was oriented with the terminator (March 22 through September 22). I believe the biggest factor causing this effect is the Sun's location in the Northern Hemisphere during these months. This results in less absorption along most of the path.

The Role of Transequatorial Propagation

Earlier I suggested the seasonal difference seen between the VK6 path and the JA path was due to the JA path crossing the equator. Let's look at this in more depth.

Refer back to Fig 4. The JA and VK6 bar graphs tell us somehow the path to JA is maintained during the summer months even though the path to VK6 usually disappears. How can this happen? A mode of propagation known as *transequatorial propagation*, TE, causes this effect.

TE is a regularly occurring phenomena at equatorial latitudes in the early evening hours (corresponding to the path between VK6 and JA around 1100 UTC). It generally is optimum around 7 PM local time in these equatorial latitudes. It usually consists of two refractions off the F layer without an intermediate ground reflection, giving a hop (commonly called a chordal hop) much greater than the normally accepted maximum of 4000 km.

TE is the result of the unique characteristics of the ionosphere in equatorial latitudes. The electron density peaks at about 15° of latitude from each side of the geomagnetic equator and the height of the electron density peaks at the geomagnetic equator. Davies' Figure 4.26, McNamara's Figure 3.16, and NM7M's Figure 21 show these characteristics.

TE is well documented. It is discussed in sections 3.4, 3.5, and 4.1.8 of NM7M's work and section 6.5 of Jacobs and Cohen's book. Other discussions are in sections 3.6 and 9.4 of McNamara's book and Davies discusses it in his section 4.6.6. Figure 4.26 of Davies' book shows an example of ray tracing providing a 5800 km transequatorial hop without an intermediate ground reflection. Villard, Stein and Yeh provide similar but more detailed analysis and ray tracings of transequatorial chordal hops of up to 7000 km in "Studies of Transequatorial Ionospheric Propagation by the Scatter-Sounding Method" in Volume 62, No 3, September 1957 of the *Journal of Geophysical Research.*

How does TE affect our K2ARO to VK6 and JA path? As stated earlier, it allows K2ARO to make JA contacts in the summer when conventional multi-hop cannot be supported. In other words, depressed areas of MUF in the summer are similar to, but more extensive than, that shown in Fig 6, which are overflown by a very long hop. I also suspect TE is the dominant mode into JA during the spring and fall months, even though Fig 6 and a similar analysis for fall months indicates conventional multi-hop could be supported.

Transequatorial propagation would explain the time shift seen in Fig 5. The path to VK6 is a multi-hop, while the path to JA relies on TE for its final segment. The multi-hop path opens to VK6, but TE isn't supported until about a half hour later to get into JA.

TE also accounts for the long-duration openings to JA (remember K2ARO's 2 hour, 42 minute opening?) These are not predicted by Fig 6. It can support frequencies about 1.4 times the conventional MUF. Thus the conventional MUF can fall to the low 20s in these equatorial latitudes while still supporting 28-MHz TE. This results in the drop out of VK6 while JA hangs in there a little longer.

Other Geographical Locations

My analysis so far has focused on the East Coast 10-m long-path to JA and VK6. However, K2ARO had numerous contacts into DU, KH6, P29, VK3, KH2, VS6, KX6, YC, HL, BV and JD1 interspersed with his JA and VK6 QSOs. What about the Midwest and the West Coast? Do they have early morning 10-m long-path propagation to JA, VK6 and similar areas?

Based on my skeds with JH3DPB and QSOs with VK6s, it is obvious the Midwest has openings to JA, VK6 and similar areas. The sun rises about an hour later in the Midwest and so openings from this area will start about an hour later than from the East Coast. Thus, the window of opportunity for openings from the Midwest will be about an hour less (since the other end of the path is the same).

The later sunrise on the West Coast makes me believe there are no long-path 10-m JA and VK6 openings from this area. By the time the sun rose in W6, it would be past midnight in JA-land. During the peak of a good sunspot cycle, it's possible the F-layer could still support propagation at this time in the vicinity of JA, but then availability of JAs becomes a consideration due to the lateness of the hour. I would be interested in hearing from any W6 (or W7) who has had any experiences in this area.

If W6s do not have early-morning 10-m long-path openings to JA and VK6, do they have long-path openings to other locations? Looking back to Fig 2 and shifting the K2ARO end of the path to W6 suggests W6s have an opening to Eastern Europe and Western and Central Asia. N6AV, for one, does confirm that W6s have QSOs on this path.

The paths examined up until now have the western end at sunrise (K2ARO and W6) and the eastern end near or after sunset (JA, VK6, and Central Asia). What about the flip of this, with the western end after sunset (K2ARO and W6) and the eastern end at sunrise? N6AV reports this is a good path for W6 into the Mideast. I'm sure the East Coast has good openings after sunset, too. But is anybody there on the other end?

The twice-a-day availability of 10-m long-path openings is supported by NT5C's log data from March 1989 through April 1992. During this period he worked BY, JA, KH6, P2, S2, VK, VS6 and YB just after his local sunrise. He worked A9, DL, EA, F, FR, HA, HB, HZ, I, LY, OK, ST0, UA, UA9, UB, UH, UI, UL, VQ9, VU, YU, Z2, ZS, 3B8, 5H, 5Z and 9K starting several hours after his local sunset until around midnight. This several hour delay is possible because the F layer goes away after sunset much slower than it forms after sunrise. I examined the log data and it was quite impressive. It made me wish I was back in W5-land to take advantage of the best of both coasts!

Guidelines and Indicators for 10-m Long-Path

In this section I'll try to pull together all the observations and analyses to give general guidelines and indicators for 10-m long-path. These are not the only guidelines and indicators. That's what makes HF propagation interesting—the unexpected.

I will begin by saying 10-m long-path is much more dependent on the sunspot cycle and daylight considerations than is long path on the lower frequency bands. I found it very instructive to look at known 10-m long-paths in relation to daylight/darkness areas. From this certain trends become obvious.

On a broad scale, 10-m long-path is possible during any portion of the sunspot cycle when the smoothed sunspot number is above about 70. When I took a long term look at the path early in the article, I stated 10-m long-path is over for Cycle 22. When will it be back? It will be back when the next cycle, Cycle 23, ascends through a smoothed sunspot number of 70 or so. Based on long-range predictions and statistics on an average sunspot cycle, 10-m long-path won't be back until the turn of the century.

Most of the openings occur from April through October. Don't rule out February/March and November—it's a possibility, although rare. Look for early morning openings generally from southeast through south just after sunrise for several hours, with the summer months opening a little earlier than the spring and fall months (because of an earlier sunrise). Don't be surprised if a heading a little west of south produces results, too. Look for late-evening openings from south through southwest for several hours.

During spring and fall months, the early-morning long-path can be available into locations both north and south of the equator. In the summer, expect the locations south of the equator to disappear, but locations north of the equator can remain, compliments of transequatorial propagation.

On a given day, the best bet for an opening occurs when the Boulder A-index is 7 and below (quiet conditions). This will give the longest-duration openings. Don't count out A-indices of 15 and less (unsettled conditions). Since the A-index is reported on WWV for the previous day, you need to follow the trend for the past several days to estimate what it is on the day in question. An alternative is to use the coarser K-index transmitted at 3-hour intervals throughout the day. An A-index of 7 corresponds to a K-index of 2, and an A-index of 15 corresponds to a K-index of 3. The NOAA/SESC BBS (1-303-497-5000) provides a three-day prediction of the planetary A-index.

Take advantage of the 10-m beacons. *The ARRL Operating Manual* has an extensive list of them. I have personally used the VK6RWA beacon on 28264 kHz as an indication of both VK6 and JA longpath.

Finally, don't be afraid to call "CQ Long Path" with target areas identified. It just may attract someone's attention.

Final Comment

Although I tried to present historical data to take the mystery out of 10-m long-path, there are still surprises. For example, NM7M and GM4IHJ report several instances of 29.4 MHz signals from the RS-12 satellite coming across the dark polar cap. It appears they were able to see propagation off drifting patches of ionization at F-layer heights. As Bob puts it, "By all that is holy and quite conventional, that should not happen!"

Acknowledgments

I would like to thank K2ARO and JH3DPB for their efforts in this study, and to NM7M for his patient help and direction. I would also like to thank NT5C for his log data, and N6AV for his further comments on the W6 10-m long path.

Notes and Bibliography

The following material was used to prepare this article:

J. Hagen's article, "10-Meter Long Path," appeared in *The DX Magazine*. Contact POB 50, Fulton CA 95439 for more information.

Copies of NM7M's book *Long-Path Propagation—A Study of Long-Path Propagation in Cycle 22*, are available from the author for $10 postpaid. Write to Bob Brown, 504 Channel View Drive, Anacortes, WA 98221. Bob is also the author of the *ULTMTLP* and *PRTYBIRD* software. Write him for further information.

The Shortwave Propagation Handbook was written by G. Jacobs, W3ASK, and T. Cohen, N4XX (CQ Publishing, Port Washington, NY, 1979).

L. McNamara, *The Radio Amateur's Guide to the Ionosphere* (Malabar, FL: Kreiger Publishing Co.)

K. Davies, *Ionospheric Radio Propagation* (US Government Printing Office, Washington D.C. 1965.)

R. Healy, NJ2L, "Propagation Broadcasts and Forecasts Demystified," *QST,* Nov 1991, p 20.

The *DXAID* software used is available for $25 postpaid. Write to Peter Oldfield, 251 Chemin Beaulne, Piedmont, Quebec J0R 1K0, Canada.

You can find additional information on smoothed sunspot numbers on page 23-13 of *The ARRL Antenna Book,* 17th Edition (1994). Smoothed numbers can also be found in *QST* each month below the band opening charts. They are also included in W3ASK's monthly propagation column in *CQ* (CQ Communications, 76 North Broadway, Hicksville NY 11801).

The publication *Ionospheric Predictions* was used extensively in the work described in this article. It was originally published by the Office of Telecommunications as *Telecommunications Research and Engineering Report 13* (Boulder, CO 1971). As noted in the text, it consists of four volumes. Volume 1 shows how to use the data, and Volumes 2, 3 and 4 consist of constant contour lines of median MUF(ZERO)F2, MUF(4000)F2 and MUF(2000)E for smoothed sunspot numbers of 10, 110 and 160, respectively. Unfortunately it is out of print. Certain portions may still be available by contacting the National Technical Information Service, U.S. Department of Commerce, Springfield, VA 22161. If you need some selected information please contact Carl, K9LA, at the address at the beginning of this article and he will try to help.

For a detailed explanation of these maps and the terminology used in *Ionospheric Prediction,* refer to "High-Frequency Propagation Estimations for the Radio Amateur," by Jerry Hall, K1PLP (now K1TD) in Mar 1972 *QST,* p 14.

E. Pocock, W3EP, "Predicting Transatlantic 50-MHz F-Layer Propagation," *QST,* Mar 1993, p 32.

QRZ DX is one source of past values of geomagnetic indices based on my groupings of the Boulder A-Index. For further information write to POB 832205, Richardson, TX 75083-2205.

Villard, Stein, and Yeh, "Studies of Transequatorial Ionospheric Propagation by the Scatter-Sounding Method," *Journal of Geophysical Research,* Volume 62, No. 3, Sep 1957.

US Army Signal Radio Propagation Agency Technical Report No. 9, *Analysis and Prediction of Sky-Wave Intensities in the High Frequency Band* is an out-of-print publication used by the author. Unfortunately no source of this report is currently known.

The ARRL Repeater Directory and *The ARRL Operating Manual* include a list of 10-m beacon stations. Although there are frequent changes, they are a useful guide to stations you might check to find open propagation paths.

Computing the Radio Horizon

By Julian "Mac" Pike, W7SDS
5822 Stonebridge Lane
Waxhaw, NC 28173

VHF and UHF propagation are only line of sight? Not really. This program predicts the length of the path and the effects of obstacles.

The year was 1933. In the *Proceedings of the Institute of Radio Engineers* a paper by Schelleng, Burrows and Ferrel appeared, entitled "Ultra-Short-Wave Propagation."[1] This paper has had a long-term significant impact on VHF propagation engineering. It set forth what is known as the *Effective Earth's Radius Model* of atmospheric refractive index for radio waves. This was the basis for a practical way of plotting VHF coverage through use of a fictitious Earth having a radius of 1.33 or 4/3 times the radius of the real Earth.

The 4/3 Earth radius geometry permitted plotting radio wave paths as straight lines. The height of a radio wave plotted as a straight line over the fictitious Earth would thus be the same as the height of a real radio wave above the real Earth.

In the days when there were no computers, engineers highly valued graphical methods for solving problems. The 4/3 Earth model, sometimes also called the *linear* model, gives rise directly to the radio horizon formula:

$$d = \sqrt{2h} \qquad \text{(Eq 1)}$$

where

d = the distance to the horizon, in statute miles

h = the height of the center of the antenna, in feet.

However, while it was insightful in 1933 and is still useful today, Eq 1 has serious shortcomings describing the real structure of radio refractivity in the troposphere. In this article, I discuss some fundamental ideas concerning the bending of VHF and higher-frequency waves in the troposphere and apply a better model to the computation of the radio horizon. A goal of this presentation is to increase understanding of propagation and make computations as realistic as possible.

HORIZON.EXE[2] is my computer program using this new model. The program includes an example of propagation in mountainous terrain to demonstrate performance.

Refractivity

Radio waves travel at the velocity of light, which is fastest in a vacuum. In any other medium they slow down. Upon entering water, light travels at about 3/4 its speed in a vacuum. The reciprocal of this speed ratio is 4/3, or 1.33. This is the index of refraction of water. Radio waves traveling in the atmosphere are slowed down only a tiny amount, since air is not very dense. The index of refraction of radio waves in the troposphere (where our weather takes place) has a typical value of about 1.000325 near the surface.

This very slight departure from 1.000000 accounts for the radio horizon being farther away than the geometric horizon. When the index of refraction increases to more than its typical value, it often causes exciting events, such as VHF DX. Since this number is so close to 1 and only changes by small amounts, it is hard to visualize. Therefore, we usually subtract 1 from it and then multiply the remainder by one million to obtain the refractivity measured in *N-units*. The typical refractivity of the troposphere near the Earth's surface is thus around 325 N-units.

When a wave encounters a change in refractivity, it is bent; that is, its direction of travel is altered. A pencil, half submerged in a glass of water, appears to be bent. This is because the refractivities of air and water are different. While radio waves can encounter regions having a boundary between air masses of different refractivities, we are concerned mainly about the change in refractivity as we move higher in the atmosphere. Intuitively, the atmosphere ends in the vacuum of space. Therefore, the refractivity has decreased from around 325 near the surface, to zero at the top of the atmo-

```
RADIO HORIZON CALCULATIONS
  Station elevation, feet         (      7836.0000 )
  Obstruction distance, miles     (        98.1000 )
  Obstruction elevation, ft       (      4400.0000 )
  Repeater distance, mi           (       180.2000 )
  Repeater elevation, ft          (      7000.0000 )
    The obstruction projects into the path by 1476.0 ft
    The radio wave radius of curvature is 4.9 earth radii
    The refractivity gradient is -31.9 N-units/km
    The gradient required to clear the obstruction is -83.4 N-units/km
                  <Enter> or <Q>uit
```

Fig 1—Screen print of *HORIZON.EXE* with the initial values for the example in the text.

sphere. Much of the time this decrease is fairly gradual.

We often picture radio waves as wave fronts, similar to ripples on a pond. A radio wave propagating horizontally does have some vertical extent. Since the upper portion of the wave is in a region of lesser refractivity than the lower, the upper portion of the wave travels slightly faster than the lower, and the wave is bent toward the Earth. Many studies of the actual refractivity of the atmosphere confirm that the values of refractivity known to the pioneers in 1933 were quite correct near the surface.

The *Effective Earth's Radius Model* is based on a decrease in refractivity of 39 N-units between 0 and 1 km elevation. The word *gradient* is used to denote a change with distance, and a negative gradient denotes a decrease with height. This value, −39 N-units per km, has become known as the standard gradient of refractivity, or *standard refraction*.

In their equation 3.18, Bean and Dutton[3] give the relationship between the refractivity gradient and the radius of curvature of the radio wave. In slightly simplified form, the radius = 1,000,000 / (gradient of N). For standard refraction, the radius is 25,641 km, or 4.02 Earth radii. Turning the equation around and making the radius equal to the Earth's actual radius yields a gradient of −157 N-units/km. This is the refractivity needed to bend a radio wave sufficiently to make it follow the surface of the Earth. This situation is called *ducting*. Widespread refractivity of this magnitude tends to occur more frequently over water, where distant contacts often occur, such as from California to Hawaii. Calculations with the improved refractivity model will yield a refractivity gradient of around −35 N-units/km at a 6000 foot elevation, with a radius of 4.5 Earth radii.

The Improved Model

In Chapter 8 of their book, Bean and Dutton present the *Bi-Exponential Model of Radio Refractivity*. This sounds awesome, but is really quite simple when one realizes that refractivity can be separately expressed in terms of a dry air component and a moisture component—hence the meaning of Bi (two). The Effective Earth's Radius model assumed that the refractivity decreased 39 N-units in each kilometer. This could not be true, since starting with 325 units at the surface would cause the refractivity to run out after about 8 km.

Therefore, you can see decreasing 39 units in each kilometer means a linear decrease. This gives rise to the alternate name *linear* for the old model. Actual measurements in the atmosphere show values of refractivity actually decrease in an exponential fashion as you go upwards in height. While the change may start at −39 in the first kilometer, the change decreases constantly for each higher 1 km layer. From this, a mathematical function can be created to provide a very accurate representation of the average real atmospheric refraction.

It is beyond the scope of this paper to present the mathematical details of the Bi-Exponential Model. It can be "tuned" to arctic, temperate or tropical climates, and can be further tuned to locations based on regional weather characteristics during particular seasons of the year. There are maps in Bean and Dutton for various seasons.

For the computer routine, I chose values for temperate climates and average annual values for the 48 contiguous states. I use it to compute values of refractivity at increasing elevations in the troposphere. Armed with this data, I then found the radius of curvature of a radio wave at each of these points. Rather than use this whole computational process each time, I summarized the relationship between elevation and radius of curvature as a best-fitting straight-line plot through a procedure known as *linear regression*.

Thus, from the average elevation of two station locations, the radius of curvature of the radio wave is calculated to determine the radio horizon information. Remember, this will be for average conditions and will not apply to conditions of unusual propagation. However, it will give insight into the role refractivity plays in propagation.

Computing the Radio Horizon

The computational model in the software uses geometric and trigonometric procedures to derive the working equations for the program. One source of input data to the program is topographic information about the path between two stations. I find a sectional aeronautical chart will quickly give me the big picture on longer paths. It is then possible to move to a 1:250,000-scale topographic map for more detailed altitude information, or even to the smaller-scale topographic maps.

The computer asks for the elevation of the reference station, the distance to the second station and its elevation, and then for the distance and elevation of an obstruction lying along the path between them. **Fig 1** is a print of the computer screen. The results will indicate if the obstruction lies below the propagation path, or projects into the path, and by how much. The program also displays the value of refractivity gradient used, the radius of the wave path, and what value of refractivity is required to clear the obstruction.

An Actual Example

To illustrate, we serve as Table Rock Forest Service Fire Lookout in east-central Oregon during the summer. The Timberline repeater on Mt Hood, about 45 miles east of Portland, can be heard quite well 180 miles away at Table Rock. This is particularly exciting for us since my wife Lola, WDØBAA, visits her family in Portland on occasion, and we are able to keep contact on 2 meters. We are using ICOM 229H rigs with 50 W available, and at the lookout, a Cushcraft A147-4 four-element beam. I scarcely remember any time the Timberline signal was not full quieting!

Much of the time, however, I cannot put a readable signal into the machine with my solar-powered station. The computer tells me Rancheria Rock, at a distance of 98 miles, projects into the propagation path by 1476 feet. The software includes this example and the first time you run the program it puts the values for this example on the screen.

Two other elevations also project into the path, one at 1463 feet, another at 1951 feet. The continual presence of a just-above-marginal signal may be due to diffraction at these obstacles, but might also be a reflection from the higher Strawberry mountains, 20 miles to the west, from which you can see Mt Hood. On one occasion, I was in the valley bottom at my brother's ranch near Mt Vernon and heard the Timberline machine with my beam pointed east toward the Strawberries, so reflection clearly takes place. Someday, I hope to get some fox-hunt capability just to see where the Timberline signal is coming from.

As you can see from Fig 1, the gradient

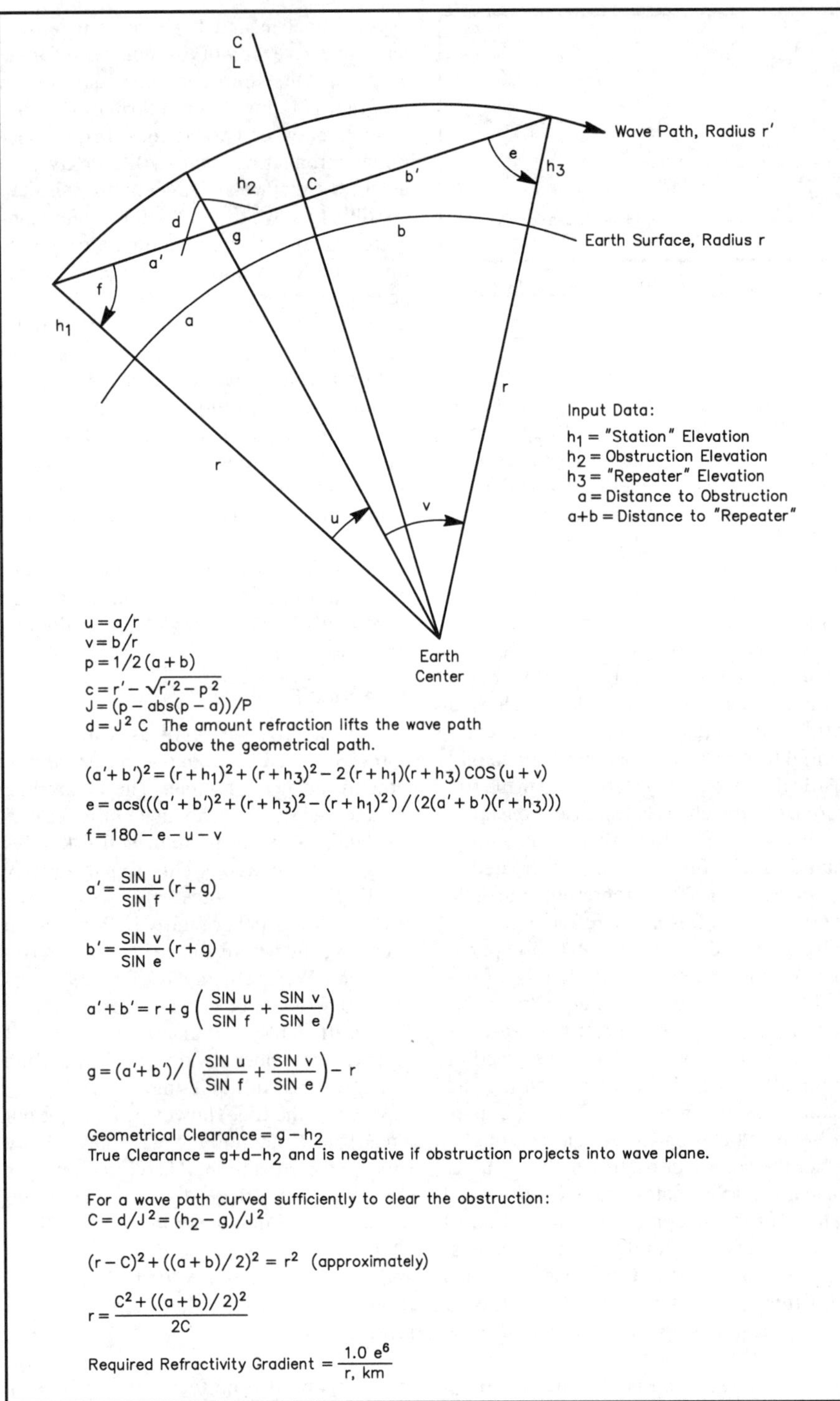

Fig 2—The software is based on this geometry and set of equations.

required to clear the obstacles is –83.4 N-units/km, a greater number than the normally existing value of –31.9 N-units/km. So how do I keep contact with WDØBAA? We use the morning and evening enhancements! As the sun warms up the air in the morning, air closer to the surface warms more slowly and this denser air below causes an increase in the refractivity gradient. At night, the air near the surface cools more rapidly than the air above, causing an enhancement. These can be very strong. In one instance I was able to call up the Timberline machine with my shirt-pocket handheld on low power (150 mW), rubber duck and all! The enhanced signals are usually quite variable, and may come and go over periods of a few minutes. But for an overall distance of around 230 miles, why complain?

With the lookout antenna at 7836 feet, we have a lot of exposure to distant signals. Other repeaters interfere. This includes the Payette machine on the Snake River off the back of the beam and the Lind, Washington, machine to the side. I have been offered a spot on the new Forest Service antenna tower for a bigger beam; that could increase our chances of contact and help reduce interference.

Our local John Day repeater is on the same frequency as the machine out on Mt Hebo in the Oregon Coast Range. Applying the program to this situation says the Cascade Range, 185 miles away, projects some 10,375 feet into the propagation path. A closer obstruction at 75 miles projects over 8000 feet. I have heard the Mt Hebo repeater (273 miles away) a number of times in the noise; it has come in readable enough to make no mistake about its identity.

Summary

I am inquisitive enough to enjoy knowing how I made the contact as well as enjoying the contact itself. Perhaps being able to go back and analyze an unusual VHF/UHF contact in hilly or mountainous terrain will shed some light on how it happened. The program (based on the geometry shown in **Fig 2**) does two things that the simple radio horizon formula fails to do. First it uses a value of atmospheric refractivity closer to the real values found in the troposphere. Second, it examines an obstruction in the path, taking into account the topography along the path.

Further examples of the program with different input values are listed in **Fig 3.** The large negative gradient required to clear the obstacle in Fig 3A suggests this path is impractical under normal circumstances. A 10.9-foot projection into the path in Fig 3B still led to a fairly high gradient requirement (–156.8 N-units/km). However, in Fig 3C, the path of a 1000-foot high repeater was above the obstruction.

The program can, of course, be applied more than once to more than one obstruction on a long path. In this way, it accomplishes what the Effective Earth's Radius Model did when propagation engineers used the special 4/3 Earth radius graph paper to plot the topography and the wave path along with it. Using the Bi-Exponential Model refractivity will make the computations come somewhat closer than the older methods. In a look-forward mode, computations may be of use in determining the feasibility of communicating between two points, and estimating the chances of getting a good enhancement to make it really happen.

References and Notes

1 J. C. Schelleng, C. R. Burrows, and E. B. Ferrel (Mar 1933), "Ultra-Short-Wave Propagation," *Proc. IRE* 21, pp 427-463.

2 The software program *HORIZON.EXE* is

```
RADIO HORIZON CALCULATIONS
  Station elevation, feet       (     7836.0000 ) 200
  Obstruction distance, miles   (       98.1000 ) 5
  Obstruction elevation, ft     (     4400.0000 ) 300          (A)
  Repeater distance, mi         (      180.2000 ) 40
  Repeater elevation, ft        (     7000.0000 ) 500
    The obstruction projects into the path by 173.3 ft
    The radio wave radius of curvature is 2.8 earth radii
    The refractivity gradient is -55.3 N-units/km
    The gradient required to clear the obstruction is -1685.4 N-units/km
                 <Enter> or <Q>uit

  Station elevation, feet       (      200.0000 ) 200
  Obstruction distance, miles   (        5.0000 ) 5
  Obstruction elevation, ft     (      300.0000 ) 200
  Repeater distance, mi         (       40.0000 ) 40                (B)
  Repeater elevation, ft        (     1500.0000 ) 1000
    The obstruction projects into the path by 10.9 ft
    The radio wave radius of curvature is 2.9 earth radii
    The refractivity gradient is -53.9 N-units/km
    The gradient required to clear the obstruction is -156.8 N-units/km
                 <Enter> or <Q>uit

  Station elevation, feet       (      200.0000 ) 200
  Obstruction distance, miles   (        5.0000 ) 5
  Obstruction elevation, ft     (      200.0000 ) 200               (C)
  Repeater distance, mi         (       40.0000 ) 20
  Repeater elevation, ft        (     1000.0000 )
    The path lies above the obstruction by 155.7 ft
    The radio wave radius of curvature is 2.9 earth radii
    The refractivity gradient is -53.9 N-units/km
                 <Enter> or <Q>uit
```

Fig 3—New values are entered to the right of each parentheses in turn. At A, the high obstruction forces an unrealistic requirement for the gradient. At B, the obstruction is not too bad (10.9-feet) but the repeater is perhaps too low to be accessed from 40 miles away. At C, the obstruction does not interfere with the path.

shareware and may be used and copied for Amateur Radio applications. For commercial applications contact the author at the address found at the beginning of this article.

3 B. R. Bean, and E. J. Dutton, *Radio Meteorology*, National Bureau of Standards Monograph 92, U.S. Government Printing Office, 1966.

Receiving Antennas

Ungrounded Beverage Antennas

By Brian Beezley, K6STI
3532 Linda Vista Dr
San Marcos, CA 92069

So you've heard how Beverages can help you hear better on the lower bands, but you're still not convinced they're worth the bother. K6STI analyzes the Beverage, especially an easy-to-construct "ungrounded" version.

Beverages are amazing receiving antennas. A good Beverage can turn a signal barely audible on your transmit antenna into solid, Q5 copy. A Beverage antenna (named after 1920s inventor H. H. Beverage) is a long, low wire used for receiving. They're most popular on the 160, 80, and 40-meter bands. Although occasionally pressed into service, Beverages are much too lossy for regular use as transmitting antennas.

Conventional Beverages are fed against ground. Although very long wires may develop a few dB of front-to-back ratio, unterminated Beverages are essentially bidirectional. Terminating a Beverage through a resistance to ground at the far end creates a highly unidirectional pattern. Beverages typically are 1 to 15 feet high and 1 to 4 wavelengths long.

Beverages are broadband, traveling-wave antennas. Power traveling toward the feedpoint is absorbed by the receiver, while that traveling in the opposite direction is absorbed by the termination resistance. Standing waves can't develop. Because waves traveling toward the feedpoint accumulate substantially in phase, while those arriving from other directions tend to cancel, the azimuth pattern of a long Beverage is highly directive. Beverages are nonresonant antennas and can be used over a wide frequency range.

A very useful Beverage variation uses two parallel wires, with one grounded at the far end. Waves traveling toward the far end reflect and return to the feedpoint differentially as transmission-line currents. With suitable termination and switching circuitry at the feedpoint, you can receive signals from either direction. A two-wire Beverage thus can replace two single-wire antennas. For simplicity, I'll consider just single-wire Beverages here.

Eliminating Ground Connections

Beverages normally connect to ground at the feedpoint and far end. See **Fig 1A**. However, obtaining a good, reliable ground can be difficult. A short ground rod may exhibit thousands of ohms resistance to ground when installed in dry soil or when corroded. Since Beverages require a termination resistance of just several hundred ohms, a poor far-end ground can grossly misterminate an antenna and destroy its pattern. A poor feedpoint ground can reduce signal levels and encourage coax-shield pickup. You can use multiple ground rods, a ground screen, or a ground-radial system to lower ground resistance, but this greatly increases the work required to install a Beverage.

A ground connection isn't essential for Beverage operation. You can place the termination resistance about one-quarter wavelength from the far end of the wire. Since the impedance of a quarter-wave wire is low and resistive, the remainder of the wire just sees the termination resistor in series with a low resistance. Similarly, you can feed a Beverage about one-quarter wavelength from the near end. Both ends of the wire are left unconnected. See Fig 1B.

Ungrounded Beverages are simple to install and have stable properties. Their radiation patterns often are superior to those of their grounded counterparts because they don't suffer unwanted response from vertical wires. (This response can be reduced by lowering wire height. One way to do this is to slant the feed and termination wires over a long distance.)

The only real disadvantage of ungrounded

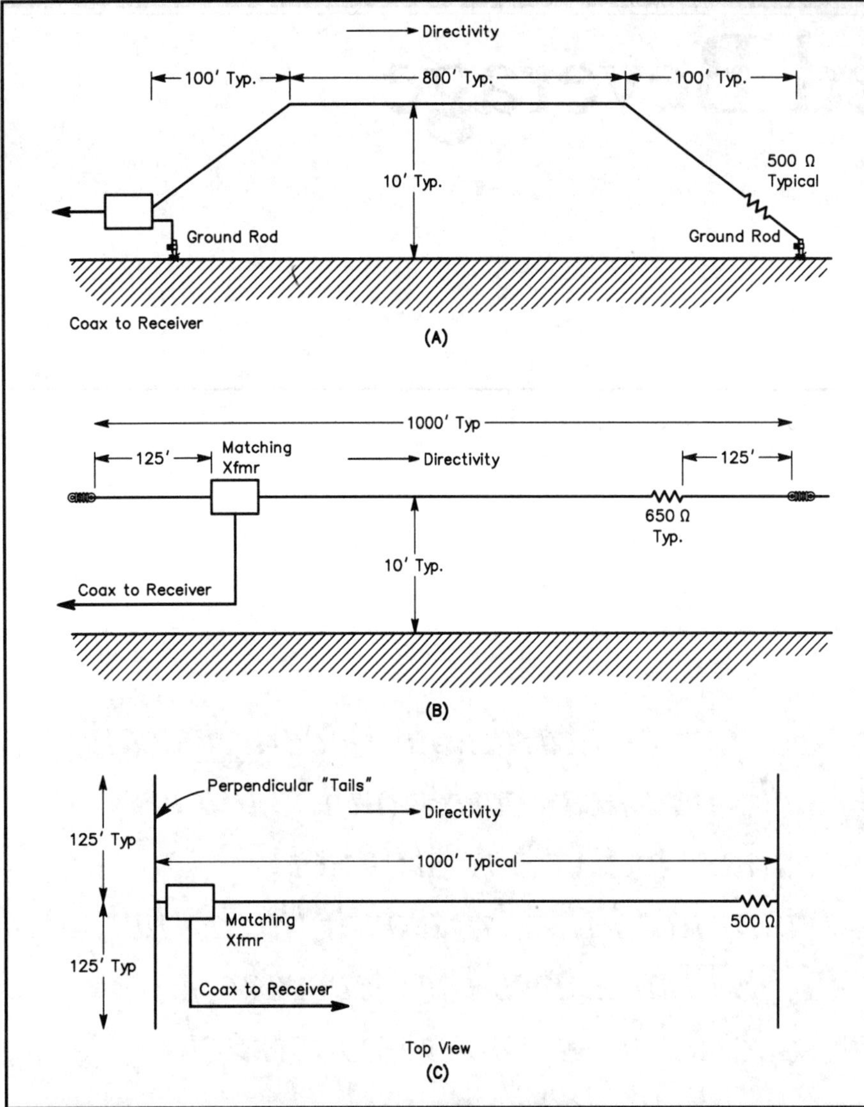

Fig 1—Single-wire Beverage antenna feeding and termination configurations. At A, conventional method is shown using slanted-wire terminations. At B, ungrounded Beverage is shown with matching transformer and terminating resistor inserted λ/4 in from either end of wire. At C, ungrounded Beverage using full span of wire by employing perpendicular "tails" at both ends. The tails function as elevated counterpoise radials at 160 meters.

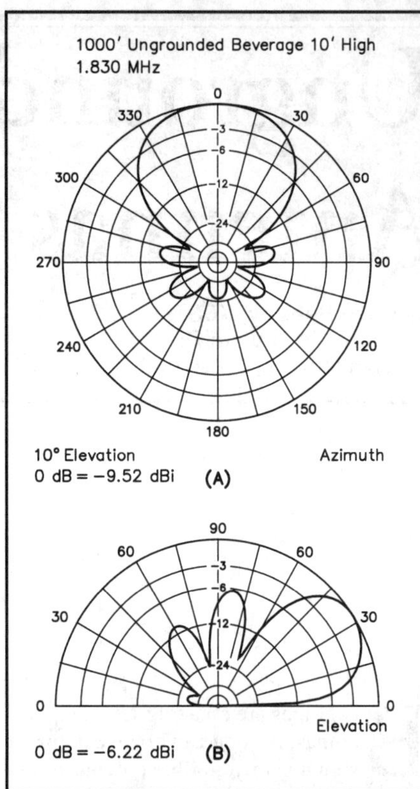

Fig 2—Azimuth and elevation patterns at 1.83 MHz for 1000-foot long ungrounded Beverage mounted 10 feet off flat ground with average conductivity and dielectric constant. The worst-case rearward lobes are better than 18 dB down compared to peak response. Note that the gain is only –9.52 dBi at 10° elevation, peaking at –6.22 dBi at about 30° elevation. If the transmission line to the receiver is very long, a preamplifier may be needed.

Beverages is that they're frequency-dependent. For example, the low-impedance, quarter-wave wire sections become high-impedance, half-wave sections at twice the design frequency. I'll show you how to multiband an ungrounded Beverage later.

Antenna Modeling

While you can use a *MININEC*-based antenna-analysis program to model Beverage antennas, you'll have to settle for approximate results at best. The *MININEC* algorithm uses perfect-conductivity ground during current calculations. It takes lossy ground into account only later when calculating radiation patterns. Because Beverages are so close to ground, their current distribution is strongly affected by ground characteristics. *NEC*'s Sommerfeld-Norton ground model takes ground dielectric constant and conductivity into account when calculating wire currents. This lets you accurately analyze antennas in close proximity to earth. I used the *NEC/Wires 1.0* program to obtain all results presented here. I modeled antennas over average-quality earth (dielectric constant = 13, conductivity = 5 mS/m).

Wire Height

Fig 2A and **2B** shows the azimuth and elevation patterns for a 1000-foot long, 10-foot high, ungrounded Beverage on 160 meters. The antenna uses #18 wire and a 650-Ω termination resistance. It's fed and terminated 120 feet from each end. The re-

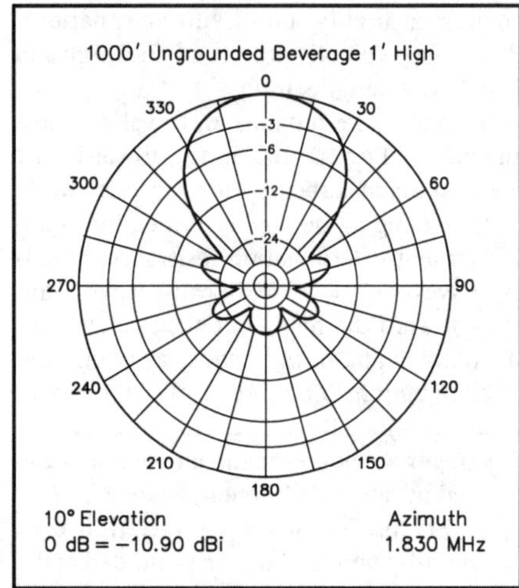

Fig 3—Azimuth response for 10° elevation angle at 1.83 MHz for 1000-foot long ungrounded Beverage mounted only 1 foot off flat ground. Compared to the higher antenna in Fig 2, this low Beverage has about 1 dB less gain, but the worst-case backlobes are suppressed even better.

markable pattern is better than that of many HF Yagis.

Fig 3 shows the azimuth pattern of a similar Beverage 1 foot above ground. The termination resistance for this antenna is 750 Ω and it's fed and terminated 125 feet from the wire ends. The pattern is even sharper than that of Fig 2. Output is about 1 dB lower.

If you're tempted as I was to just lay a Beverage on the ground, take a close look at **Fig 4**. This is the Beverage of Fig 3 but just 1 inch off ground (*NEC* provides accurate results for wire heights down to within several wire radii of ground). While still useful, the pattern has degraded considerably. Signal output is quite a bit lower. (Nevertheless, some users report excellent results from on-the-ground Beverages. Don't hesitate to try one if it's your only option.)

In the West you can just lay an ungrounded Beverage in the chaparral a few feet off ground. In the East you can unroll one on top of a foot or two of snow (snow has very low dielectric constant and conductivity and should be transparent to RF at low frequencies). Beverages like these are very easy to install (no support poles or ground rods). You can roll one out just before a contest.

Occasionally a Beverage may have to clear a height-sensitive area. **Fig 5** shows what can happen when wire height varies. The first 200 feet of this antenna is 3 feet high (chaparral height). Next, a 50-foot section slopes from 3 to 15 feet. This is followed by a 200-foot span 15 feet high, a 50-foot downslope, and finally the remainder of the wire at 3 feet. This model is similar to a Beverage used by K6TQ which must clear an agricultural area with produce trucks. While the rear part of the pattern is relatively unaffected, the height variation causes a large sidelobe bulge. It doesn't matter much if your Beverage is high or low, but try to keep the height as constant as possible.

Wire Length

Fig 6 shows the azimuth pattern of a 2000-foot ungrounded Beverage. The termination resistance is 550 Ω. The pattern of a very long Beverage like this may be too narrow for complete coverage of a general direction.

Fig 7 shows the pattern of a 500-foot ungrounded Beverage. The termination resistance is 1100 Ω. While much broader than the patterns of longer antennas, it's still quite directive and can provide a substantial receive improvement.

Feeding Ungrounded Beverages

Conventional Beverages usually are fed with an autotransformer having a 3:1 turns ratio. This matches a 450-Ω impedance to 50 Ω. Ungrounded Beverages typically have an input resistance of 500 to 1000 Ω (roughly equal to the termination resistance) and an input reactance of up to 100 Ω. An exact impedance match isn't important; a better match just delivers a higher signal. Although usually not required to improve signal-to-noise ratio, a receive preamplifier can equalize Beverage and transmit-antenna signals for easier direct comparisons.

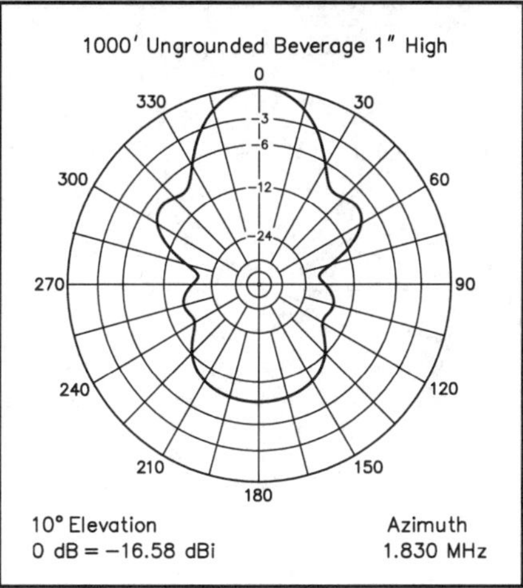

Fig 4—Azimuth response for same antenna as in Fig 3, except that the height is now only 1 inch off ground. The rearward pattern suffers and the gain fails about 5.4 dB compared to the Beverages in Figs 2 and 3.

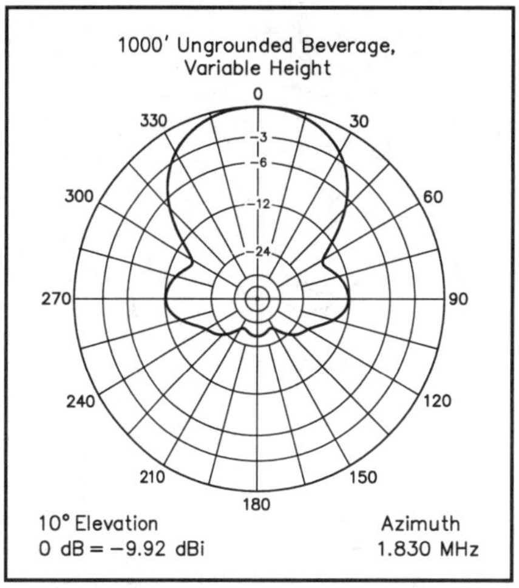

Fig 5—Azimuth response at 1.83 MHz for 1000-foot long, variable height ungrounded Beverage. The sidelobes at 90° and 270° are larger than the antenna in Fig 2, but the pattern and gain are still quite acceptable.

Both grounded and ungrounded Beverages can be sensitive to unwanted coupling at the feedpoint. If you connect coax through an autotransformer to a Beverage, the outside of the coax shield can become part of the antenna system if it exhibits an impedance similar to or less than whatever connects to it. It's the same thing that can happen with a coax-fed dipole. To prevent this, use a matching transformer with separate primary and secondary windings and low interwinding capacitance. This will isolate the outside of the coax shield from the antenna. Alternatively, you can use an autotransformer followed by a current-type balun. (Because the feed tail of an ungrounded Beverage is likely to provide a lower-impedance termination than a typical ground rod, these Beverages should be less sensitive to coax-shield pickup.)

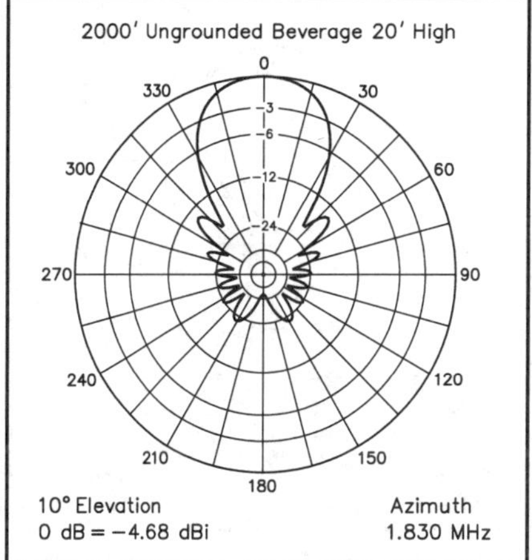

Fig 6—Azimuth response at 1.83 MHz for 2000-foot long ungrounded Beverage mounted 10 feet over flat ground. The gain is almost 5 dB higher than that of Fig 2, and the backlobes are all suppressed in excess of 23 dB. This is a great 160-meter receiving antenna, but must be oriented carefully at desired receiving bearing because of narrow frontal lobe.

Terminating Ungrounded Beverages

Optimal termination resistance depends on wire length, height, diameter, and ground constants. The termination resistance normally just affects the rear part of the pattern. It's easy to find the optimal value by modeling the exact geometry of your antenna with *NEC*. If you use long spans of wire with considerable sag, model the sag using multiple wires.

You can determine optimal termination resistance experimentally using a signal source to the rear of the antenna. Simply adjust the termination resistance for minimum signal. (Actually, minimum response directly to the rear isn't quite optimal for most designs. This condition usually causes the pattern to degrade slightly over a broad region elsewhere. Nevertheless, adjusting for minimum response is close enough for all but the most fanatical Beverage enthusiast.)

Multibanding Ungrounded Beverages

If you provide a termination resistance one-quarter wavelength from the far end for each band of interest, you can use a Beverage on several bands. At the near end use a

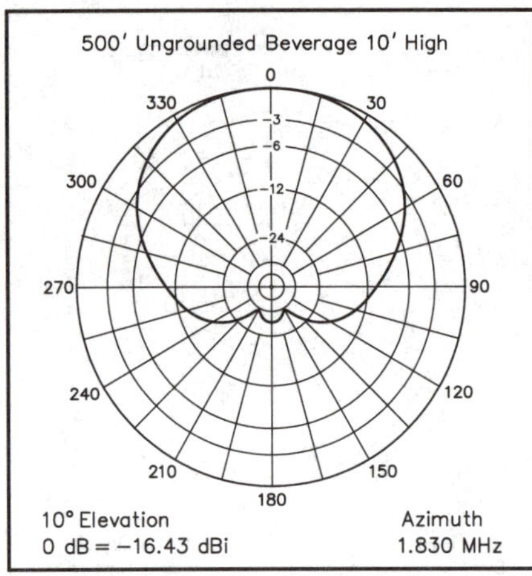

Fig 7—Azimuth response at 1.83 MHz for 500-foot long ungrounded Beverage mounted 10 feet over average ground. The gain is almost 7 dB less than Fig 2 at 10° elevation. The sides lobes are large, even though the rear lobes are well suppressed.

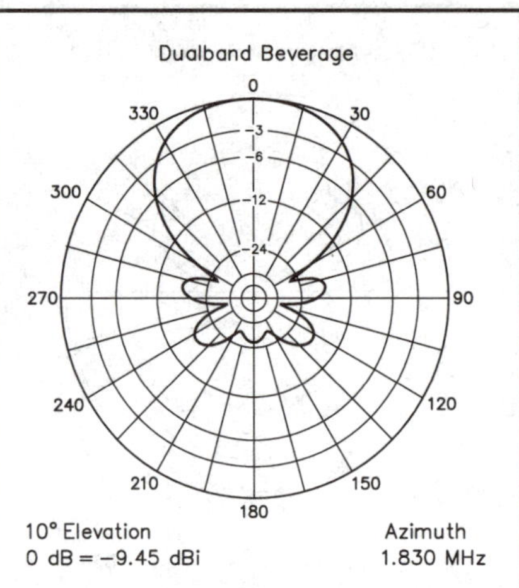

Fig 8—Dualband Beverage on 1.83 MHz, using a 700-Ω resistor 63 feet from the far end, and a 350-Ω resistor 125 feet out.

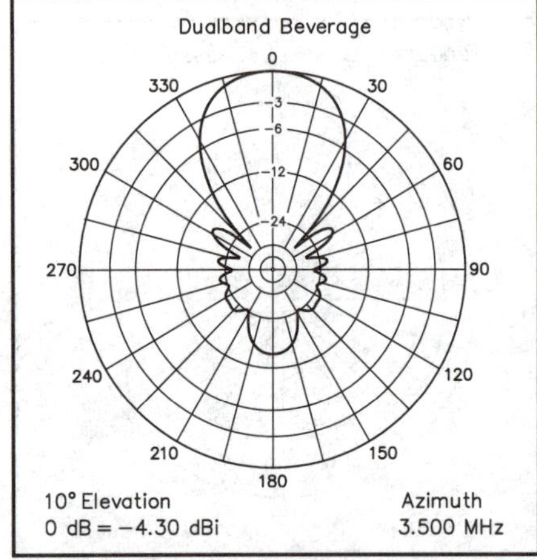

Fig 9—Dualband Beverage on 3.5 MHz, using same terminations as in Fig 7. The termination is not optimal for this band, but still yields desirable directivity.

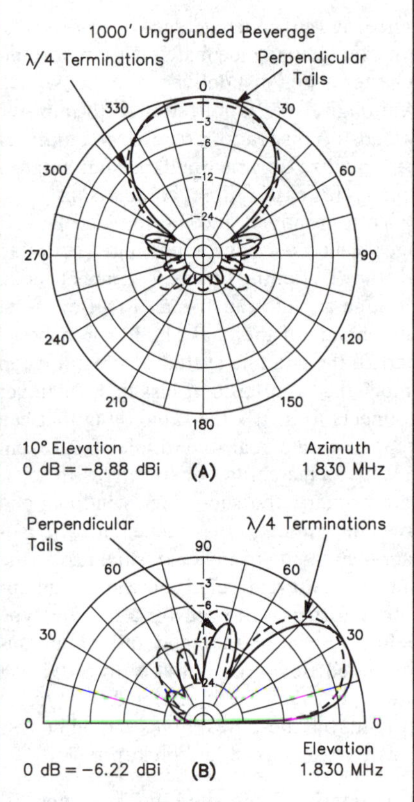

Fig 10—Ungrounded Beverage, 1000 feet long with two λ/4 perpendicular termination "tails." Pattern from ungrounded Beverage in Fig 2 is overlaid for comparison. At A, the azimuthal patterns at 10° elevation are shown; at B, the elevation patterns are shown.

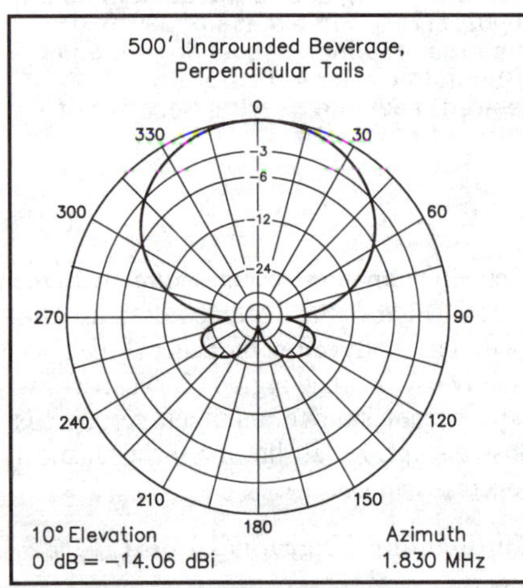

Fig 11—Azimuthal pattern for ungrounded Beverage, 500 feet long with two λ/4 perpendicular termination tails.

separate quarter-wave tail for each band and connect them in parallel at the feedpoint.

Fig 8 and **Fig 9** show 160-meter and 80-meter patterns of a 1000-foot-long, 10-foot-high, dualband Beverage. This antenna uses a 700-Ω resistor 63 feet from the far end and a 350-Ω resistor 125 feet out. The two termination resistances provide near-optimum performance on 160. The backlobe of the 80-meter pattern is suboptimal but the sidelobes are still well down. I didn't experiment much with termination-resistance values or positions. I'll bet both can be improved.

Another alternative is to use a single termination resistance with multiple quarter-wave tails. The resistor value will be optimal on just one band since antenna length isn't constant in wavelengths. But the value won't be far off on any band if the wire is long. For best performance on all bands, use a separate resistor for each tail.

Perpendicular Tails

The ungrounded Beverages described so far really don't make optimal use of the entire antenna length. Although they contribute power, the near- and far-end quarter-wave tails don't support traveling waves. If space permits you can feed and terminate against a pair of quarter-wave wires perpendicular to each end. See Fig 1C. This effectively lengthens the Beverage by one-half wavelength. Essentially the perpendicular tails form two, two-wire, elevated radial systems.

Fig 10A and **10B** shows the azimuth and elevation patterns for a 1000-foot-long, 10-foot-high Beverage with perpendicular tails, with the patterns from Fig 2 overlaid for comparison. This antenna uses a 500-Ω termination. The pattern is somewhat narrower than that of Fig 2 and the backlobes are even smaller. The elevation plot shows that overhead response is about 10 dB less with perpendicular tails and that the forward lobe cants at a lower angle.

Since the quarter-wave tails of a 500-foot Beverage comprise half its length, you might expect a more dramatic improvement for perpendicular tails on this antenna. **Fig 11** shows the pattern of a 500-foot-long, 10-foot-high Beverage with perpendicular tails. This antenna also uses a 500-Ω termination. Side response is down quite a bit from Fig 7 and forward response is 2 dB greater.

While they provide better performance, designs with perpendicular tails are more complex and occupy additional space. You can make better use of the additional space by phasing two Beverages with in-line tails to obtain a much better azimuth pattern...but that's another story.

Beverages Simplified

By Tom Russell, N4KG
ARRL Technical Advisor
29836 Country Lane
Harvest, AL 35749

You gotta hear 'em to work 'em. N4KG is an avid DX and contest operator, with only 3 countries left on 40, 20 on 80 meters, and more than 200 worked on 160 meters. He gives some practical advice on how to improve your low-frequency receiving capabilities using Beverage antennas.

Have you noticed in many of the articles on Beverage receiving antennas the emphasis on making a "perfect" installation? Sloping feeds and termination, matched termination resistors, radials at the termination and feedpoints, perfect matching transformers, tight and straight wires, and elaborate switching boxes are all emphasized. Photographs make you think they were laid out with the aid of a transit, in unobstructed fields covering tens of acres.

That's nice of course, but it's not all necessary. The bare necessities include a reasonably straight run of wire from 250 to 1000 feet long, a matching transformer and a ground (earth) connection at the feedpoint. Dog legs of 30° or less can be tolerated for geographical reasons. Notice that I did not include terminating resistors on this short list.

The objective with any low-band receiving antenna is to improve detected signal-to-noise ratio. While "perfect" installations do produce excellent pattern rejection in the undesired directions, simple unterminated wires with a matching transformer and ground stake, house electrical ground, grounded metal fence, or grounded tower leg can be utilized to make easily installed and effective receiving antennas.

I have eight such simple Beverages and have given up trying to maintain terminating resistors in the woods. The patterns are sufficiently good to allow me to distinguish between continents by merely turning a coax switch. The Beverages show a 10 to 20 dB improvement in SNR over my transmitting antennas, often allowing me to hear stations that are not even perceptible on the transmit antennas. What more does one need?

Well, on 160 meters some longer wires might be helpful. A Beverage can be thought of as an electrical telescope. The longer the wire, the narrower the view, both in the vertical and horizontal directions. A one-wavelength wire "sees" a 60° wide pattern, about equivalent to the beamwidth of a 3-element Yagi. Longer wires produce narrower patterns down to about 30°. After this, additional length provides no benefit. Short Beverages are useful—longer are better, sometimes.

Let me cite some experiences with different length Beverages. Toward Europe, I have two wires—my original 500 footer at 60° azimuth and a 1000 footer at 45°. On 160 meters, the 1000 footer shows as much improvement over the 500 footer as the 500 footer shows over my vertical. On 80 meters, however, the situation is a little more complex. At sunset the 500 footer is usually best. There have been times when I ran outside to see if a tree limb had fallen on the long Beverage or if the matching transformer connections were broken. I simply could not hear signals on the longer wires that I was hearing on the shorter one! Finally, around midnight the longer wire produced equal signals.

It's a matter of wave angles. Wave angles

are higher at sunset and take some time after darkness to fall below 30°. I simply do not believe the maximum wave angle numbers published for years in antenna books. These numbers were generated years ago from one single study of a single path 60 years ago without reference to the sunspot level. [The material in *The ARRL Antenna Book* prior to the 17th edition came from data measured in 1934, the low point of that Solar Cycle. —*Ed.*] The signals I'm hearing are arriving at much higher angles than the books would have you believe, but that's another subject.

A 2-wavelength long 80-meter Beverage (560 feet long) makes an excellent receiving antenna for 80 and 40 meters and while not perfect on 160 meters, it is still better than most transmit antennas. I have a short (240 foot) wire to the East, which is often effective on 40 meters. For 80 and 160 meters, my 800 foot East Beverage hears signals in East Africa and the Indian Ocean that are not even perceptible on any other antenna. To the Southeast I am limited to 400 feet but that wire has pulled out such delectables as VK9YM, 9M2AX and 9V1YC on 80 meter long-path at my sunset. I even use it to hear JAs on 160 meters, but only when the owners clear their vacant lot across the street.

OK, my secret is out. My Beverages, like my dogs, don't pay much attention to property lines! How much wire can you run on 4 acres, after all?

For you city dwellers, there is still hope. You can make up a "temporary" Beverage kit and string it out (during the night) on the ground in whatever direction you need for the latest DXpedition. In the morning (or after you work the station you need) you can roll it up again! Use insulated wire with a ground rod on your property and a matching transformer.

I like stranded insulated wire (#20 or larger) for Beverages, especially when running in the woods, because tree limbs are always falling on the wires, necessitating field repairs. Stranded wire is easy to tie back together, stripping the ends and soldering with a portable butane soldering iron. Another feature of insulated wire fitting in with my simplified Beverage philosophy is that you can simply take a heavy weighted object, tie the wire to it, and throw it over tree limbs for as far as you want to go! Leave a little slack in the wire for repairs. It helps to live near woods.

Matching transformers are a must (unless you live next to a 50 kW AM broadcast transmitter, which will overload them). The mismatch loss to a 50-Ω input is about 10 dB. Most Beverages are already about 10 dB below the signal level recovered by your transmit antenna, and another 10 dB is too high a price to pay. You either use a preamp or a matching transformer.

I like simple broadband 9:1 autotransformers. A high permeability core is a must, and "good" RF core material does not necessarily make a good, reliable broadband transformer. I used to use small cores and small wires—until I got tired of replacing them after nearby lightning strikes, the same reason that I stopped bothering with terminating resistors. I often used to find a piece of perforated insulation with no wire inside protruding from my small cores after a strike! Now I use 1-inch cores and #22 wire. (I'll gladly send you one for $20.) Keeping things simple, I just solder the transformer to the Beverage and coax—no boxes, no connectors.

Beverage receiving antennas are a boon to the low-band DXer. Everybody should own some!

Rotators and Measurements

Plotting Antenna Polar Diagrams Using an IBM-Compatible Computer

By Peter Dodd, G3LDO
37 The Ridings
East Preston
West Sussex BN16 2TW
United Kingdom

After automating the design procedure for a new antenna, you can now automate the measurement of your new creation.

Introduction

For many years I have experimented with antennas and manually plotted the polar diagrams of many configurations. More recently I automated the plotting of these diagrams using a BBC computer.[1] This article describes ways of using the more available IBM or IBM-compatible computer to produce these antenna directional patterns. It also discusses equipment and software requirements as well as plotting scales: linear, log and ARRL-style polar plots.

The *MININEC* program and its derivatives, such as *MN* and *ELNEC,* use mathematical models of antennas for design. While these programs are excellent they do have algorithm limitations. These limitations may be significant depending on the type of antenna being investigated.[2] Any antenna designed using a mathematical model has to be physically constructed to complete the study. Plotting the actual polar diagram of new design will then complement the mathematical model.

VHF Modeling of HF Antennas

HF antenna configuration experiments are often difficult since even a compact configuration is rather large. Because size decreases with frequency any beam antenna for 28 MHz is half the size of one for 14 MHz, yet its performance is the same. This is because wavelengths, capacitances and inductances are reduced in proportion to the linear dimensions, while gains, impedances, dielectric constants and permeabilities are unchanged. Therefore you can go one step further and study the performance of a model of a proposed antenna at VHF, where the antenna test range is far more manageable. This practice is known as *VHF modeling* and has been used for many years.

The 2-meter band is very practical for experimenting with antennas. On the 70-cm band the antenna physical construction is rather small to accurately model a 14-MHz antenna. Modeling can be performed on the 50-MHz band but more space is required. Thus the 2-meter band is a good compromise.

Measuring Equipment and the Antenna Test Range

The professionals measure antenna performance by exciting a source antenna to flood the antenna test area with RF. A receiver or field-strength meter measures the received signal strength from the *Antenna Under Test* (AUT) as the antenna is rotated.[3]

Because of the reciprocity relation of an antenna (the transmitting and receiving directivity characteristics are usually the same) it is not important which antenna is energized and which is used as the receiver antenna. When making measurements of Amateur Radio antennas I prefer to use the test equipment the other way around and energize the AUT. I connect a simple antenna to the receiver or field-strength meter. There are two reasons for this: first the SWR of the AUT can be checked to ensure it is correctly matched to the feeder. Antenna comparisons lose meaning unless the SWR is controlled; preferably better than 1.5:1. Next the SWR meter can be used as a sensitive indication of relative transmitter power output if the transmitter is being used to energize the AUT.

You require several items to plot antenna polar diagrams using a computer. Details of the different items of equipment are described later.

- A field-strength meter (FSM) with remote analog, digital field-strength level indicator or a receiver with calibrated signal level indicator.
- Loop or dipole antenna for field-strength meter.
- Signal generator with calibrated variable RF output, or a transmitter with variable power output.

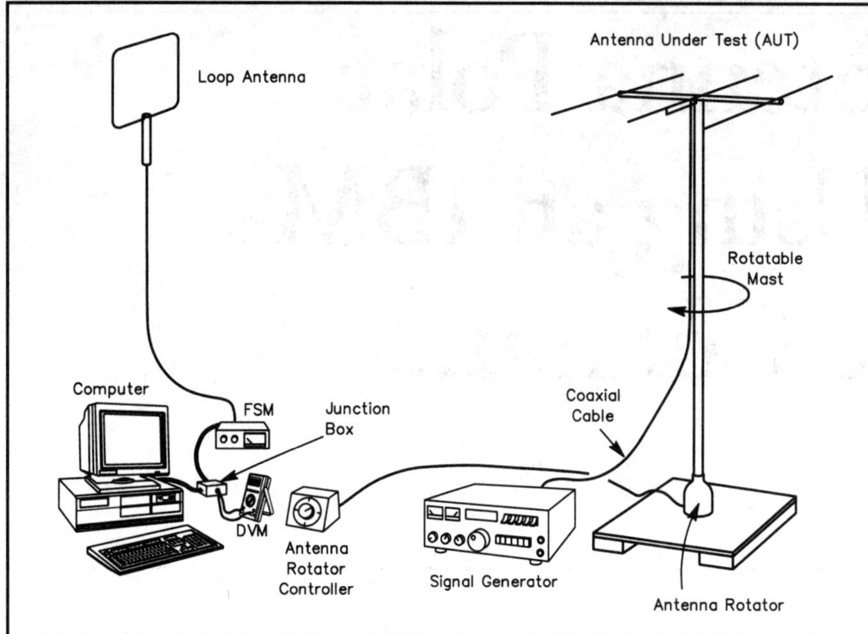

Fig 1—Equipment required to produce antenna polar diagrams using a computer. The AUT is energized by the signal generator. The computer takes analog data from the FSM and displays it on the screen as an evolving polar diagram while the AUT is rotated.

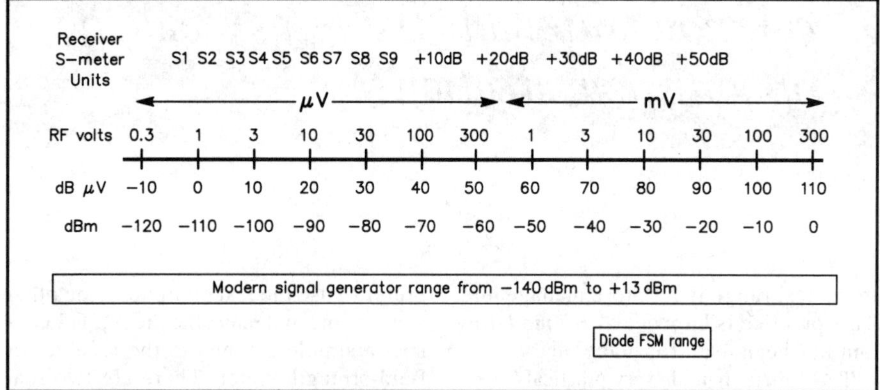

Fig 2—Units of signal strength and approximate comparisons. Examples of some of the units in use are μV, mV (RMS or peak), dBμV and dBm. The range of signal levels shown are those often encountered while experimenting with an antenna. Note the usable range of a typical unamplified diode FSM.

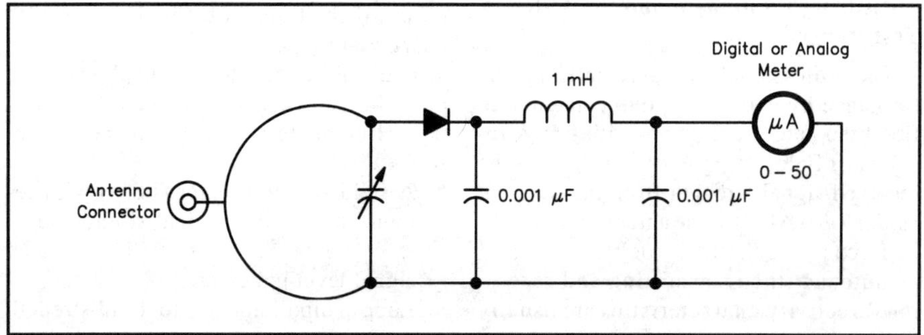

Fig 3—VHF diode field-strength meter. The coil is just over one turn of #18 SWG wire. The capacitor value is not critical; my MK1 FSM used a small air-spaced variable of unknown capacity. The antenna is tapped one third of a turn from the bottom of the coil and the diode one third of a turn from the top.

- Antenna Under Test (AUT).
- AUT rotator mechanism, mast and position indicator. The position indicator can be a piece of white cardboard held to the mast by a clip or clothes peg.
- SWR meter, coaxial cable and connectors.
- IBM computer
- Analog-to Digital Converter (ADC)
- Digital voltmeter to monitor the signal level at the input to the ADC.
- Software to enable the data to be displayed as a polar diagram and to save and retrieve data from files.

The equipment is connected together as shown in **Fig 1**.

Measuring Equipment Considerations

Field-Strength Meters/Signal Generators or Transmitters.

My early polar diagram measurements were performed using a transmitter to energize the AUT antenna. A simple diode FSM measured the signal strength. Volts rather than millivolts are required from the FSM to drive the ADC. The ADC used has a range of 0 to 5 V. The output range of the FSM and the ADC must match. If the input signal to the ADC is only 0 to 2.5 V for a 0 to 5 V ADC the converter errors are effectively doubled.

Fig 2 lists signal strength levels and units of RF field strength measurement. You should use it as a guide to decide what equipment to use.

The following methods can be used to obtain enough volts to drive the computer:

Method 1—Use a simple diode FSM and a transmitter with a power output of up to 10 W. This option has the advantage of simplicity. The transmitter should be stable, free of spurious outputs and have a controllable power output level.

The FSM circuit most often described in literature for antenna adjustment is the simple diode FSM. A circuit is shown in **Fig 3**.

The size and shape of the enclosure are unimportant but should be made of metal to prevent RF energy from entering the tuned circuit by any path other than the pick-up antenna. You can calibrate the instrument using a GDO or signal generator. The meter leads are filtered as shown in the schematic.

Method 2—Use a more complex FSM with stages of gain, and a low-power transmitter. The circuit in **Fig 4** is based on a design by Doug DeMaw, W1FB,[4] and should be suitable with many low power sources.

Reference 5 describes a test instrument including attenuator, amplifier and detector. It includes an amplifier with a gain of 40 dB and bandwidth of 65 MHz.

Method 3—Use a modified receiver or commercial FSM and a signal generator to energize the AUT.

I use this approach because I have found

suitable equipment. The signal generator should have an output of about 100 mV and the level should be calibrated, either on a meter or a dial. Sensitive field-strength meters are not plentiful but they can be obtained at flea markets.

If a sensitive FSM is not available, then a communications receiver can be used. It should be tunable so the AUT performance over a range of frequencies can be examined. The detector and amplifier, shown in Fig 4, can be tagged on to the end of the IF chain with a suitable coupling circuit to drive the computer ADC.

A receiver is rather sensitive as an FSM and a switched attenuator may be necessary. This is connected into the feed line between the pickup antenna and the receiver antenna connector. Using a switched attenuator has a further advantage. If you wish to make relative gain comparisons in dB the S-meter calibration is unimportant. It is used only to set a reference level indicator. The relative dB levels are then indicated by the switch settings on the attenuator. A suitable attenuator is described in *The ARRL Handbook*.[6]

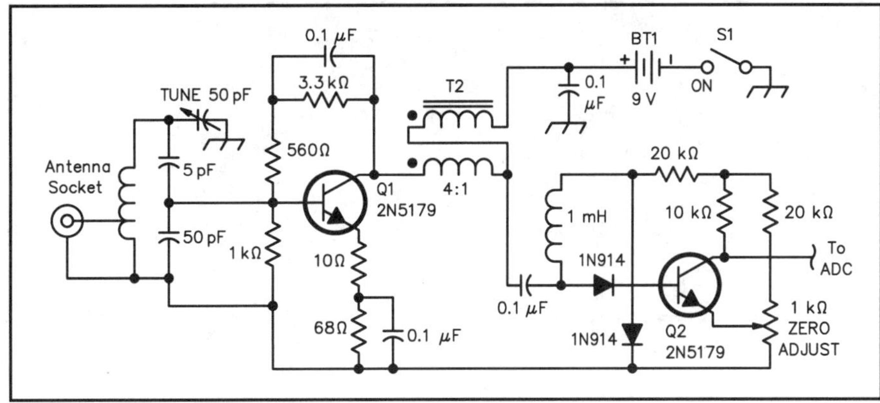

Fig 4—Field-strength meter for use with low-power transmitter. A simple diode FSM with RF amplification. The RF amplifier Q1 operates as a Class-A linear amplifier. T2 is a 4:1 broadband transformer using 15 turns of #28 enamel wire (bifilar wound) on an Amidon FT50-43 toroid core. The 1-kΩ potentiometer in the emitter of Q2 should be adjusted for zero volts under no-signal conditions.

Analog to Digital Converter (ADC)

A method of converting the analog data from the FSM to a digital form is required. All my original work was done using a BBC computer containing an ADC as standard equipment and using a special BASIC language command called ADVAL.

The IBM computer is not fitted with an ADC so some extra hardware is required. There is a whole range of usable ADCs. Some I/O boards plug into the expansion slot. Others connect through the RS232 serial port or the printer port. The one I have used is based on a simple, commercially available device called the Pico ADC-10, designed to plug into the printer port. It has a range of 0 to 5 V. Availability of other units is given in the Appendix.

As noted before the output-voltage range of the FSM and the input-voltage range of the ADC should match.

AUT Rotator

The AUT can be rotated by a light-duty rotor such as a TV antenna rotator. However, this could have a couple of disadvantages. The rate of rotation is rather low and since it is not normally continuously rated may get rather hot after a couple of measurements. I use a motor and gear box from a previous industrial application. It has a rotation speed of about 15 seconds. It is convenient for the rotator to be able to rotate in either direction to prevent the feed line from getting wrapped around the mast during successive measurements.

Software

The program performs the following tasks:

1. Displays a graticule with heading information to provide a reference for the polar plot.
2. Samples analog data from the FSM and plots it as a polar diagram, over the graticule as the AUT is rotated.
3. Saves antenna data to a file and redisplays it when required.

In addition, provision is made for adjusting parameters, such as aligning the plotting time to the AUT rotation speed and adjusting the sampled values for the required polar display.

The software is written in BASIC. It will run in GW-BASIC, QBASIC and QuickBASIC and is included on the disk accompanying this book under the filename G3LDOPLT.BAS. The program is menu driven and straightforward to use. To recall the file use the full file name (such as *BEAM1.DAT)* including extension.

To keep the system as simple as possible there is no provision for the computer to read the AUT heading. The large number of IBM and IBM-compatible computers all operate at different speeds. Therefore plotting time and the AUT rotation time must be adjusted to be identical. This adjustment is selected at the menu and has a default time of 30. The default can be modified (line 1610).

A gain factor is used to modify the plot with reference to the graticule and is selected from within the program. A default of 1.5 is used if a scale factor is not entered for the gain factor. This default can be modified at line 1200.

The plot scale can be modified to display data, previously stored on disk, as Linear, Log or ARRL-log periodic. The significance of these scales is described later.

The subroutine for acquiring the analog data from the FSM at line 760 will have to be modified if an ADC other than the ADC-10 is used.

The number of circles and spokes in the graticule can be altered and the number of dots in each of the lines of the graticule can be changed (graticule subroutine at line 260). Generally the program will execute faster with a lower number of dots. A higher number of dots will look better if the plot is printed.

The plotting points in the signal-level lines are actually circles. The thickness of the line can be altered by changing the diameter of the circle, lines 1010 and 1410.

The display can be sent to the printer using the PRINT SCREEN key. Normally, this will print only ASCII characters, but can be changed to print graphics by executing GRAPHICS.EXE before GWBASIC.EXE. The program GRAPHICS.EXE is usually supplied on the DOS source disk for your computer.

Measurement Procedure

This measurement procedure assumes a signal generator and a sensitive FSM or receiver is being used. First check to be sure the AUT is matched to the coax feeder using a transmitter and an SWR meter.

Next connect the equipment as shown in Fig 1. The connection to the computer is made via the ADC (not shown). When making measurements on the 2-meter band I place the AUT about 40 inches high. The measurement antenna is placed 12 to 15 feet high, and the two antennas are spaced about 40 feet apart. For more details of setting up an antenna range see *The ARRL Antenna Book*.[7]

Place all operating equipment, such as the computer, the FSM output voltmeter, the FSM frequency control and the AUT rotator so they can be operated or seen from an operating position. The AUT mast and position indicator must also be visible from the operating position.

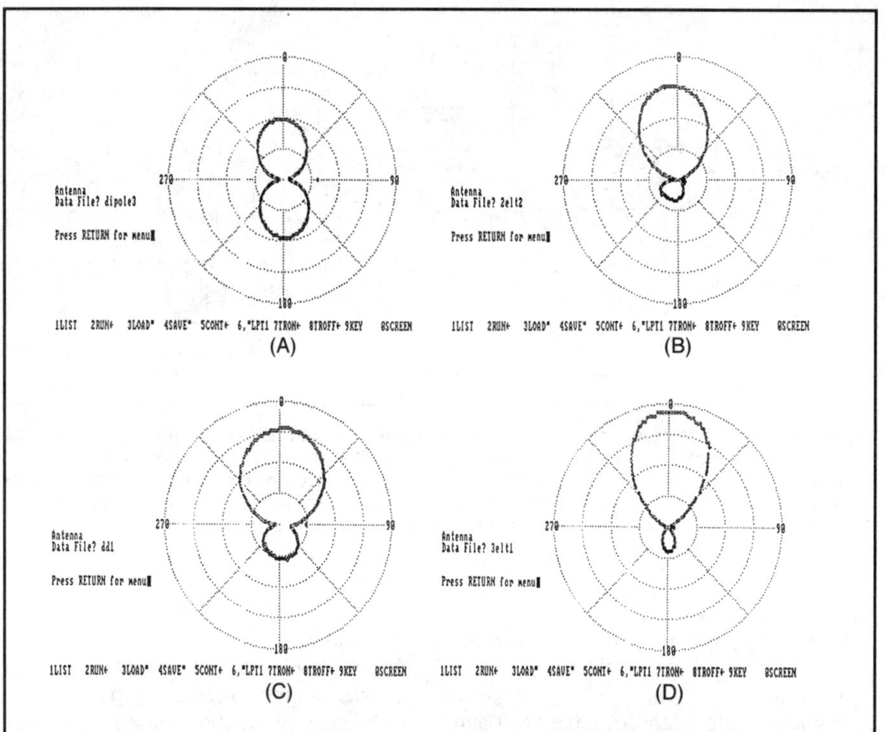

Fig 5—Polar diagrams: At A, a dipole; at B, a 2-element beam; at C, a Double-D and at D, a 3-element beam compared. Diagrams plotted using a linear scale.

Point the AUT at the FSM antenna. Adjust the signal generator and FSM to the measurement frequency. Adjust the signal generator level and the FSM attenuator or gain control so the output is appropriate to the ADC in use.

Rotate the antenna through 360° while watching the FSM signal level. In this way you can gain a mental picture of the AUT pattern. It is useful to make this preliminary check to ensure the antenna is working reasonably well before going through the routine of plotting the data.

Measure the AUT rotation time using a watch with a seconds display. Select T at the menu and adjust the plotting time. The program inserts a signal to plot a circle.

The program starts at 90° and rotates counterclockwise. Rotate the AUT to be 90° clockwise from the FSM antenna.

Select option P at the menu. Start the AUT rotation and the software plotting simultaneously.

Check gain level for a correct setting. As an example suppose you are making comparative plots of a dipole and a 3-element beam. The gain should be set to plot the polar diagram of the antenna with the greatest gain within the graticule.

When the polar plot is complete check for correct orientation (relative to the graticule). If it is not, alter the position of the marker on the AUT mast and repeat the plot.

When you are satisfied with the plot the data can be stored on the disk by returning to the menu and selecting option S.

Scale Considerations

The polar diagrams of four sample antennas plotted using a linear scale are shown in **Fig 5**. A linear FSM (without an attenuator) can only show features of the polar diagram from maximum signal strength down about 15 to 20 dB or so. This gives a very optimistic picture of directivity and suppresses the back lobes. The linear scale does have the advantage of making the instrument very sensitive to changes in level in the region of maximum signal strength. This sensitivity to gain differences is useful for making gain comparisons between different antennas.

Other scales can be used. For example, **Fig 6A** shows a polar plot of a dipole and an array comprising four 19-element Yagis, plotted on a linear scale. Most antenna manufacturers' polar plots use this scale.

If either of these antennas were connected to a receiver the plot of Fig 6B would result. The receiver's S meter generally has a logarithmic (linear dB) scale and will display a signal-level range of at least 50 dB. This scale gives a much more accurate picture of how the antenna will perform when connected to the station receiver, and explains why the directivity of a given antenna is often not as good as its published polar diagram might suggest.

The ARRL has devised a unique polar diagram having a log periodic scale as shown in Fig 6C. Instead of the graduations varying linearly with the log of the field

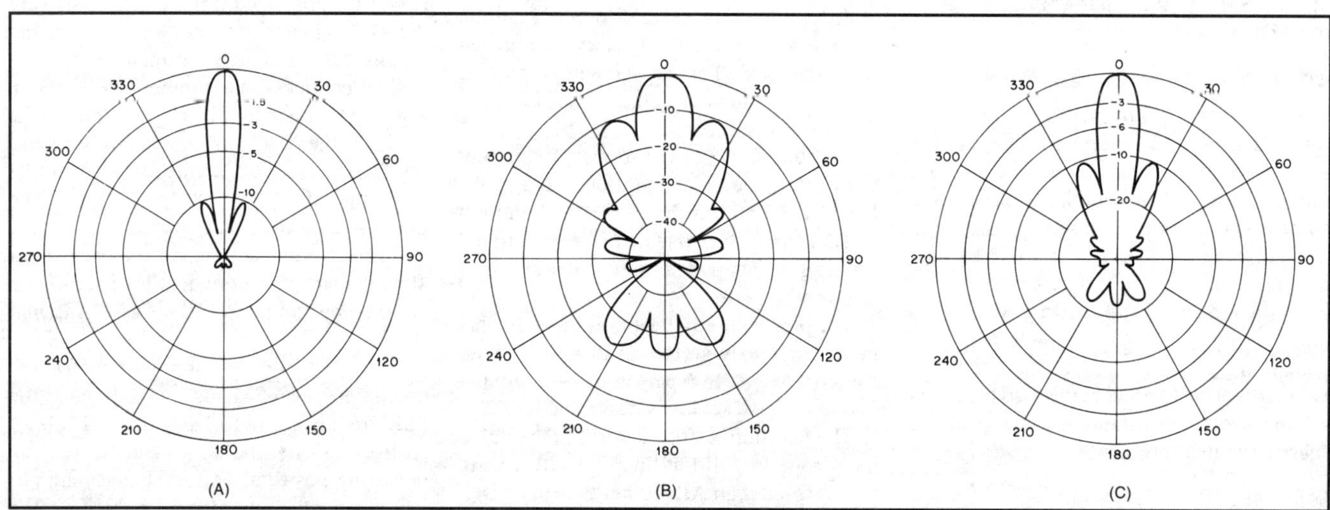

Fig 6—Antenna polar diagrams using various scales: Linear at A, Log at B and ARRL at C.

intensity, they vary periodically. The constant of this periodicity is 0.89 for 2 dB intervals. This represents a compromise between the linear and logarithmic scales just described. It possesses good sensitivity to small changes in maximum signal while at the same time is able to display the minor lobes.

The program described plots the stored data in the scale selected. A typical output plot as seen on the computer monitor is shown in **Fig 7**.

Fig. 7—Photo of IBM-compatible computer with 3-element beam displayed on the monitor. To the left of the computer is the AUT control box and on the right a commercial FSM. The AUT, rotator and signal generator can be seen through the open window.

Appendix

The Pico ADC-10 used in this project is the size of a printer cable connector and is self contained. It plugs into the parallel printer port and requires no external power. The resolution is 8 bits and the input voltage range is 0-5 V. A standard BNC connects the input signal to the ADC-10. It is supplied with software routines in C, PASCAL and BASIC. The manufacturer is Pico Technology Ltd, Broadway House, 149-151 St Neots Road, Hardwick, Cambridge CB3 7QJ, United Kingdom

Tel: +44-954-211716
Fax: +44-954-211880

The price £49.00 Sterling includes postage and packing. Pico Technology accepts credit cards, Sterling checks and bankers drafts.

A similar ADC is described in an article "High Speed Data Acquisition" by Mike Gray N8KDD.[8] It also interfaces the IBM via the parallel printer port. A printed-circuit board is available and parts are available from Radio Shack. Software information (BASIC) is also supplied.

Another solution is given in *Linear Technology Design Notes*, June 1990. This circuit uses the LTC1290 ADC and interfaces via the serial COM port. Software information (BASIC) is also supplied.

The address is:

Linear Technology Corporation
1630 McCarthy Blvd.
Milpitas, CA 95035-7487
Tel 408-432-1900

The subject of signal/IBM computer interfaces is described in *Interfacing Sensors to the IBM PC* by W.J. Tompkins, published by Prentice Hall. It also describes a low cost data acquisition system (ADC) that interfaces with the expansion bus. A circuit is given together with a circuit and a BASIC program.

References

1Peter Dodd, G3LDO, *The Antenna Experimenter Guide.* $15, shipping $5, airmail. 37 The Ridings, East Preston, West Sussex, BN16 2TW, UK. Tel 903-7770804. Credit cards accepted.

2Roy Lewallen, W7EL, "MININEC, The Other Edge of the Sword," *QST,* Feb 1991, p 18.

3*Antenna/RCS Measurement Products*, Hewlett Packard publication 5958-0396.

4Doug DeMaw, W1FB, "Field-Strength Meters," *QST*, Mar 1985, p 26.

5*Solid State Design for the Radio Amateur* (Newington CT, ARRL, 1977), p 148.

6*The 1994 ARRL Handbook for Radio Amateurs*, p 25-37.

7*The ARRL Antenna Book*, 17th Edition, pp 27-43 to 27-48.

8Mike Gray, N8KDD, "High Speed Data Acquisition," *73 Amateur Radio Today,* Aug 1991, p 28.

An Automatic Position Control for Your Beam–Updating the CDE/Telex Rotator Controller

By John Svoboda, W6MIT
2261 Peaceful Garden Way
Rescue, CA 95672

W6MIT continues the popular tradition creating useful modifications for the Ham-M rotator.

The CDE/Telex Ham-M or Tailtwister rotator is probably the most popular one around and has changed little over the years. This article offers a simple servo solution to automate the infamous three buttons on the front panel.

What's in the Standard Control Box?

The unmodified rotator controller box consists of two very basic power-supply circuits. One circuit powers the motor and the other operates the indicator. See **Fig 1**. The brake-release pushbutton S1 supplies ac power to motor transformer T2, which applies 30 V ac to the brake solenoid (located inside the rotator housing) as well as to the direction switches. The direction switches S2 and S3 route power to one side or the other of the motor winding. A thermal circuit breaker CB is built into T2, since it was designed to be small and yet provide plenty of power for short periods of time. It was not designed for continuous operation—it will overheat. The line fuse is to protect transformer T2 in the event of a cable short. Fuse F2, located inside the housing, protects the indicator supply from a cable short. The 130 µF non-polarized motor-run capacitor C1 is located in the box for convenience; it could be located elsewhere.[1]

Making It Automatic

By adding a simple servo amplifier we can eliminate the manual direction buttons and turn the rotator to the desired heading by merely setting a pointer knob. **Fig 2** shows the circuit. The servo idea is not altogether new; it is based on an article that appeared in June 1973 *QST*.

Circuit operation is straightforward: the heading control is turned to the desired position and the START button S1 (see Fig 2B) is pressed. Comparator U4 is unbalanced, causing optoisolator U5 to conduct. This in turn triggers triac Q3 and power is applied to T2. Pulses via D5 and D6 prevent U5 from timing-out until the system reaches the desired heading. U1 now has power supplied to it by means of rectifiers D1 and D2, and produces an error signal. Optoisolators U2 and U3 are wired so that one or the other will conduct when the signal swings from plus to minus. Both are off at null. In turn, they trigger two triacs, Q1 and Q2 feeding ac to the appropriate motor winding. The position pot located inside the rotator moves to bring the system gradually back into balance. The motor then stops and the brake times out.[2]

Resistors R7 and R8 establish the system gain and deadband. Initial tests with an R7 value of 470 kΩ resulted in endless "hunting" when tested with an inertial load. With the values shown, the system will null properly.

Selecting a Heading Control

The heading control R2 is basically an azimuth dial and ideally requires a 360° device. A position potentiometer is ideal for this application. A position potentiometer is a special form of linear pot that turns through 360° and that only have a dead space encompassing 2° or 3° of rotation. In this application the dead space should be placed South—as this is where the rotator limit stops are located. Such a position potentiometer is best found in a surplus store. Values from 5 to 20 kΩ are suitable.

Some provision must be made, however, to prevent the arm of such a pot from accidentally landing in that dead space and creating an open circuit. An open circuit will confuse the servo amp. Ra and Rb shown in the schematic will cause the system to head toward North rather than toward either limit switch should such an event occur. Ra and Rb should be approximately ten times the resistance value of R2. They are not required if a mechanical detent is used that precludes stopping on "open." Of course, an ordinary 270° linear pot can be used instead of a special pot, but one loses the flavor of a 360° azimuth control.

Another solution to the position control problem is to use a rotary switch. A 24 point (15°) detent switch will provide adequate resolution for 3 and 4-element beams, whose beamwidths are pretty wide. These are not as

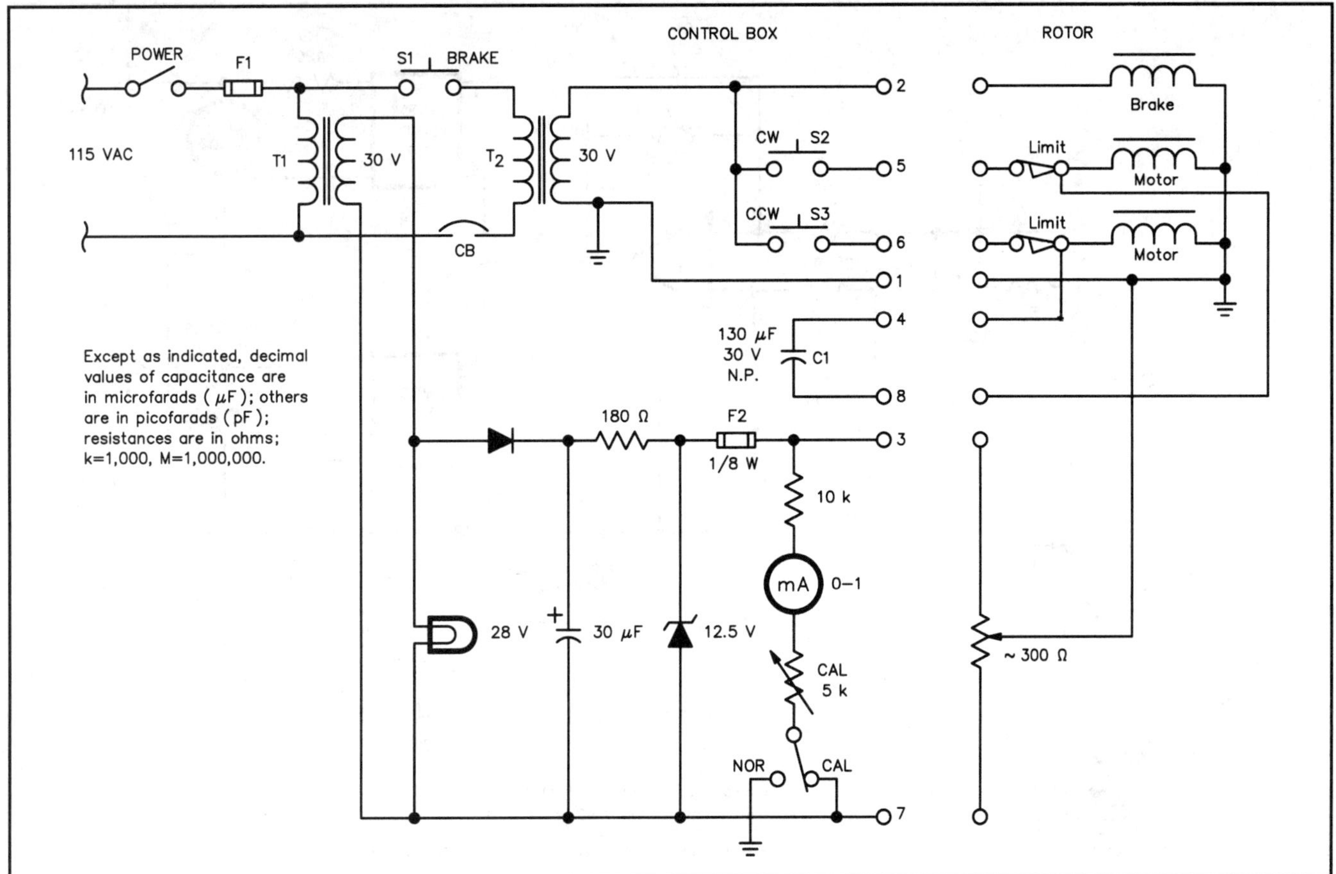

Fig 1—Schematic of unmodified Ham-M or Tailtwister rotator box.

common, however, as the 30° variety. I found some switches at the local surplus store that had only 16 contacts in place. Fortunately these had the arm contact located on the back of the switch deck. It was clear that they could be modified to offer 24 active positions. I bought two switches and removed the contacts from one deck—very carefully—and installed them onto the other. I also found that removing one of the detent ball bearings improved switch operation—it made it not so stiff, more fluid. The resistors used on the switch should be 1% or better, and all 24 must be the same. The value can be anything from 400 to 900 Ω. My switch happened to end up with 681 Ω 1 % resistors. Best buys again are surplus.

Avoiding the Limit Switches

Resistors R1 and R3 of the divider in Fig 2 are designed to prevent the motor from running to the limits. To say it another way; they make sure that the null occurs before a limit switch opens. If U1 doesn't null, the brake will stay released (power on) forever, or until the transformer overheats and the circuit breaker opens! The value of these resistors will depend on the resistance of the heading pot/switch. My tests showed that 5° of rotation was enough to clear the limit switch. Using either a pot or switch; you can calculate your R1 and R3 values as follows:

24 steps × 681 Ω = 16344 total resistance.
16344 / 360° = 45.4 Ω/°
45.4 × 5° = 227 Ω (R1 and R3). Standard value = 220 Ω

Because the bottom resistor (R2-24) of the switch string and R3 are in series, the two can be combined into a single resistor or trimpot (681+220 = 801 Ω). See Fig 2.

Construction

The controller board layout is not critical. I used a 4.5×5.0 inch piece of perfboard. Terminals were placed so that jumpers could be installed after individual circuits were tested before tying the system together. The board was mounted in the bottom of the controller chassis on ½-inch standoffs. The op amp U1's OFFSET control R12 should be adjusted to zero volts, measured from R6 to ground, before installing U1. R12 is then readjusted for best null after the circuit is operational. All the triacs should be the same type (200 V, 6 A). Although the motor operates at only 30 V, it does create much higher voltage peaks due to mutual and self inductance.

I replaced the standard thin outer panel in my control box. See the photos in **Figs 3** and **4**. The original CAL pot and switch were discarded and replaced with smaller parts

→

Fig 2—Schematic for new rotator control box with servo directional control.

mounted on a bracket inside the cover. The CAL pot permits compensating for cable resistance/length, so chances are very good that it won't need adjustment too often. I used an illuminated switch for my START button to indicate that the brake is released. It is powered from the 30 V ac brake line (terminal 2) through a diode to dim it. I also moved the motor-run capacitor from under the chassis to the top to make room for the board. Power for the brake delay timer, U4, was obtained by replacing the original indicator control transformer with one with two secondary windings.[3]

Testing It Out

The board can be checked out before tearing into your controller as follows: Connect 30 V ac to terminal 2 of the board (see Fig 2). Connect 28 V bulbs (to substitute for the motor) between terminals 5 and 6 and transformer common, terminal 1. Connect the heading pot/switch temporarily across the +9 V and –9 V supplies at Zeners D3 and D4, instead of the normal connection to ter-

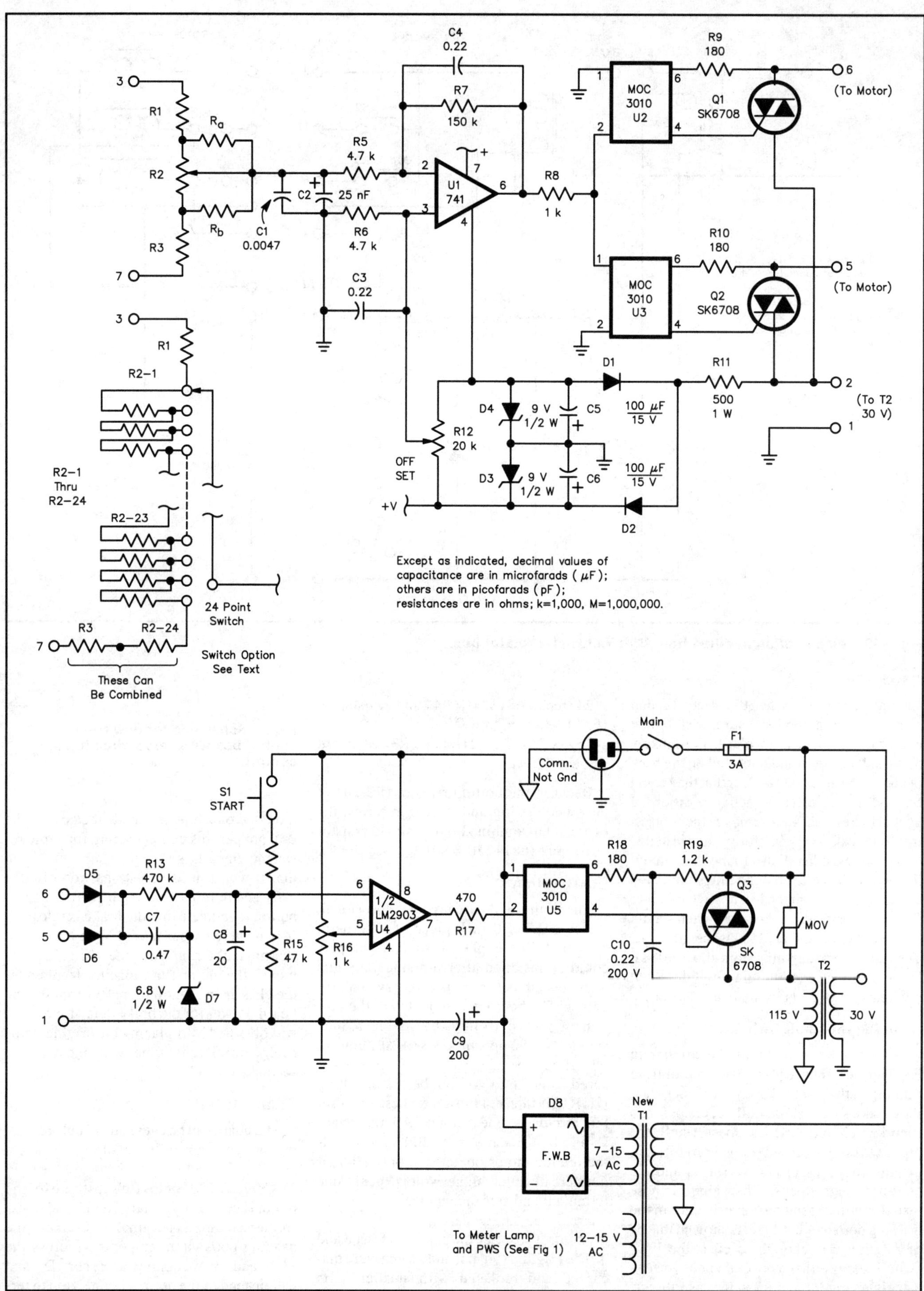
C4
0.22
R7
150 k
R9
180
R1
Ra
R2
R3
Rb
C1
0.0047
C2
25 nF
R5
4.7 k
R6
4.7 k
U1
741
R8
1 k
MOC
3010
U2
Q1
SK6708
(To Motor)
R10
180
MOC
3010
U3
Q2
SK6708
C3
0.22
D1
R11
500
1 W
(To T2
30 V)
D4
9 V
1/2 W
C5
100 µF
15 V
R12
20 k
OFF
SET
D3
C6
D2
+V
R2–1
R2–1
Thru
R2–24
R2–23
R2–24
24 Point
Switch
Switch Option
See Text
These Can
Be Combined
Except as indicated, decimal values of
capacitance are in microfarads (µF);
others are in picofarads (pF);
resistances are in ohms; k=1,000, M=1,000,000.
Main
F1
3A
Comn.
Not Gnd
S1
START
R13
470 k
D5
D6
C7
0.47
C8
20
R15
47 k
R16
1 k
1/2
LM2903
U4
470
R17
6.8 V
1/2 W
D7
C9
200
R18
180
R19
1.2 k
MOC
3010
U5
Q3
SK
6708
C10
0.22
200 V
MOV
T2
115 V
30 V
D8
F.W.B
New
T1
7–15
V AC
12–15 V
AC
To Meter Lamp
and PWS (See Fig 1)

minals 3 and 7. Chances are, one or the other lamps will illuminate—if not, change the setting of the heading control. Adjust it so that both lamps go out. At null, measure U1's output at pin 6 and adjust the OFFSET control for minimum output. When using a switch, a null should occur at one detent position. If not, adjust the OFFSET control slightly. Switching to an adjacent position should cause one of the lamps to light. When using a pot, there should be some small degree of rotation where both lamps are off.

The brake circuit may also be tested at this time. Provide 7 to 12 V ac to the full-wave bridge D8. Adjust R16 for 2 V at U4 pin 5. Connect a 115 V lamp in series with the brake triac and the line. With op amp U1 at null (motor lamps off), the brake lamp should also be off. When the pot is moved and a motor lamp comes on, so should the brake lamp. Then move the position pot/switch back to null. The brake lamp should stay on for another 3 to 4 seconds and then go out. Adjust R16 to change the delay time. Too long a delay may result in unnecessary "hunting."

Fig 3—Photo of front of author's control box, using 24-position switch.

When all is assembled, if the motor chooses to run counter to the direction that the switch or pot is turned, simply reverse the leads of the heading control (terminals 3 and 7). The motor must run toward the knob so there is some hope for it to find the null before encountering a limit switch.

In the End

The overall accuracy of the system is about 6°. My tests, with a 35-foot-pound inertial load, resulted in a clockwise error of 1.7° and a clockwise error of 4.2°. The difference is probably due to the age and wear of my rotator. The slack in my rotator is about 1.5°. During testing, I noted a positional error near the 270° heading. As it turned out, one of the 1% resistors on my heading switch was out of spec. It might be a good idea to use 0.1% resistors for R2 if you find them.

The use of a START button has several advantages; the beam is not affected by accidental knob movement, since power is off most of the time. This also reduces the opportunity for RFI problems.

From the standpoint of operating convenience; the servo positioning system is great. I look at my Great Circle wall map and turn the knob to what seems right. Once set, the START button is pushed, and it's automatic from there! Use of a bar-type knob also clearly shows the reciprocal heading.

Fig 4—Photo of inside of control box, viewed from bottom.

Notes

[1]The numbered terminals on Figs 1 and 2 refer to the numbers of the barrier strip on the back of the controller box (as well as the one on the rotator).

[2]The sequence of nulling is somewhat more complicated than the explanation just offered. Wind loading and inertia have an important effect on the outcome. As the rotator approaches null, the motor slows (providing braking action) and at the null the motor stops. Since the brake delay circuit is no longer receiving triggers, it begins to time out. However, inertia and wind forces may be adding (or subtracting), causing the rotator to continue to rotate (motor coasting). As the rotation continues, an error signal may again be detected. Power will be applied to the opposite motor winding so as to pull the rotor back to null. The process can then repeat itself. Hopefully, because the brake timer circuit has received fewer triggers, it will continue to time-out. "Hunting" is common in all servo systems to some degree. This also explains why the brake indicator may seem to stay on longer for some headings than for others. In very windy conditions, it may be necessary to turn off the power switch for a few seconds to ensure the brake is engaged.

[3]Models have changed but the transformers seem to be the same. Early indicator circuits used a 20 V dc unregulated supply. Newer units have a 12 V Zener regulated circuit using the same 20 V transformer. Therefore, almost any transformer that can supply somewhat more than 12 V ac at approximately 300 mA. will work. The transformer I used had 15 and 7 V windings and was rated about 500 mA.

Transmatches and Transformers

A Balanced-Feed Transmatch and Antenna System

By Harry R. Hyder, W7IV
1638 West Inverness Drive
Tempe, AZ 85282

This system keeps the RF out of the shack and sends it where it belongs —to the antenna.

I had a problem. My lot is only 65 feet square and I wanted to operate on all bands—10 through 160 meters. Here is my solution. You probably will not want to duplicate it exactly, but it might give you some ideas towards solving your own antenna problems.

A trap vertical on the roof of my house took care of 10, 12, 15, 17 and 20 meters. The real problem occurred on 30 through 160 meters. For these bands an inverted-V antenna was erected with a 450-Ω open-wire feed line.

The shack is in front of the house and the antenna (naturally) is on the other side in the rear of the house. Draping open-wire line across the roof presented a number of problems. In addition, I don't like to bring open-wire line directly into a shack because the end of the line can be quite hot. [With a short dipole and 100 W on 160 meters the voltage at the end of the open-wire line can exceed 10,000 V!—*Ed.*]

Transmatch Design

My solution was to locate the tuner or Transmatch on a shelf above the rear patio, protected from the weather. Since I did not want to go out and twist knobs on the tuner whenever I changed bands, I designed and built the tuner to provide a match on six frequencies. A switch selects 10.125, 7.15, 3.55, 3.9 and 1.84 MHz.

I decided to use an L-network, since in my experience this configuration will match almost anything with the fewest components—one L and one C for each frequency. **Fig 1** is a schematic of the configuration. The balanced line requires half of the inductance in each leg of the network. My rig is connected through coax cable to the low-impedance side of the 4:1 balun.[1] The high-impedance side of the balun goes to the matching network. While

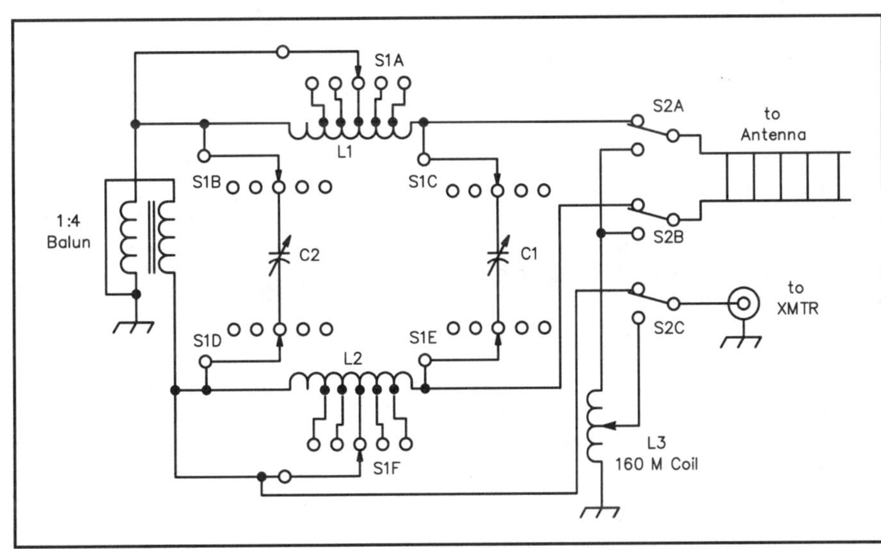

Fig 1—Schematic of the tuner. S1 is a 4-section switch. Two sections select taps on L1 and L2. The other two sections select capacitors. W7IV measured the following values for his installation:

Frequency MHz	*C1 pF*	*C2 pF*	*Turns for L1 and L2 each coil*
10.125	0	150	3
7.15	0	27	12
3.55	1000	0	10
3.8	1000	0	8
3.9	1200	0	7

Capacitance values are approximate.

Fig 2—Top view of the tuner. The balanced feed line is connected to the terminals on the left of the unit. The large coil is the 160-meter loading coil, and S1 is in the center.

Fig 3—Coil L1 is in the lower center of this side view. Several tuning capacitors (C1 and C2) are visible in the center of the top deck.

the schematic shows capacitors on both sides of the inductor only, a single capacitor is needed on any one frequency. The switch-selected capacitors are never needed on both sides of the inductor simultaneously.

Initially I breadboarded the tuner from junkbox parts to arrive at approximate component values. I used a low-power signal source in the shack and inserted an SWR bridge in the coax line at the tuner. After determining the values of capacitance needed I measured them with a capacitance meter. The values in Fig 1 are those I use. Your antenna and surroundings will no doubt require slightly different values, but the ones listed in the figure will give you an idea of the range of values needed.

Be careful when you select the capacitor types. Some older units do very well in RF service with high current capability while others will produce smoke. Some junk box dipped-mica capacitors date back to the mid 1940s. Their original capabilities, even if known, may have changed with age.

Fig 4—The complete unit in its protective box. The top can be lifted off without disturbing the knob connected to S1.

Fig 2 and **Fig 3** are a top view and side view of my unit. The small coils, L1 and L2, (one is visible in the lower center of Fig 3) are used on all bands. *Mini clips* connect the coil taps. These clips, as well as many of the other parts, are often available at flea markets.

160-Meter Loading

On 160 meters I use the antenna as a top-loaded vertical rather than a very short dipole. Since the apex of the antenna is only 26 feet high (a limit caused by the available guys and supports in my yard) and the dipole length is $^1/_8\ \lambda$, I feel the vertical antenna approach radiates more power horizontally.

The vertical consists of the two conductors of the open-wire transmission line tied together. Switch S2 bypasses the tuner and the transmission line is fed through the loading coil on the right of Fig 1.

The transmission line is only 35 feet long but it apparently gets some top loading from the dipole and works very well. I use a relay controlled from the shack to change from tuner to loading coil configuration. A switch could have just as easily been used in place of the relay (as shown in the schematic) but the original idea of the tuner was to make the whole thing controllable from the shack. At first I was planning to use a surplus motor-driven switch and indexing mechanism for the tuner switching. Later I found it just as easy in my case to open the patio door and throw a switch by hand.

I built the housing for the tuner from half-inch particle board. The top can be lifted off without disturbing any components or settings. A liberal treatment of water seal and a few coats of exterior paint provide protection from the elements. **Fig 4** is a view of the closed unit. The knob protruding is connected to S1 in the schematic (Fig 1) and used to select the frequency.

Performance

On the narrow 30-meter band the SWR is low across the entire band. On 40 meters the tuner is centered on 7.15 MHz and the SWR is only slightly above 2:1 on 7.0 and 7.3 MHz. Tuning is naturally more restricted on the 75 and 80-meter bands. Each of the switch positions for these bands have ± 25 kHz range before the SWR exceeds 2:1. Larger frequency excursions on these bands as well as on 160 meters are handled by a Transmatch in the shack.

I have used the tuner and antenna at a power of 100 W for about 7 years with good results. All common grounds in the tuner are connected to a garden hose connection a few feet from the tuner. Although the antenna has been down a few times to replace a broken halyard, the tuner has never needed any maintenance or retuning.

Note

[1] *The ARRL Antenna Book*, 17th (1994) Edition, pp 26-10 and 26-11 discusses balun design and construction. Additional information is also found in *The 1994 ARRL Handbook*, on pp 11-12, 11-13, 16-8 and 16-9.

A 10 to 80-MHz Broadband 75 to 50-Ω Transformer

By Tom Rehm, K9PIQ
11653 N Pinehurst Circle
Mequon, WI 53092

K9PIQ explains how to build an efficient transformer to match 75-Ω Hardline to 50-Ω at 6 meters.

I wanted to develop a ferrite-core transmission-line transformer to match 75-Ω Hardline to 50-Ω RG-8 coax. The goal was to achieve a broadband impedance match on 6 meters. I was concerned that transmission-line losses would be high for a 375 foot cable run between my transmitter and antenna using standard 50-Ω coaxial cable. To make matters worse, 85% of the cable was outside and a good part of that had to be buried, so there was concern about long-term cable losses caused by contamination.

The choice became clear—use the 75-Ω CATV Hardline that was coiled up and collecting dust in the garage. I intended to use the Hardline from the top of the tower to the house and then transform it to 50 Ω for standard coaxial feed to the transmitter. At the antenna side of the installation, I elected to modify the antenna to create a 75-Ω match directly to the Hardline rather than inserting another transformer.

Several iterations of design were attempted in an effort to make the transformer bandwidth as large as possible, while keeping it centered around the 6-meter band. It would be nice too, if its bandwidth was wide enough to cover the upper HF frequencies for future antenna projects. The final design provided a flat SWR from 10 to 80 MHz. and could easily handle my 300 W transmitter output.

Design and construction ideas were derived from the book *Transmission Line Transformers*, by Jerry Sevick, W2FMI,[1] and from *The ARRL Handbook*.[2] These are excellent reference materials and offer many transformer designs covering the entire HF spectrum.

My transformer is shown schematically in **Fig 1**, with the physical winding distribution shown in **Fig 2**. One word of caution: Transmission-line transformers should be terminated in their design impedance for peak efficiency—in this case, 50 and 75 Ω, respectively. Try to avoid using them simply to match arbitrary impedances. When properly terminated, this transformer is extremely efficient and should readily handle power levels of 500 W.

Insertion-loss tests were performed using the method of back-to-back transformers suggested by Sevick. Results indicate that the transformer loss is less than 0.1 dB for each transformer on all amateur bands between 10 and 80 MHz. Measured loss was only 0.07 dB at 52 MHz. One final test was conducted at 52 MHz—to compare receive input sensitivity both with and without the back-to-back transformers connected. Test results showed no measurable difference in the minimum discernible signal.

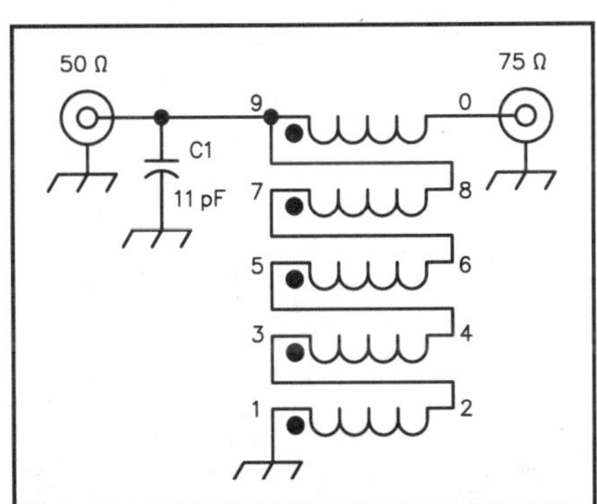

Fig 1—Schematic of K9PIQ broadband 75 to 50-Ω transformer.

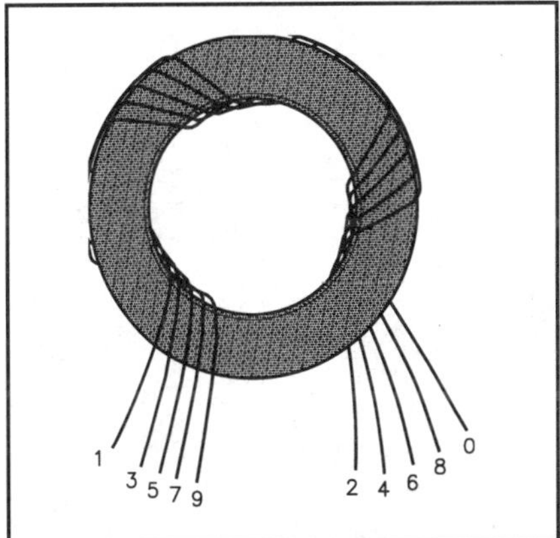

Fig 2—Drawing showing winding details for transformer.

Material

Wire:	#14 Thermaleze (less than 5 feet total)
Core:	Toroidal ferrite core #67; 1.4-inch OD, Amidon FT-140-67
Tape:	1/2-inch glass cloth tape, Scotch no. 27
C1:	11 pF cap., silver mica, 500 V
Box:	Aluminum, 1 1/4 × 2 1/2 × 3 1/4 inches, Ten-Tec TP-13
Connectors:	SO-239, chassis mount, Amphenol

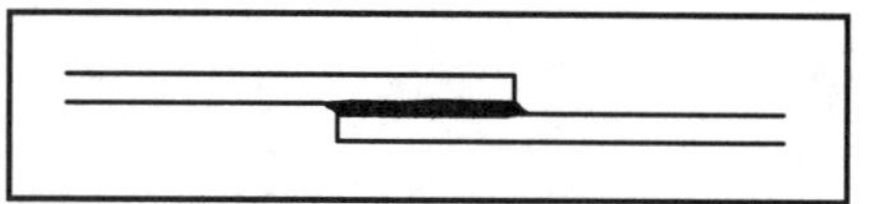

Fig 3—Wire-end bonding detail.

Construction

The five coils shown in Figs 1 and 2 represent the five adjacent turns of a coil. The five individual wires should be placed flat and grouped side-by-side to a length of about 10 inches before winding. The group can be held together with two wraps of glass tape placed every 1 1/2 inches. In total there are 3 group-turns through the center of the core. Normally core preparation is not required. However, I found it a bit easier to wind if it is first covered with a single layer of glass tape.

Wire ends 2-3, 4-5, 6-7 and 8-9 should be kept short and routed to continue the circular winding pattern. I soldered along the length of the wire ends rather than twisting them together as you would do for a wire nut. See **Fig 3** for recommended end-turn bonding. Make very sure that you identify the individual wire ends and connect them together properly. If you follow Fig 2 carefully you should have no problem.

I put the completed assembly in an aluminum box fitted with two closely spaced SO-259 chassis connectors. You may want to add a ground lug to the outside of the box as a termination point for system grounding.

Notes

[1]J. Sevick, *Transmission Line Transformers* (Newington, CT: ARRL, 1990).

[2]*The 1994 ARRL Handbook* (Newington, CT: ARRL).

Transmission-Line Measurements and Accessories

Measuring RF Impedance Using the Three-Meter Method and a Computer

By Peter Dodd, G3LDO
37 The Ridings
East Preston
West Sussex BN16 2TW
United Kingdom

G3LDO updates a clever method for impedance measurement developed originally by W8CGD. Don't be misled by the name—it does not involve a new amateur frequency band; three dc meter readings are needed!

Introduction

The ability to measure impedance at RF is particularly useful when experimenting with antennas or making antenna adjustments. An impedance measurement says more about an antenna than SWR ever can.

The most common method of amateur impedance measurement found in Amateur Radio journals and handbooks[1] is the Rx noise bridge. It has an upper frequency limit between 30 to 50 MHz, depending on the bridge construction. The noise bridge has the advantage that the variable bridge component dials can be made direct reading. The disadvantages are the need for a detailed calibration of the adjustable standard resistance components and that the reactive bridge component can only be calibrated for one frequency.

The Three-Meter Method of Measuring Impedance

A simple, yet reasonably accurate, method of measuring impedance using three meters was described by W8CGD.[2] Although it is an easy instrument to construct, it can produce results comparable to a laboratory instrument, provided that sound construction and operation practices are followed.

This method compares the unknown impedance with a *fixed* standard impedance, and the ratio is indicated by the reading of a meter. Because the meter can only indicate a scalar relationship, provisions for making more than one comparison are required so that the resistive and reactive parts of the unknown can be determined. See **Fig 1**. Three readings are required, suggesting the name for the technique—the Three-Meter Method. The main features of this method for measuring RF impedance are:

- The fixed reference components do not have such large values of stray impedances as the variable type. This enables the instrument to be used at a much higher frequency. I have used the Three-Meter bridge on 2 meters with good results.

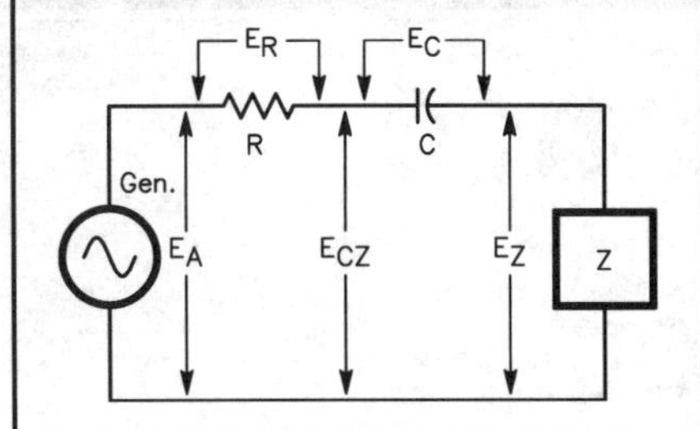

Fig 1—Three-Meter RF bridge using a fixed standard reference.

- Provided the fixed reference component values are known, the need for calibration is not so great.
- The main disadvantage of the method is that it is not direct reading. W8CGD used a graphic method, using a pair of compasses and a ruler, to extract the impedance data from the Three-Meter data.

An alternative method of extracting the impedance value is to use a computer with appropriate software. A method of achieving this is described in *QEX*[3] and in my book.[4] This article expands on this work and corrects some of the errors in the original *QEX* article.[5,6]

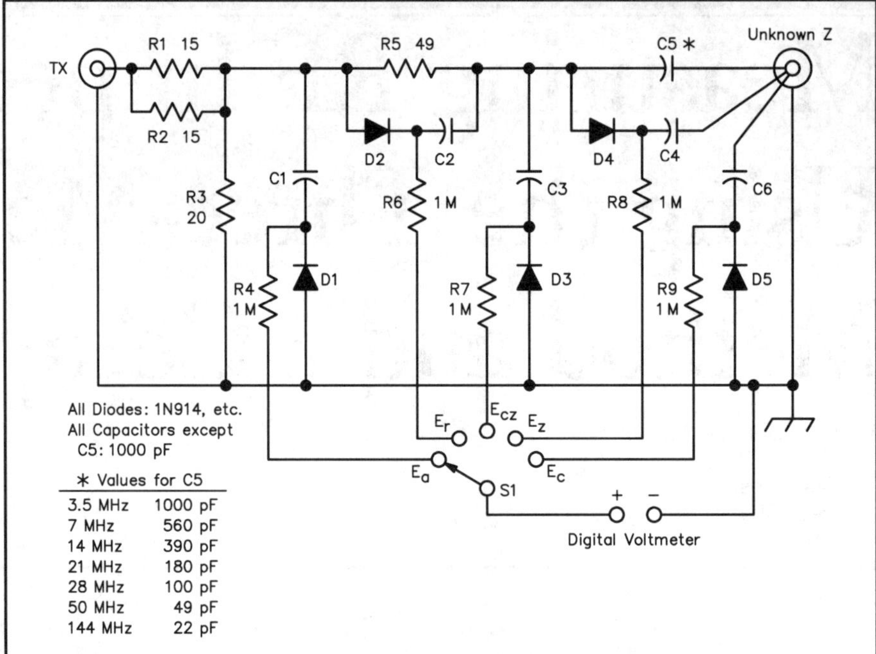

Fig 2—Circuit diagram of Three-Meter box.

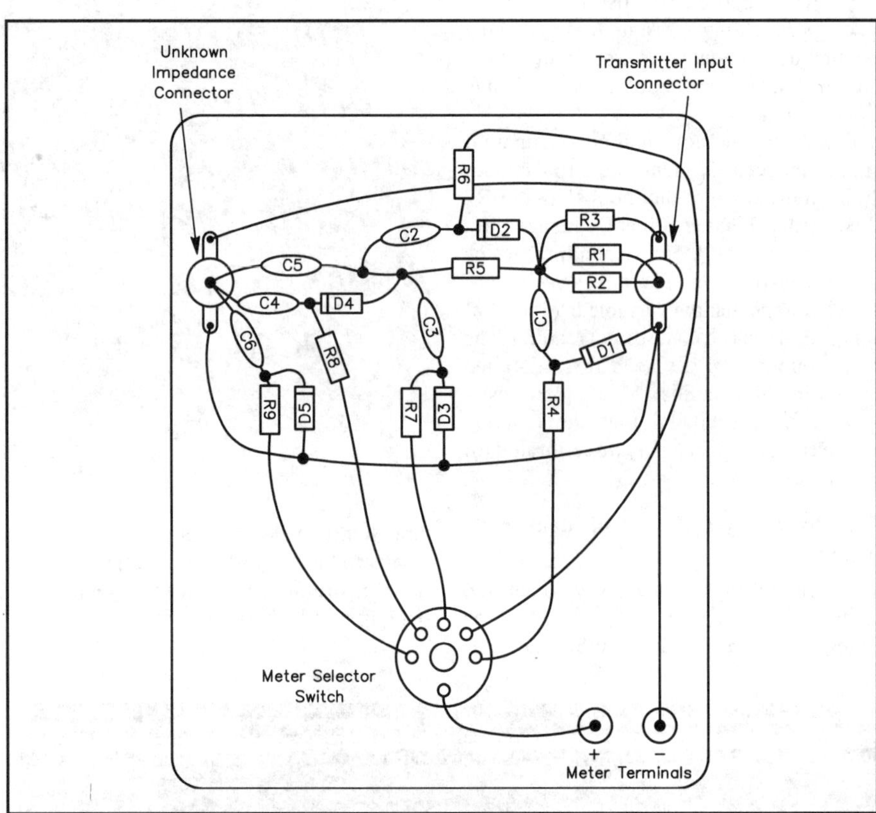

Fig 3—Layout diagram of Three-Meter box.

Three-Meter Box, Circuit Description and Construction

The circuit and the layout of the impedance measuring box are shown in **Figs 2** and **3** respectively. The unit is built in a die-cast aluminum box with the components mounted on the inside of the lid. In order to achieve the greatest accuracy, you should change the reference capacitance for each frequency band—some thought should be given to making the replacement of this component fairly easy. The original article suggested a reactance value of between 25 and 50 Ω for reference reactance C. Although values of up to 100 Ω give reasonable results, the capacitor value really should be changed for different bands for best accuracy.

The area around the unknown Z port is quite sensitive to component lead length. On the other hand, the lead for the reference capacitor (next to the reference resistor) appears as a small component of negative capacitance and is accommodated by the voltage measurements. Nevertheless, component leads should be kept as short as possible for reference capacitor C.

A low-power transmitter with a variable power output is used as an excitation source. It is very important that the excitation source have a low harmonic content. Erratic measurements may be caused by harmonics. A half-wave filter, or similar, between the transmitter and the Three-Meter box may be required.

An attenuator is used at the input so the transmitter is isolated from the variations of unknown impedance as the transmitter frequency is varied during a series of measurements. These attenuator resistors must be capable of dissipating the transmitter power without overheating. Otherwise the voltage readings will drift during the measurement. One measurement takes around 30 seconds, so the attenuator resistors do not have to be continuously rated.

Power stability of the excitation source is also important—that is, E_R in Fig 2 should remain constant while the other parameters are being measured. (Note: In the following, I use two designations for the measured voltages; that is, E_a = A, E_r = B, E_{cz} = C, E_c = D and E_z = E. This is because these voltages are used as variables in the computer programs, and some BASIC interpreters, such as the Casio PB110, use only single letter variables.)

Recording Three-Meter Impedance Data

Connect the unknown impedance, excitation source and digital voltmeter to the test box, as shown in **Fig 4**.

1. Set the switch to read E_r (B).
2. Set the transmitter to the lowest power level and switch it on.
3. Increase the power until the voltmeter reads 5 V; record this on a notepad as "50." Note the voltages at the other switch positions (for example, 12 V being recorded as "120" and 6.3 V as "63"), then check the stability of the transmitter power output by checking position E_r (B) again.

I tried using a lower voltage for E_r (B), as suggested in *QEX* correspondence.[5] However this resulted in an increase in data errors.

Processing the Three-Meter Data Using a Pair of Compasses and a Ruler

Although this article is primarily about processing Three-Meter data using a computer, I feel it is necessary to describe the method originally used by W8CGD. The program algorithm uses the same technique. Also, if a lot of errors are being reported, the graphic method may reveal the cause of the problem. All that is required is a sheet of linear graph paper, a pencil and a pair of compasses. The method is illustrated in **Fig 5**.

1. Draw a horizontal line, with length equal to E_r, on the graph paper .
2. Draw a vertical line, whose length is equal to E_c, down from the right-hand side of line E_r.
3. Place the point of the compasses on [1] and draw an arc whose radius is equal to E_a. Repeat for [2] and [3] for radius values E_{cz} and E_z respectively.
4. Mark the point at which the arcs intersect. An exact intersection of all three arcs is not always possible due to the errors in the data.

A horizontal line from the reference point to the arc intersection gives the resistive value of impedance. A vertical displacement up or down from the resistive line gives the value of the inductive or capacitive reactance respectively. If the arcs do not intersect, use the center of the triangle formed by the non-intersecting arcs as the impedance point.

Fig 6 illustrates a case using the graphic solution where the E_{cz} measurement is in error for some reason. An increase in E_{cz} results in a proper intersection.

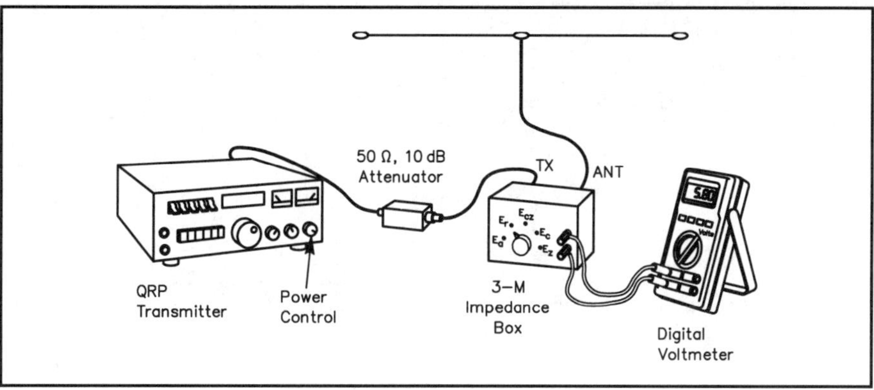

Fig 4—Typical Three-Meter measurement test set-up.

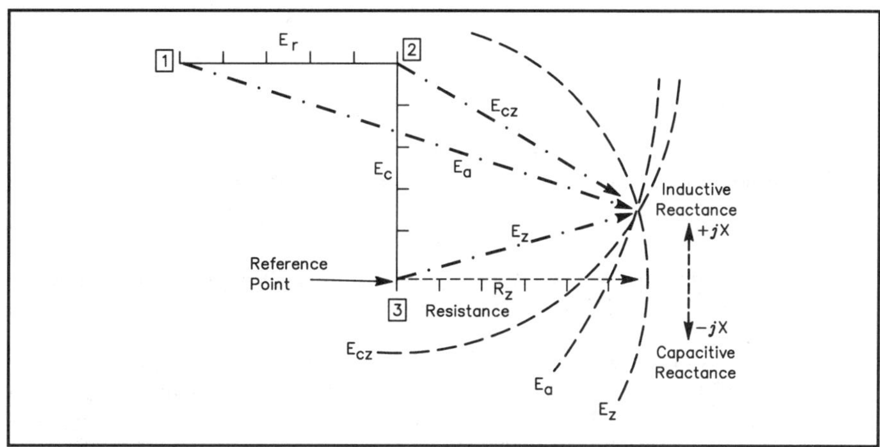

Fig 5—W8CGD method to extract impedance from Three-Meter data, using linear graph paper.

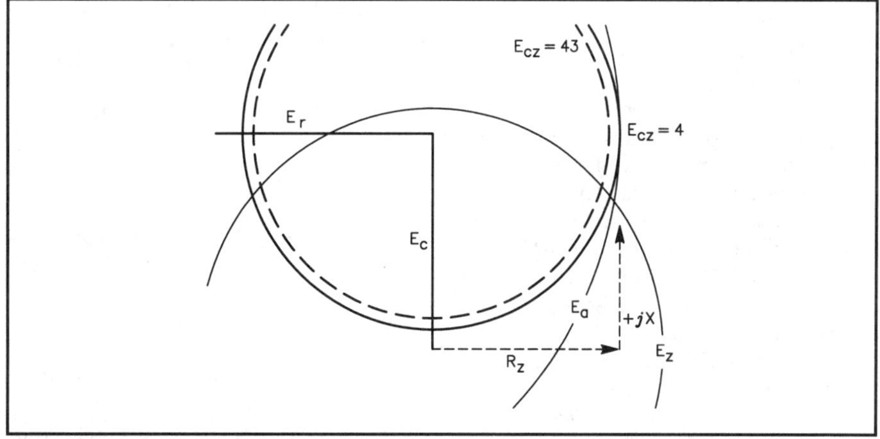

Fig 6—Graphic solution showing an error in E_{cz} (dotted) resulting in a non-intersection. An increase in E_{cz} (solid line) shows an intersection.

Extracting Impedance Data from Three-Meter Data Using a Computer

The following two BASIC program listings are included on the diskette accompanying this book:

1. *ZSMALL* is used with a BASIC program-

```
        Ea (A)     95
        Ecz(C)     43
        Ec (D)     50
        Ez (E)     55
   If there is more than one line of data below then the data has been corrected
   Last line indicates corrected data used to determine impedance.

   Ea   (A) = 95 ,   Ecz   (C) = 43 , Ec   (D) = 50 , Ez   (E) = 55
   Ea   (A) = 95 ,   Ecz   (C) = 44 , Ec   (D) = 50 , Ez   (E) = 55
   Ea   (A) = 95 ,   Ecz   (C) = 45 , Ec   (D) = 50 , Ez   (E) = 55

   SOLUTION
       Resistance     43.7
       Reactance    + 39.5

       ERRORS
          R= (+/-)     1.3
          X= (+/-)     10.5

   Goto Main Menu 1. Repeat this program 2. ?
```

Fig 7—Printout of screen display showing corrections made by the program to create an intersection.

mable calculator such as the Casio PB110. This listing is a derivative of program *ZCALC*. It has been designed specifically to fit into the 1K of memory of the Casio PB110, or similar, pocket computer. It performs all the calculations but does not amend faulty data.

2. A more comprehensive program, *ZCALC*. This program is written in GWBASIC in such a way that it will also work with QuickBASIC.

When either *ZCALC* or *ZSMALL* is run, you will be prompted for items of data. See the "READ.ME" file on the disk for detailed operating instructions for the programs. The exception is E_r (B), which is assumed to be 50 V. Note that entry of voltages is multiplied by ten times the actual voltage reading at each diode detector. Either program shows the solution as soon as the last item of data is entered. In addition, each program reports errors, giving an indication of data reliability.

The program will amend any data resulting in a small geometric non-intersection. It will also inform you of any change in any item of data required to effect this intersection. The program will make 10 attempts to correct a non-intersecting error. If the data cannot be corrected during these attempts the program will report this and not give a solution. **Fig 7** shows a screen printout of a case where *ZCALC* has automatically corrected the value of E_{cz} in order to achieve a proper intersection.

Development of the Program Algorithm

The algorithm was devised by the late Tom Lloyd, G3TML/9L1TL. The following description is based on notes written by Tom.

Not only does this algorithm yield a solution but it will also work with imperfect data and gives some idea of its reliability. The geometry of the vector diagrams, shown in the graphic solution in Fig 5, calls for a trigonometric approach. The use of angular functions in the algorithm could have been avoided by using geometrically derived equations, as suggested in Reference 5, where a solution is given by:

$$R_{es} = \frac{E_a^2 - E_r^2 - E_{cz}}{2E_r} \qquad \text{(Eq 1)}$$

$$jX = \frac{E_{cz}^2 - E_c^2 - E_z^2}{2E_c} \qquad \text{(Eq 2)}$$

Here the real part of the solution (R_{es}) is a function of E_a, E_r and E_{cz} only. Hence any change in the values of E_c or E_z will not affect the value of R_{es}. Similarly, the imaginary part of the solution (jX) is a function of E_{cz}, E_c and E_z only. This means that jX does not respond to changes in the values of E_a and/or E_r. Consequently, an accurate solution using these equations is possible only when ideal data are used, and in no case is an indication of the reliability of the solution given. Also the solution is in error disproportionately to the increase in data error.

It is rare that "ideal" data are available; that is, the three arcs with radii E_a, E_{cz} and E_z intersect in a common point. Generally, however, data are "well behaved." The three arcs intersect to give six intersections, some of which may coincide. Three of these intersections will lie close to the true solution and form a small error triangle. It is the accepted practice to take a solution to lie at the "center of gravity" of this triangle.

Provided the size of the error triangle is small, its sides may be approximated by straight lines without introducing significant errors. This algorithm uses a procedure to find a solution together with attendant errors, the latter giving an indication of its reliability.

There is a further class of data that may or may not have significant errors, causing one of the smaller arcs not to intersect the other two. This data is designated "acceptable faulty data" and the program *ZCALC* incorporates a routine for dealing with it. However, it must be stressed that, in practice, most data are well behaved.

Very occasionally arc intersections fall within one of two narrow sections of the vector diagram where the program has the effect of calculating a solution lying to one side, or the other, of the true solution. Happily, this only occurs rarely and the true result lies within the "error" limits accompanying it.

Additional Software

I have developed additional software to produce tables and graphic displays of antenna or component impedance signatures over a range of frequencies. The programs required to produce these tables and graphics are also located on the diskette bundled with this book. They are called *TODISK*, which allows the operator to enter and store to disk up to nine sets of data at different frequencies for a particular load; *TABLE*, which converts the data stored by *TODISK* into a tabular listing, which in turn can be printed or input to *GRAPH*, which produces an on-screen graph of the impedance characteristics versus frequency. The three programs can be run from a "shell" *MENU* program or individually.

Fig 8 shows a graph of the measured impedance versus frequency for an 80-meter mobile whip. It's not too hard to see that the resonant resistance (which also includes ground losses) would result in a best-case SWR of about 2:1. The values at 3.55 and 3.62 MHz are slightly out of line with the rest of the data on the graph, indicating small errors associated with the measurements for these data items.

Table 1 illustrates how the Three-Meter Method can be very useful for determining the electrical length of a transmission line. It does so with greater accuracy than other methods tried. The far end of the transmission line must be terminated with a resistor that is approximately twice (or half) the value of the transmission line characteristic impedance—the value is not critical. Resonance is indicated by zero measured reactance.

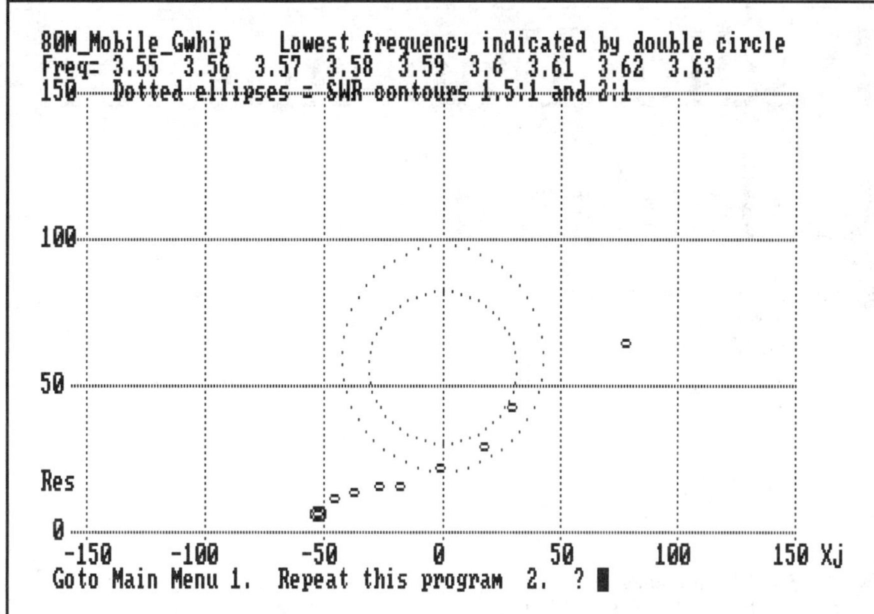

Fig 8—Measurements on a short 80-meter mobile whip, illustrating why the best SWR obtained was 2:1. Note that the measurements at 3550 and 3620 kHz are slightly out of line with the other data on the graph and show probable errors.

Future Software Design Improvements

It would be useful, and quite feasible, to have a routine in this software that calculates the impedance transformation effect of the transmission line between the antenna and the Three-Meter box. This would allow the impedance of an antenna to be measured, regardless of the transmission line length. An additional routine could be used to calculate the electrical length of the transmission line with a known resistive termination. A small program, called *TXFORM*, useful for calculating the impedance-transforming effect of transmission line, is included on the disk.

Table 1
Coax, Electrical Length

Freq (MHz)	*A*	*C*	*D*	*E*	*Results (Ω) Res*	*jX*	*± Errors (Ω) Res*	*jX*
20.93	85.0	48.0	36.4	23.7	23.1	−6.5	1.1	2.2
21.00	83.7	46.5	37.0	23.5	23.3	−3.3	0.1	0.2
21.1	82.6	45.0	36.4	23.0	22.9	−2.4	0.1	0.1
21.2	81.6	43.0	36.4	23.0	23.0	0	0.1	0.1
21.3	82.0	42.0	36.6	23.2	23.6	+1.31	1.0	1.6
21.4	80.4	40.0	36.3	23.2	23.2	+3.4	0.4	0.7
21.5	80.2	38.5	36.0	23.7	23.7	+5.1	0.8	1.4

Note: line is terminated with resistor equal to twice Z_0.

Future Three-Meter Hardware Design Improvements

This instrument appears to give very good results when built with standard components and is accurate enough for most antenna adjustment purposes. However, WD8KBW[8] suggests the following if greater accuracy is desired:

1. The power attenuator should be located outside the test box. This reduces the chance of coupling between the high voltages and currents in the attenuator and the voltmeters.
2. Provide shielding between the voltmeter circuits and the RF line.
3. Calibrate the voltmeter diode probes on an ac voltage (between 5 and 10 V); they should read within 2% of each other.
4. Calibrate the reference resistance using a laboratory-quality ohmmeter.

Notes and References

[1]A Noise Bridge for 1.8 through 30 MHz, *The ARRL Antenna Book*, 17th Edition (Newington: ARRL, 1994), pp 27-23 to 27-26.

[2]D. Strandlund, W8CGD, "Measurement of R + *j*X," *QST*, Jun 1965, p 24.

[3]Dodd and Lloyd, "Measurement of Antenna Impedance," *QEX*, Nov 1987.

[4]P. Dodd, *The Antenna Experimenter's Guide*, $15, shipping $5 air mail. Credit card. 37 The Ridings, East Preston, West Sussex BN16 2TW, UK, 903-770804.

[5]Technical Correspondence, *QEX*, May 1988, p 3.

[6]P. Dodd, "The Mobile Roof-Rack Antenna," *QST*, Nov 1988, p 29.

[7]J. Grebenkemper, KI6WX, "Improving and Using the RX Noise Bridge," *QST*, Aug 1989, p 27; Feedback, Jan 1990, p 27.

[8]A. Weller, WD8KBW, "Variations on the Three Meter Method of Measuring Impedance," unpublished article.

Radio Transmission-Line Calculator

By John R. O'Donnell, N1AQ
2144 Cemetery Road
Fabius, NY 13063

N1AQ updates the classic Smith Chart Calculator for ease of use.

When I was manager for a company called *FXR*[1] back in the 1950s, I gave away hundreds of Smith Chart Calculators as a sales promotion. It was a good gimmick and everybody wanted one. However, there were mistakes in the printing. The rotary radial arm had all the "Radially Scaled Parameters" printed on it, making it nearly impossible to read anything underneath it. The slider fixed to the radial arm continually slipped or came off and got lost. When everything did go right you still had to swing the half-length arm through 180° (which had to be measured somehow) to change impedance points to admittance points.

As a result of the problems I never used mine, but I did keep it for sentimental reasons. The other day I came across the calculator, and decided the basic idea was good. With some changes I decided the calculator would be extremely useful solving Smith Chart problems. I formulated the following criteria for a new calculator:

- Move any impedance point any wavelength distance, without making an off-line calculation.
- Be able to go from admittance to impedance points at a glance.
- Read electrical degrees without calculation.
- Solve a stub-matching problem by plotting just one point on the Smith Chart.
- Follow impedance/admittance plots along constant resistance/conductance circles so that the final location intersects a R = 1.0 or G = 1.0 line.
- Nothing should be allowed to interfere with seeing the chart.

To accomplish these goals the following were necessary:

- The outer "Wavelengths Toward Load" and "Wavelength Toward Generator" scales must be rotatable, just like the old *FXR* unit.
- The "Radially Scaled Parameters" should be separate from the calculator. This information is readily transferred to the Smith Chart using a set of dividers.
- In place of the opaque radial arm (covering the radius of the chart) a clear plastic ruler should be used to cover the diameter of the chart.
- The plastic ruler should have equal graduations on either side of center and should, of course, be rotatable around the center of the chart.
- Some means for reading electrical degrees should be provided.
- Anything placed over the chart should be of clear plastic and easily removed.

The calculator was built by making two copies of the Smith Charts located at the back of *Antenna Impedance Matching* by Wilfred Caron, published by the ARRL.[2] Most good copiers have an expander mode and the charts can be adjusted to a desirable size—a bonus to those of us in need of longer arms. I made mine approximately 10 inches in diameter.

The charts are pasted to poster board and later covered with plastic laminate for protection. One of the charts is trimmed with a pair of scissors around its outermost circle. The second chart is trimmed around the outer circle containing the scale labeled "Angle of Reflection Coefficient In Degrees." The "Radially Scaled Parameters" scale is also copied and attached to poster board.

Before covering the second (outermost rotating) circle with the plastic laminate, use a compass to add a circle centered at 0.333 on the resistance axis. Make the radius equal to the distance from 1.0 to 3.0 on this same axis. This added circle is the equivalent "overlay" of the constant resistance circle R = 1 and will simplify the solution of some problems. (See Example 3.)

Since the charts are covered with laminate, a marking type pen can be used to plot points on it. The ink is easily removed with ordinary rubbing alcohol.

While in search for something that would be suitable for the plastic ruler, I came upon a "Rolling Ruler" (as seen on TV), costing just a few dollars. At first I planned to modify the ruler, but the more I looked at it, the better it seemed, just as it was. These rolling rulers are graduated in cm and have a series of holes at each cm mark into which a pencil tip can be placed in normal operation. They measure 30 cm in length and pivot nicely at the 15-cm hole around a thumbtack pushed upwards through the backs of the two poster boards, through the center of the two concentric Smith Charts. I slipped a 1/2 inch length of red plastic tubing (it was part of an extension squirt tube from a can of WD-40) over the protruding thumbtack shaft to make a pivot for the ruler. See

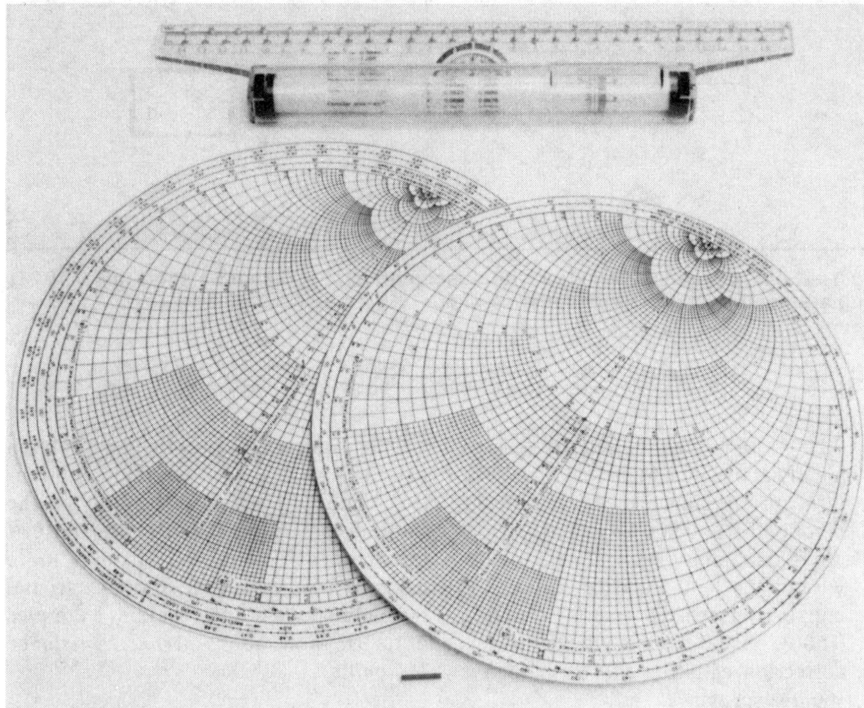

Fig 1—Photo showing components for N1AQ Smith Chart calculator. A Smith Chart from *Antenna Impedance Matching* is photo-expanded with copier machine to make two identical 10-inch diameter circles. One is cut with scissors at the outermost circle (after label "Wavelengths Toward Generator"). The other is cut at outer circle labeled "Angle of Reflection Coefficient in Degrees." Both are pasted to poster board for rigidity and are later laminated in plastic, after a circle centered on 0.333 on the resistance axis is added to smaller diameter circle. See text. A thumbtack is pushed through the back of both charts at the center of each, and a ½-inch length of plastic tube (from a can of WD-40) fits nicely on the tack shaft to act as a pivot point for the "Rolling Ruler" shown below the two chart wheels. The small plastic pivot tube is shown at the top of the photo. (*Photo courtesy Regina L. Nash, Aesthetic Prints, Oneida, NY.*)

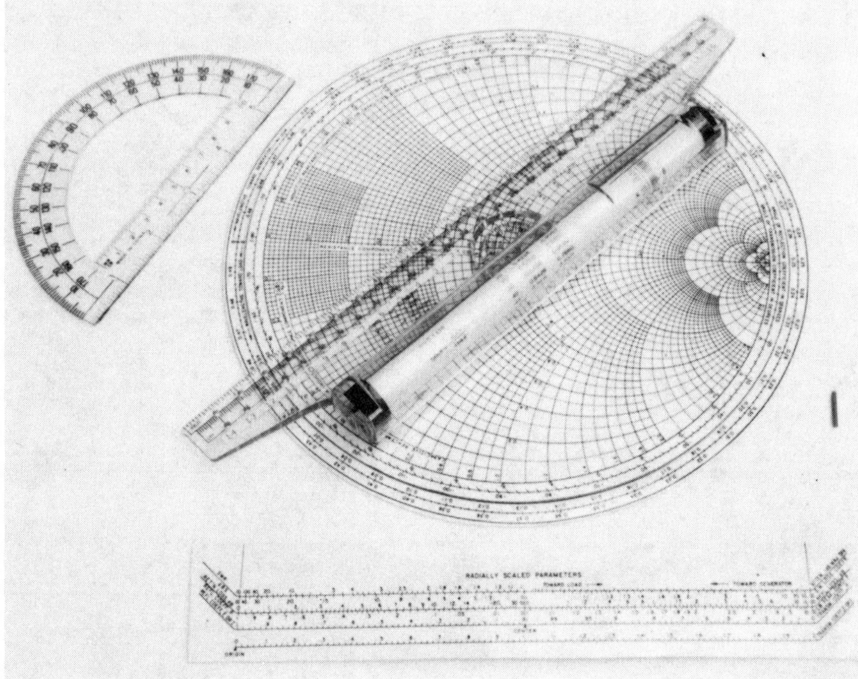

Fig 2—Photo showing complete N1AQ Smith Chart calculator, with Radially Scaled Parameters scale below and rotatable protractor to the upper left. The Rolling Ruler pivots at the tube at the center of the two concentric chart circles. (*Photo courtesy Regina L. Nash, Aesthetic Prints, Oneida, NY.)*

Fig 1, where the plastic tube is shown at the top of the photo. You may have to enlarge the hole in the rolling ruler with a no. 44 drill, as I did.

The rolling ruler also has a built-in protractor, which can be used to measure electrical degrees. However, it is easier to use a large clear plastic protractor graduated in 0° to 180° marks. Most have a hole at the proper location for use over the thumbtack pivot point. Electrical degrees will be half the marked value, or 0° to 90°. See **Fig 2**.

To facilitate using the rolling ruler transforming from impedances to admittances and vice versa, I added two scales of typewritten numbers labeled from 1 to 14 to the ruler. The zero in the middle is implied, but is not actually written—doing so would obscure the rolling ruler's protractor and make it hard to see through the ruler. The numbers are located in a line adjacent to the cm holes, starting at the 15-cm center hole and extending to both outer edges of the ruler. See Fig 1. Even when plotted holes are not lined up with a particular ruler hole, it is easy to spot the points through the unobstructed plastic, within a few mm.

Just a few examples of the new calculator's use should demonstrate its usefulness.

Example 1: Find impedance at any given point along a transmission line, when impedance at another point is known.

1. Line up the center line of the ruler so that it runs through the known impedance point. Note the plotted position by "hole" number and by deviation in mm from the center thumbtack.
2. Rotate the wavelengths scale until the zero point is at this same center line.
3. Rotate ruler only the desired wavelength distance circle and read the impedance at the position noted in Step 1.

Example 2: Match an antenna with feed-point impedance of 80 – *j*40 Ω, feeding it with a length of 50-Ω transmission line, followed by a parallel-connected stub.[3]

1. See **Fig 3**. Plot Z_a as 1.6 – *j*0.8 (normalized to 50 Ω). On a 10-inch diameter chart, this conveniently falls at the 4 cm hole.
2. Move "Wavelengths Toward Generator" zero to line up with 1.6 – *j*0.8.
3. Move the rolling ruler clockwise so that the 4 cm hole on the opposite side of the pivot point crosses the 1.0 resistance circle.
4. Read the answers.
   a. The necessary length of transmission line, 0.104 Ω toward the generator, is indicated directly on the wavelengths scale.
   b. The resulting susceptance component +*j*0.8 is read directly through the 4 cm hole where it crosses the

1.0 resistive circle. The resistive component at the input end of the line is 1.0, denormalized to 50 Ω.

Now we need to resonate the reactive component at the input of the transmission line. A parallel stub susceptance of B_p = $-j0.8$ (50) = $-j40$ will transform the admittance point Y_n to the center of the chart, where G = 1.0 and B = $j0$. The antenna will be matched as desired.

In Example 2 we used a short section of line to move an impedance or admittance point to a R = 0 or G = 1 line on the Smith Chart calculator. It is fairly easy to visualize the operation using a transmission line, since the movement is concentric with the center of the Smith Chart. However, when using a coil or a capacitor, a reactance of susceptance is involved and the move must be made along resistance/conductance circles. To simplify this operation using the new calculator we added the circle to the outermost chart before lamination. This is the same inverted constant-resistance circle R = 1 that would be seen using an overlay tracing box, as Caron describes in his book.

The following example illustrates the use of this added circle.

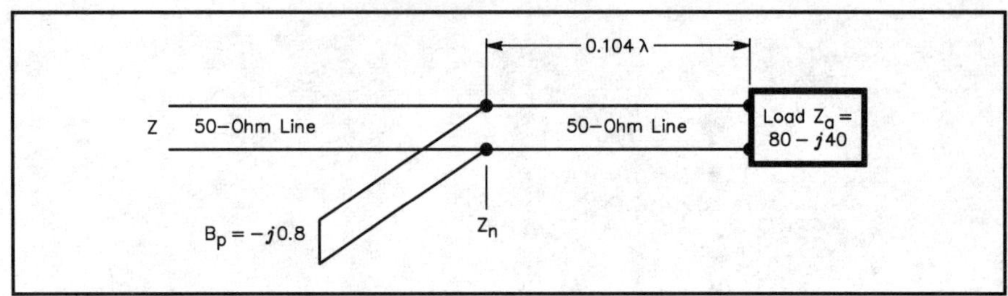

Fig 3—A parallel-connected stub with susceptance of – j0.8 and a short line 0.104 Ω long are used to match an impedance of 80 – j40 to 50 Ω.

Example 3: We measure 10 – j30 with our bridge located in the shack and want to design a simple L-network to give us a 50-Ω match.

1. Normalize 10 – j30 to 0.200 – j0.600 Ω.
2. Locate (without needing to plot it) 0.200 – j0.600 Ω. This falls conveniently at the 8 cm hole. Make a note of the 8 cm hole diametrically opposite the pivot point, at 0.5000 – j1.500.
3. Move this point by eye counter-clockwise along the 0.5 constant circle line until it intersects the "overlay" circle at 0.500 + j0.510 point, and note that this correction calls for –j0.990 inductive shunt susceptance.
4. Diametrically opposite the 0.500 + j0.510 point is the new impedance, which has been moved to 1.000 – j1.000. This point is easily seen as the intersection of the center line of the ruler and the R = 1 line. Note that a correction of +1.000 (series inductive reactance) is needed for a match. Also note that we did not have to rely on any hole notation in this last step.

This somewhat complicated problem has been solved with a simple subtraction, without plotting any points on the Smith Chart. I hope that the ease of use of this new calculator will inspire more hams to make use of the Smith Chart and take advantage of the excellent examples of its use in the book *Antenna Impedance Matching*.

Notes

[1]FXR, an early microwave company of the 1950s.

[2]W. F. Caron, *Antenna Impedance Matching* (Newington: ARRL, 1989). Charts are photo-expanded from chart toward end of book.

[3]See Reference 2, example p 4-8.

[4]See Reference 2, example p 5-13 to 5-15.

Determining Transmission-Line Voltage and Current from the Line Impedance

By Don Patterson, KK6JI
884 Rubis Drive
Sunnyvale, CA 94087

This project started when I decided to try to determine the maximum voltage and current on my tuned transmission line. I also wanted to calculate the voltage on the line at the tuner output, the power input to the antenna and the transmission-line losses.

The unexpected bonus from this study was that I could calculate and graph the voltage and current at any point along the line and readily visualize the standing-wave pattern introduced by a mismatched load. The effects of line loss on the standing-wave pattern can also be studied.

The impedance at any distance from the load can be calculated from the transmission line's characteristic impedance, velocity factor, attenuation and the impedance of the load. The power delivered to the load can be calculated by knowing the line's characterisitic impedance, total attenuation and the load impedance.

The forward power at any point on the line can be calculated by knowing the load power and the attenuation between the point and the load. The voltage and current at any point on the line can be calculated by knowing the forward power at that point and the impedance of the line.

Transmission Line Connected to a Mismatched Load

A transmission line of characteristic impedance Z_0 is driven by a tuner that provides a conjugate match[1] between the line and the transmitter. The line is connected to an antenna that has a complex load impedance, $R_L + jX_L$. **Fig 1** shows the configuration of such a system. The impedance looking toward the load at any point on the line can be calculated by[2,3]

$$Z_x = \frac{Z_0\left(Z_L + Z_0 \tanh(A_1 + jB_1)\right)}{Z_0 + Z_L \tanh(A_1 + jB_1)} = Z_{sr} + jZ_{si}$$

where

Z_x = impedance looking into the line toward the load

Z_{sr} = real or resistive component of Z_x

Z_{si} = reactive (imaginary) component of Z_x

Z_L = load impedance = $R_L + jX_L$.

A_1 = line attenuation in nepers = AT× DR

where

AT = matched-line attenuation, dB/100 feet, converted to nepers/meter by multiplying by (3.2808/100) × (Ln(10)/20)

DR = distance from the point to the load, meters

B_1 = line velocity coefficient = BP × DR, radians

where

BP = line phase coefficient, radians/meter = 2π × F / (300 × V)

F = frequency, in MHz

V = line velocity factor

As this equation was derived by considering the fields and attenuation along the line, it yields the same results as found with the Smith Chart[4] and may therefore be used to calculate all the characteristics of a transmission line, including the effects of line losses.

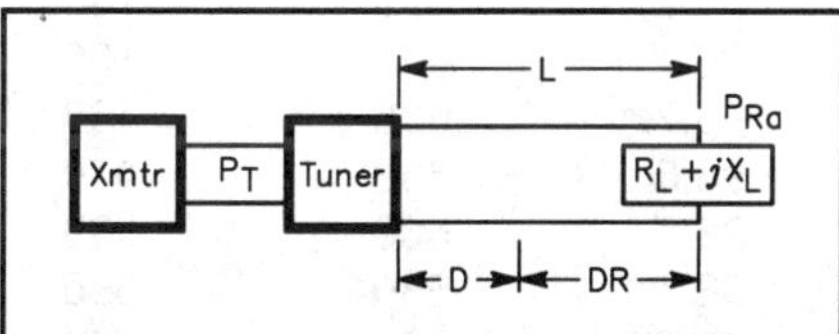

Fig 1—Block diagram of transmission-line system analyzed by KK6JI's program LNVI.BAS.

Voltage and Current at Any Point on the Line

In his classic paper, Byron Goodman, W1DX,[5] showed how to calculate the voltage and current along the transmission line. The power input to the antenna, P_{Ra}, is dissipated in the antenna's resistance. Therefore:

$$P_{Ra} = I_L^2 R_L$$

Thus, the load current is

$$I_L = \sqrt{P_{Ra} / R_L}$$

and the RMS load voltage is:

$$V_L = I_L \times |Z_L| = I_L \sqrt{R_L^2 + X_L^2}$$

The net forward power, P_1, transmitted past any point on the line is equal to the load power plus the power dissipated in the section of the line between the point and the load.

$$P_1 = P_{Ra} e^{2AT \times DR}$$

where $e^{2AT \times DR}$ is the line attenuation between the point and the load.

The line current I_1 and the line voltage V_1 can now be calculated, using the line power P_1 and the line impedance $Z_{sr} + jZ_{si}$ in the same manner that I calculated the load current and voltage.

$$I_1 = \sqrt{P_1 / Z_{sr}}$$

and

$$V_1 = I_1 \sqrt{Z_{Xr}^2 + Z_{si}^2}$$

These values can be plotted as a function of the distance from the load to give a visual presentation of the line voltage and current.

Power Delivered to the Load

When the line is driven by a tuner presenting a conjugate match to the line, and if we know the value of AT (the line attenuation factor in nepers/meter), L (the distance from the tuner to the load), P_T (the power delivered to the line from the tuner), and ρ (rho, the reflection coefficient at the load), then we can calculate P_{Ra} (the power dissipated in the load).

$$P_{Ra} = P_T \frac{(1-\rho^2)\,e^{-2AT \times DR}}{1-\rho^2\,e^{-4AT \times DR}}$$

The reflected energy, attenuated by traveling twice down the line, is combined with the power P_T at the output of the tuner and then is attenuated as it travels to the load. The load accepts the fraction $\rho^2/(1-\rho^2)$ of the power arriving at the load end of the line and reflects the remainder back to the tuner output.[6]

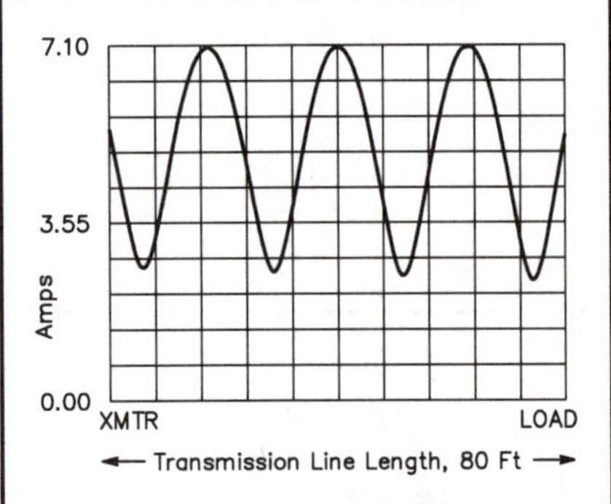

Fig 2—Example of current along RG-8 foam transmission line, which is 80 feet long at 14.25 MHz and terminated in load impedance of 30 +*j*40 Ω. Note that minimum level of current decreases from the load due to attenuation in line. This reduces the SWR at the input of the line compared to the load end.

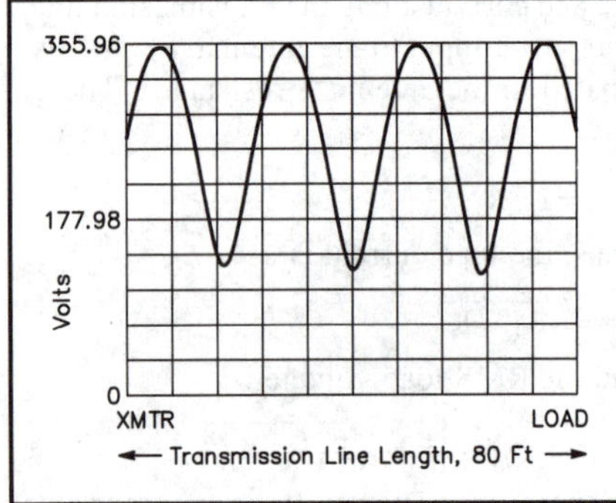

Fig 3—Example of voltage along same line as in Fig 2.

Standing Wave Ratio and Reflection Coefficient

The complex impedance of the antenna input ($R_L + jX_L$) can be determined by direct measurement[7] or calculated by an antenna modeling program. The standing wave ratio can be calculated from the equations in *The ARRL Antenna Book*.[8]

$$SWR = \frac{A+B}{A-B}$$

where

$$A = \sqrt{R_L^2 + Z_0^2 + X_L^2}$$

$$B = \sqrt{(R_L - Z_0)^2 + X_L^2}$$

The reflection coefficient can be calculated from *The ARRL Antenna Book*.[9]

$$\rho = \frac{SWR-1}{SWR+1}, \text{ load reflection coefficient.}$$

The matched-line attenuation coefficient, in dB/100 feet, can be found in the cable data in *The ARRL Antenna Book*[10] or from specifications from cable manufacturers, as can the velocity factor for a particular cable. The matched-line attenuation coefficient expressed in nepers/meter rather than dB/100 feet is:

$AT = (dB/100 \times 3.2808) \times (Ln(10)/20)$ nepers/meter.

Numerical values using the above equations are calculated in a BASIC computer program, LNVI.BAS. It prompts you to enter the required data, then calculates and displays the desired output data. For example, if the antenna input impedance is 30 + *j*40 Ω, fed by 80 feet of RG-8 foam cable, and the output power at the output of the antenna tuner is 1000 W at 14.25 MHz, the inputs to the program are:

CHARACTERISTIC IMPEDANCE, OHMS:	50
LINE ATTENUATION, dB/100 foot:	.57
LINE VELOCITY FACTOR	.66
LENGTH OF LINE, FEET	80
SERIES RESISTANCE COMPONENT OF THE LOAD, OHMS	30
SERIES REACTIVE COMPONENT OF THE LOAD, OHMS	40
OPERATING FREQUENCY, MHz	14.25
POWER INPUT TO THE LINE FROM TUNER, WATTS	1000

The program calculates and displays the following:

POWER INTO THE LOAD, WATTS	846.858
POWER DISSIPATED IN THE LINE	153.142
LOAD STANDING WAVE RATIO, SWR	3
MAX VOLTAGE	355.9617
MINIMUM VOLTAGE	121.9312
MAX CURRENT	7.103488
MIN CURRENT	2.397733
MAX V/MIN V	2.919365
MAX I/MIN I	2.962585

The program prompts you to graph the current or voltage along the line. A plot of the current is shown in **Fig 2** and the voltage waveform is shown in **Fig 3**.

Off-Center Fed Horizontal Antenna

The program can be used to calculate some of the parameters of a proposed antenna system, such as:

1. What is the power input to the antenna when the power output of the tuner is 100 W?
2. What is the maximum transmission-line voltage?
3. What is the voltage across the tuner output?

For example, I determined that I could string #16 wire 36 feet high between two short masts mounted on our garage and roof top. A convenient feed-line connection could be made 22 feet from one end. The feed line needed to be 50 feet long, as it had

Table 1
Example of 68-Foot Long Dipole, 36 Feet High, Fed 22 Feet from One End with 50 Feet of 600-Ω Open-Wire Line with 100 W

Freq MHz	*Line Loss dB/100'*	*Antenna Imped, Ω*	*SWR*	*Antenna Power, W*	*Max Line Voltage, V*	*Tuner Output Voltage, V*
28.50	0.100	156–*j*63	3.89	97.7	477	143
24.94	0.093	377–*j*1095	7.40	96.1	652	130
21.22	0.085	3169–*j*1668	6.79	96.7	625	468
18.18	0.080	588–*j*1351	7.03	96.8	639	479
14.20	0.070	121.5–*j*36.5	4.96	98.7	539	539
10.13	0.057	2757+*j*514	4.76	98.4	530	530
7.15	0.050	132+*j*65.6	4.60	96.6	521	344
3.75	0.030	11.6–*j*1117	231	71.5	2768	387
1.90	0.023	2.11–*j*2677	5944	11.4	6174	1420

to be routed through the attic into a second-story room. I wanted to operate on as many of the HF bands as possible.

To minimize feed-line losses I chose to use a 600-Ω open-wire line with a velocity factor of 0.975. I found the characteristics of this line in the 1949 *ARRL Antenna Book*. The antenna input impedance was calculated with an antenna modeling program.[11] The resistive parts of the impedances for 80 and 160 meters were corrected for errors introduced by the modeling program because the antenna height was less than 0.2 wavelength.[12]

Table 1 shows the results of this study. Feed-line losses are less than 0.2 dB on all bands between 40 and 10 meters. The losses rise to a usable 1.5 dB on 80 meters, and as expected are more than 9 dB on 160 meters.

The radiation pattern of this antenna, including the effects on unbalanced feed-line current, is a subject for future study.

Acknowledgment

I want to thank Bill Comer, K6JV, for his many helpful suggestions, and Dr Don Pettibone for reviewing this paper.

Notes and References

[1] *The ARRL Handbook* (Newington, CT: The American Radio Relay League, 1989-1994), p 16-11.

[2] Ramo, Whinnerey and Van Duser, *Fields and Waves in Communications Electronics* (John Wiley and Sons, 1967), p 46.

[3] John D. Kraus, *Antennas* (New York: McGraw-Hill, 1988), p 848.

[4] *The ARRL Antenna Book*, 15th edition (Newington, CT: The American Radio Relay League, 1988), Chapter 28.

[5] Byron Goodman, W1DX, "My Feedline Tunes My Antenna," Mar 1956 *QST*, Apr 1977 *QST* reprint.

[6] M. Walter Maxwell, W2DU, *Reflections* (Newington, CT: The American Radio Relay League, 1990), p 23-8.

[7] Reference 6, p 15-2.

[8] Reference 4, p 24-9.

[9] Reference 4, p 24-10.

[10] Reference 4, pp 24-18 to 24-19.

[11] Brian Beezley, K6STI, *MN* and *YO* antenna programs, 3532 Linda Vista Dr, San Marcos, CA 92069.

[12] John Belrose, VE2CV, "Modeling HF Antennas with MININEC," *The ARRL Antenna Compendium, Vol 3*, (Newington, CT, 1992) p 156.

VHF/UHF Antennas

A Vertical 6-Meter Wire Extended Double Zepp

By Wayde S. Bartholomew, WA3WMG
RD3 Box 3769
Pottsville, PA 17901

WA3WMG describes his simple, low-cost but very effective 6-meter wire antenna for repeater coverage.

I needed a low-cost gain antenna with vertical polarization for 52.525 MHz. The concept of an Extended Double Zepp looked interesting. I constructed a mock-up antenna at my QTH to compare it with a quarter-wave vertical at a height of 50 feet. The Extended-Double Zepp typically outperformed the vertical by two to three S units for both receive and transmit. This was with a power output of 25 W, talking with stations 30 to 100 miles distant. This encouraged me to construct the final version, which was placed at the 100-foot level on a commercial tower 1800 feet above sea level.

Construction is fairly straightforward. See **Fig 1**. The upper and lower legs of the dipole sections are supported by 1½-inch PVC pipe. The 1:1 balun and matching section are placed inside another piece of 1½-inch PVC pipe for support and weather protection.

Spacing from the tower has a definite effect on directivity. **Fig 2** shows the effect on the azimuthal and elevation patterns of spacing the antenna 5 feet away from a tower, compared to a 10-foot spacing from the tower. In this case the bottom of the antenna is at the 50-foot level on a 100 foot high tower. The front-to-back ratio is about 9 dB for the 5-foot spacing.

Although it is still distorted, the pattern for the 10-foot spacing is more omnidirectional in nature than the 5-foot spacing. Of course, the larger spacing does present more of a construction problem when using PVC pipe.

Tuning of the antenna is simple. The legs are temporarily stretched out on the tower with the bottom of the antenna a few feet off the ground. The ladder line is purposely made a little long at installation (3 feet is a good starting point), and then trimmed a little at a time for the lowest SWR. This should be close to 1:1.

Once the feed line is trimmed, the ends of the PVC pipe should be sealed with RTV sealant. A small hole should be drilled in the bottom of the PVC to drain off any condensation that may form inside. A small screen should be placed over the drain hole to keep out spiders and wasps—they seem to like antennas for homes!

Results have been impressive. During band openings the antenna is an excellent performer, and during marginal band conditions it has the gain to still be effective. In three months time from November 1991 to January 1992, 25 states were worked. Three other Extended Double Zepps were constructed for repeater use, with excellent results.

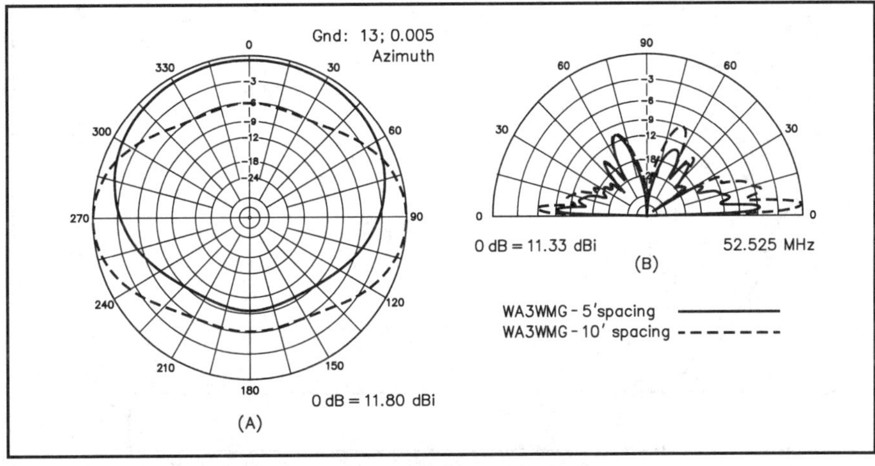

Fig 2—Patterns for vertical 6-meter EDZ mounted parallel to metal tower. At A is a comparison of computed azimuthal patterns for 5- and 10-foot spacings from tower; at B is a comparison of elevation patterns. The larger spacing yields coverage that is more omnidirectional.

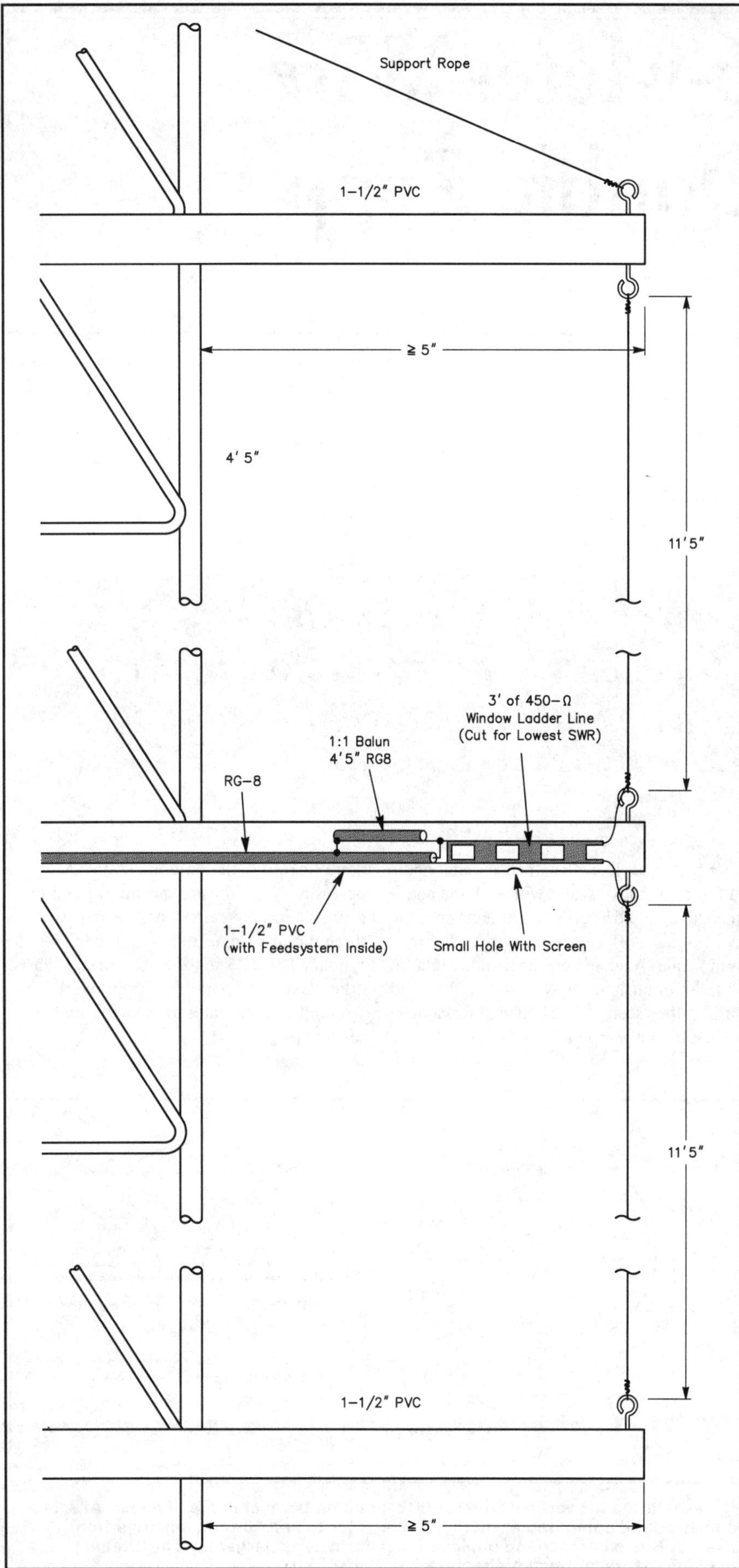

Fig 1—Mounting dimensions for 6-meter Extended Double Zepp antenna. The 450-Ω "window" ladder line is cut long initially at 3 feet and trimmed for best SWR at installation. The feed system, including the 1:1 coax balun and window line, are placed inside 1½-inch PVC pipe for support and weather protection.

An (Almost) Invisible, (Almost) No Mounting Holes 2-Meter Mobile Antenna

By David B. Brown, K8AX
5538 Naiche Road
Columbus, OH 43213

Ever lose your mobile 2-meter rig to thieves? K8AX has a solution to throw them off the track.

Every ham has memories of radio experiences that will last forever—that very first contact, a home brew transmitter that actually worked, five band WAS; the list goes on and on. My list contains one experience that stands apart from all others. If you haven't had a similar experience, you most likely know someone who has.

After an enjoyable evening of bowling, I returned to my car to find that the front door was unlocked. I never leave the car unlocked, especially in places like dimly lit parking lots. A second later, it hit me like a ton of bricks—my 2-meter transceiver was gone! Never have I felt so violated. How did the thief know I had that rig? Did he know me personally? Days later, after regaining my composure, I surmised that the trunk-mounted $^5/_8$-λ antenna must have been the giveaway.

That experience profoundly affected my mobile operating. After several months of not having a rig in the car, an old crystal-controlled transceiver was installed out of sight under the front seat. That got me back on the air, but was inconvenient and inflexible. Eventually, a new synthesized transceiver was purchased and mounted back under the dash. However, being the paranoid person that I had become, the rig was rarely left unattended in its mounting bracket. Although inconvenient, it was usually stowed in the trunk or carried into the house when not in use.

In fact, having to remove the rig evolved into the practice of simply leaving it at home most of the time. So there I was, mobiling again, but usually without a 2-meter rig. Finally, the antenna described in this article was conceived. With an (almost) invisible antenna that doesn't advertise the presence of a radio, the rig was once again permanently mounted under the dash.

The (almost) invisible antenna is simply a λ/4 whip made from a piece of 0.032 inch diameter steel piano wire purchased at a local hobby shop. Piano wire is available in diameters of 0.025, 0.032 and 0.047 inch. I found the 0.025 inch diameter wire too flimsy and had a hard time choosing between 0.032 and 0.047. The 0.032 is easier to work with and somewhat less visible than the 0.047. However, 0.047 wire will stay upright a little better at 55 MPH.

The piano-wire whip is attached to a banana plug that plugs into a banana jack mounted to the car body just under the trunk lid. Two fiber washers insulate the banana jack from the car body. In **Fig 1**, my daughter Emily poses just behind the antenna. While the antenna is not invisible, it is hardly noticeable at a quick glance.

Fig 1—K8AX's daughter poses just behind the (almost) invisible 2-meter mobile antenna.

Fig 2 shows a close-up view of the antenna

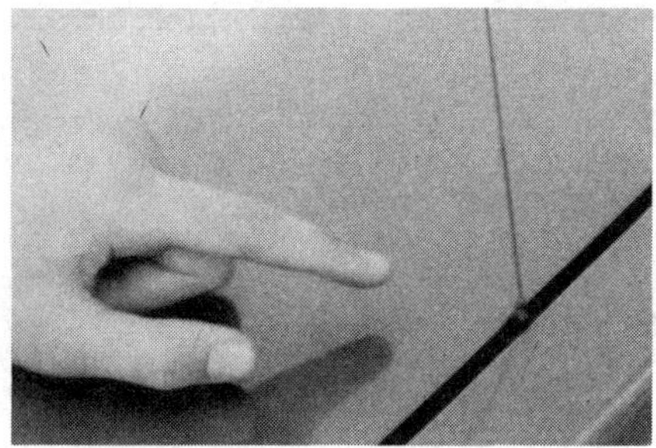

Fig 2—View of the antenna emerging from the gap between the trunk lid and fender.

emerging from the gap between the trunk lid and fender. **Fig 3** reveals the antenna mounting after opening the trunk. Two small holes are drilled in the car body under the trunk lid. The larger of the two holes mounts the banana jack, while the smaller one grounds the outer shield of the coaxial feed line. These out-of-sight holes will certainly not affect the car's resale value. Note also from Fig 3 that the piano wire is bent to match the car-body contour between the jack and fender top. The antenna measures 19 inches from the point where it emerges out of the trunk to its tip. A loop, approximately 3/8-inch in diameter, is formed at the tip to eliminate the danger of being poked by the sharp wire.

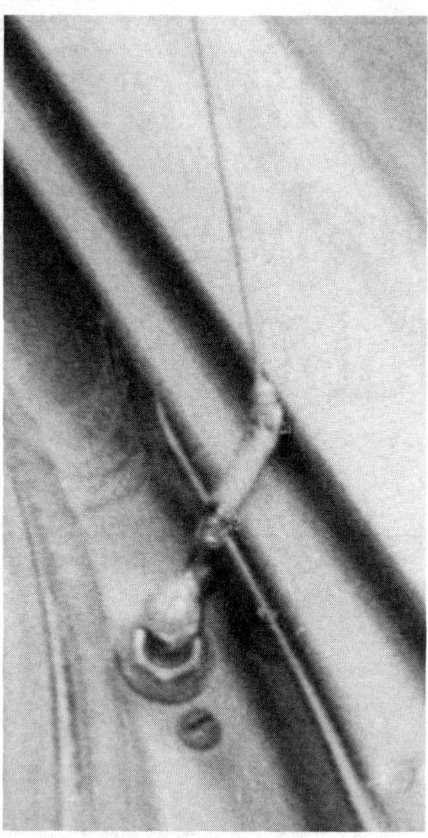

Fig 3—View with trunk lid open, showing banana-jack mounting.

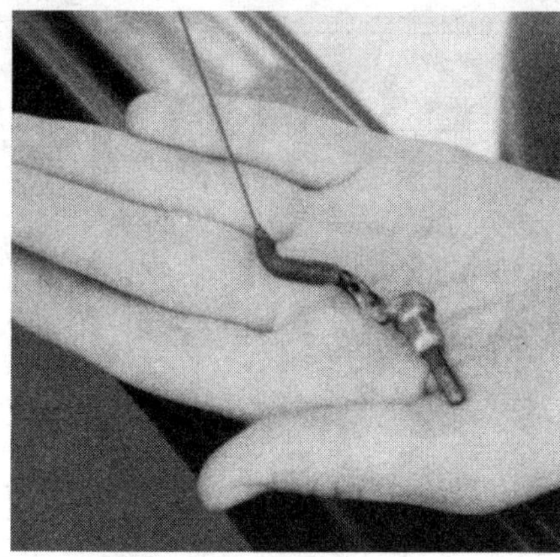

Fig 4—Here, the antenna has been removed from the banana jack to show the shaping of the piano wire and insulating details.

The antenna is removed from its jack in **Fig 4**. Here you can see that the piano wire has been thickened with plastic electrical tape and heat shrink tubing between the top of the banana plug and the point where it emerges from the trunk. In turn, this is sealed with a coating of clear silicon rubber. This is an important step to prevent rust and subsequent breakage where the whip attaches to the banana plug. Not only does this provide the necessary insulation between the wire and the car body, it also makes for a snug fit between the trunk lid and fender. Finally, the antenna is coated with gray Krylon spray enamel to retard rust. In this case, the color matches that of the gray Honda.

This antenna has been in use for approximately six years and is all that is needed to operate any of the repeaters here in Columbus, Ohio. If everything is purchased new, the total cost is less than $12. **Table 1** lists the parts list and estimated prices. A good source for the banana plug/jack components is an old plug-in antenna tuner coil assembly, readily available at hamfest flea markets.

This antenna really is invisible in dimly lit bowling alley parking lots. Try it—it's cheap; it works well and it is a significant step towards reducing the possibility of mobile radio theft.

Table 1
Parts List for the (Almost) Invisible 2-Meter Mobile Antenna

Item	Price
36 inch length 0.032 inch diameter steel piano wire	$ .15
Banana plug	.40
Banana jack	.40
2 fiber washers	.20
6-32 screw/nut/ground lug	.06
10 feet RG-58C coaxial feed line	1.70
PL-259 connector with reducing adapter	1.80
Clear Silicon Rubber Sealant (Dow Corning or GE)	3.95
Krylon spray enamel	2.67
Total:	$11.33

Vertical Extended Double Zepp for 2 Meters

By Andrew S. Griffith, W4ULD
203 Lord Granville Drive, Route 2
Morehead City, NC 28557

Here is a vertically polarized 2-meter gain antenna, designed to be hung unobtrusively in the trees and constructed like a battleship!

Two important factors needed to achieve wide coverage with a 2-meter vertical antenna are height and gain. However, many of us who live in restricted neighborhoods cannot achieve height with masts and towers but we can hide very effective antennas in tall trees, such as pines. The 2-meter antenna described here has gain and is designed to be suspended from a tree limb. However, the principles employed could be applied to a free-standing system. The calculated gain[1] of the antenna is about 2.6 dB over a vertical dipole.

The antenna is constructed mainly from copper pipe, brass "hobby" tubing, and PVC pipe, all of which are readily available at reasonable cost. Construction is quite simple and requires an electric drill with a sanding disk, a hacksaw, and a propane torch in addition to common tools found around the ham shack. The total cost should be about $20.

Description

The antenna is an extended double Zepp fed in the center with a transmission line formed by the lower element. The total length is about 12 feet. See **Fig 1A**. The radiating portion of the antenna consists of two collinear elements that are 48 inches long, or about 0.65 λ long electrically. The top and lower elements are constructed from nominal 1/2-inch copper pipe (5/8-inch OD) with 0.025 inch wall thickness. The lower element is actually 85 inches long but is decoupled from the bottom part of the antenna by a λ/4 open-ended sleeve. The open end of the sleeve is located 48 inches from the top of the lower element. The sleeve is made from nominal 1-inch copper pipe (1 1/8 inch OD). The lower element extends through the sleeve to a matching section at the bottom of the antenna. A #12 bare copper wire runs through the center of the lower element, forming a 117-Ω low-loss transmission line. The wire is centered in the lower element by plastic spacers every 6 1/16 inches along the wire. The stub matching section consists of an adjustable length of transmission line and an adjustable parallel stub fabricated from a copper tee, copper pipe, and hobby brass.

The length of the transmission line from the feedpoint (between the top and lower elements) is 1 λ electrically, plus a stub matching section[2] to transform the feedpoint impedance of about $106 - j323$ Ω to 50 Ω. The inductive reactance of the closed-end stub is about 27 Ω.

The center and end (top) insulators are made from 1/2-inch PVC water pipe.

Construction Notes

Construction of this antenna involves cutting metal and heating components to high temperature for soldering. During these operations you should wear leather gloves and eye protection.

Hard drawn 1/2-inch copper pipe or tubing is usually available with a wall thickness of 0.025 or 0.035 inch. It is very important that 0.025 inch thick pipe be used for the lower element, because the inside diameter of the element determines the impedance of the transmission line formed by the bottom element and also establishes the length required for matching to 50 Ω. Further, the thin wall pipe is much cheaper and lighter.

As shown in Fig 1A the lower element is 85 inches long. The ends should be reamed and dressed to remove any burrs. The #12 bare copper wire through the center of the element should be cut about 90 inches long to leave some to work with on each end. The wire should be straightened by pulling it between a couple of clamped wooden blocks. The spacers used by the author to support the center wire were made from Lucite rod dressed down to about 9/16 inch diameter and sawed to 3/16 inch thickness. The center was drilled 5/64 inch diameter to make a tight fit over the #12 wire. The spacers are placed each 6 1/16 inches along

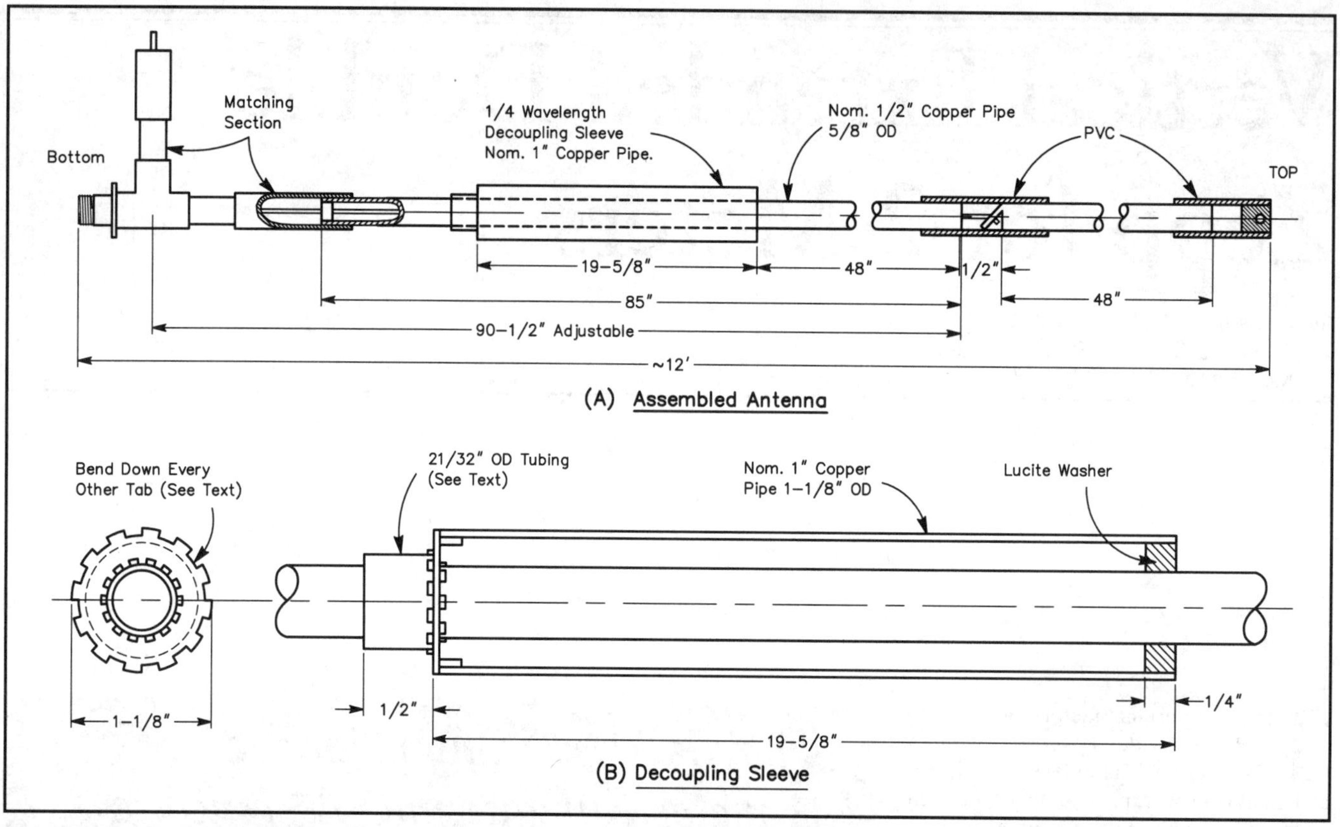

Fig 1—At A, drawing of W4ULD antenna. At B, details of decoupling sleeve.

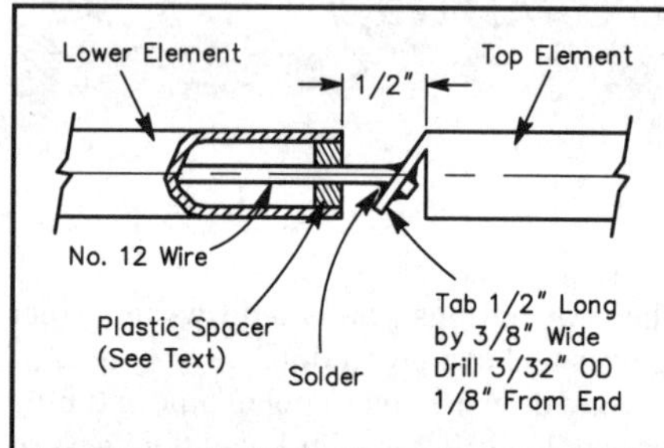

Fig 3—Details of feedpoint for W4ULD antenna.

the wire with a spacer just inside each end of the element. Fifteen spacers are required. The author later found that cheap and easy spacers could be made from six-pointed plastic stars purchased in the sewing section of a local department store. The stars are about 3/4 inch diameter and 0.3 inch thick, with a 0.07 inch diameter center hole. Drill the stars with a 5/64 inch bit and sand them down to 9/16 inch diameter. Cement the spacers in place along the wire with epoxy or household cement. When the cement has set, slide the wire and spacers into the lower element.

The top element has a 1/2 × 3/8 inch tab on its lower end, as shown in **Fig 3**. The tab was sawed out with a hacksaw and drilled 3/32 inch to accommodate the #12 wire from the lower element.

The decoupling sleeve is shown in Fig 1B. It is fabricated from nominal 1-inch copper pipe (1 1/8 inch OD). Lay out the plastic washer with a compass. Drill or drill and file the inner hole to 5/8 inch ID so that it will fit tightly over the bottom element. Drilling a large hole in Lucite requires slow drill speed and extreme care to prevent a catastrophic breakthrough. Either start with a 1/4-inch hole and gradually increase the bit size to 5/8 inch, or file the hole with a round file. Finally, cut out the washer and sand it down until it slides easily into the decoupling sleeve. The end cap at the bottom of the decoupling sleeve is made of copper (0.020 inch thick common) or hobby brass (0.015 inch thick) sheet. Lay out the end cap with a compass. When drilling thin stock, wear gloves and hold the stock with pliers. Drill and file the center hole until a tight fit is obtained over 21/32-inch diameter brass hobby tubing. Cut out the end cap and make about 24 1/16-inch slits around the circumference with tin snips. Cut 20 1/8-inch slits in the end of a piece of 21/32-inch tubing. Place the tubing over a piece of 1/2-inch pipe and bend every other tab outward. The pipe keeps the tubing from collapsing below 5/8 inch ID.

Remove the pipe and slightly bend the remaining tabs inward. Slide the end cap over these tabs, restraighten the tabs and replace the pipe. Bend the straight tabs outward locking the end cap in place. Squeeze the inner and outer tabs together with needle nose pliers for a rigid mechanical connection and solder the connection. Cut off the 21/32-inch tubing about 1/2 inch from the connection. Bend every other tab on the outside rim of the end cap so that the bent tabs fit snugly into the end of the decoupling sleeve. Clean the end cap and the end of the sleeve with fine steel wool and solder the end cap onto the sleeve using a propane torch. Leave a couple of small areas unsoldered so that moisture can escape from the sleeve.

The matching section is shown in **Fig 2**. Follow the dimensions carefully. The first step is to saw four 5/16-inch long slits 90° apart in one end of the 1/2-inch copper tee for later clamping the assembly together. If an SO-239 connector is available with hard brown or yellow insulation, the SO-239 can be later soldered directly to the tee. If not, the next step is to make a mounting flange for the connector. Drill and file a hole in a piece of copper or brass sheet to fit snugly over the end of the tee opposite the slits. Cut the stock to 1 inch square as shown in Fig 2 and slide it over the tee. Solder it flush with the end of the tee. Place the SO-239 in the end of the tee and clamp it in place even with the edges of the flange. Drill through one hole in the SO-239 and through the flange with a 1/8-inch bit, using a small block of wood to back up the flange. Screw the

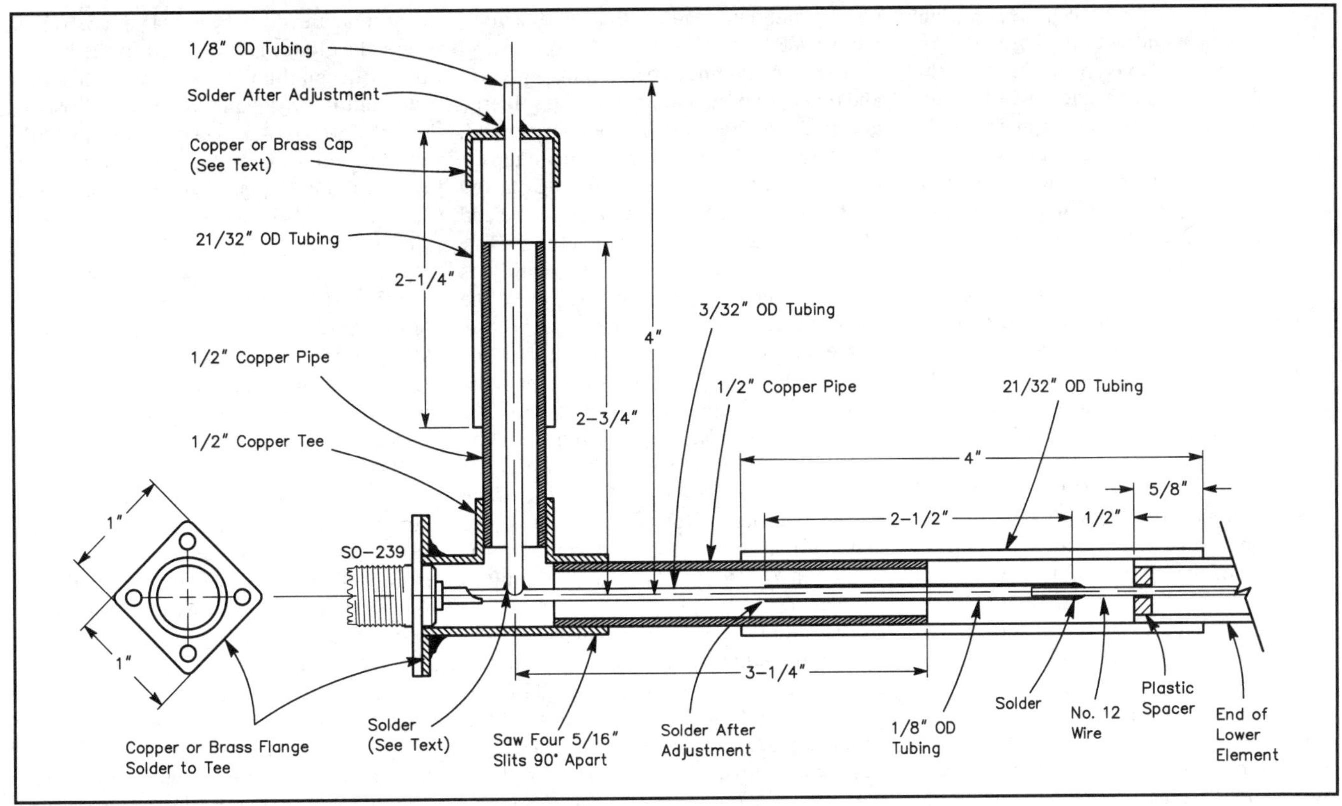

Fig 2—Details of matching section for W4ULD antenna.

SO-239 to the flange with a 4-40 screw and nut. Drill the other three holes.

Prepare the two lengths of $^1/_2$-inch pipe that go in the end and side outlet of the tee. Since tees differ, measure from the center of the tee as shown in the drawing. Solder a length of $^3/_{32}$-inch diameter brass tubing to the center conductor of the SO-239 making sure that it is straight in all directions. Insert the proper piece of pipe into the slotted end of the tee and clamp with a $^3/_4$-inch hose clamp. Insert the SO-239 in the other end. Cut off the end of the $^3/_{32}$-inch tubing flush with the end of the pipe and dress the end of the tubing. Cut a 4 inch length of $^1/_8$-inch brass tubing and file a saddle in one end. With the SO-239 and its tubing in place mark where the tubing is in the center of the side outlet with a pencil. Remove the SO-239 and heavily tin the center mark on the tubing with solder. Heavily tin the saddle end of the $^1/_8$-in tubing. Remount the SO-239 and lock the tubing in the center of the $^1/_2$-inch pipe. One of the plastic stars drilled with a $^3/_{32}$-inch bit will work nicely. With the solder gun passing through the side outlet of the tee solder the $^1/_8$-inch tubing to the $^3/_{32}$-inch tubing. When the solder melts, the saddle will help hold the tubing in place. Make sure the $^1/_8$-inch tubing is straight and in the center of the side outlet. Tin one end of the side outlet pipe and force it into the tee. If you choose to solder the SO-239 to the tee, do so at this point. Leave a crack for moisture to escape.

Fabricate the adjustable end section for the stub as shown in Fig 2 from $^{21}/_{32}$-inch diameter tubing. Make the end cap from brass or copper sheet. Drill the hole in the center with a $^7/_{64}$-inch bit and swage it with a nail set for a snug fit over the $^1/_8$-inch tubing. Make tabs around the circumference of the cap and bend them over the $^{21}/_{32}$-inch tubing. Solder the cap to the $^{21}/_{32}$-inch tubing. Assemble the stub. The adjustable section should slide easily but snugly in and out.

Cut and dress the 4 inch length of $^{21}/_{32}$-inch diameter tubing and the $2^1/_2$-inch length of $^1/_8$-inch diameter tubing for the adjustable main line section.

The 5 inch long center insulator, made from $^1/_2$-inch PVC water pipe, must slide easily over the top and lower elements so it must be either reamed or bored to just over $^5/_8$-inch ID, or else it must be slit its length with a hacksaw. If it is reamed, four slits 90° apart and $^5/_8$ inch long must be sawed in each end of the insulator for later clamping to the top and lower elements. If it is slit end to end, the end slits are not needed. The longitudinal slit will be sealed later with silicone caulking.

The 5 inch long top insulator is also made from $^1/_2$-inch PVC pipe. Sand down a $^5/_8$ inch diameter wooden dowel or piece of wood until it can be driven $1^1/_2$ inches into the end of the insulator. A watertight fit is required. Drill a $^1/_4$ inch hole through the PVC and wooden plug about $^3/_4$ inches from the end for the supporting rope. Saw four slits $^5/_8$ inch long and 90° apart in the other end of the insulator. File a slight bevel on the top end of the top element and drive the end insulator onto the top element until the element is at least 1 inch into the insulator. Clamp the insulator to the element with a 1 inch "all stainless" hose clamp.

Assembly

The components should be placed on a long flat surface. A picnic table will do but a pair of picnic benches end-to-end is better, since the entire antenna can be supported. Using an awl or penknife, lightly scribe a mark on the lower element exactly 48 inches from the top of the element. Slide the center insulator and two 1-inch hose clamps (all stainless) over the lower element. Cut off the wire in the top of the lower element to within 1 inch of the element. Make sure the plastic spacer is just inside the end of the element. Pass the wire through the hole in the tab on the top element. Manipulate the two elements (let the wire bend) until they are aligned and $^1/_2$ inch apart. Solder the wire in the tab and snip off any excess beyond $^1/_8$ inch as shown in Fig 3. Slide the center insulator over the two elements until it is centered over the feedpoint and clamp each end with the hose clamps.

Slide the plastic washer for the decoupling sleeve over the lower element until the edge of the washer matches the scribe mark on the element. Slide the decoupling sleeve onto the lower element until the open lip of the sleeve is also on the scribe mark and even with the top edge of the washer. Clamp the sleeve in place with a $^{3}/_{4}$-inch hose clamp over the $^{21}/_{32}$-inch nub on the bottom end cap.

Cut off the wire coming from the bottom of the lower element to $^{7}/_{8}$ inches from the end of the element. Make sure the plastic spacer is just inside the end of the element. Tin the end of the wire to within $^{1}/_{2}$ inch of the element. Slide on the $2^{1}/_{2}$ inch length of $^{1}/_{8}$-inch tubing and solder it $^{1}/_{2}$ inch from the end of the lower element. Slightly crimp the other end of the tubing so that it will slide snugly over $^{3}/_{32}$-inch tubing. Scribe a mark $^{5}/_{8}$ inches from the bottom of the lower element. Slide two $^{3}/_{4}$-inch hose clamps over the bottom of the lower element. Slide the 4 inch length of $^{21}/_{32}$-inch tubing over the bottom element until about 1 inch of the $^{1}/_{8}$-inch tubing can be seen. Insert the $^{3}/_{32}$-inch tubing in the stub matching section into the $^{1}/_{8}$-inch tubing. Slide the $^{21}/_{32}$-inch tubing down and over the pipe in the end of the tee. Adjust the $^{21}/_{32}$-inch tubing until the top edge is even with the scribe mark on the lower element. Slide the first hose clamp down and leave loose. Tighten the second clamp around the top of the $^{21}/_{32}$-inch tubing. The matching section should slide snugly up and down. Lightly tighten a hose clamp on the bottom of the $^{21}/_{32}$-inch tubing.

The antenna is now ready to be suspended and tested. Suspend the antenna so that the stub is about eye level.

Tuning and Final Assembly

Set the length of the matching section to $6^{1}/_{8}$ inches from the top of the 4-inch section of $^{21}/_{32}$-inch tubing (scribe mark) to the center of the tee. Adjust the stub to $3^{5}/_{8}$ inch from the center of the tee. Measure the resonant frequency and SWR. This can best be done at the antenna with an instrument such as the MFJ-249 Antenna Analyzer attached to the antenna with about two feet of coax. The second best is with a hand-held radio and an SWR meter. Sweep the band to determine the frequency of minimum SWR. The SWR should be quite low. At the resonant frequency adjust the stub length for minimum SWR, which should be close to 1:1.

The adjustments in the matching section should allow setting the resonant frequency from about 145 to 147 MHz. The antenna's 2:1 SWR bandwidth will be about ± 1 MHz. To decrease the frequency, lengthen the main line portion of the matching section. Set the antenna to the desired center frequency and then readjust the stub for minimum SWR.

After the antenna is performing properly and tuned, two final joints must be soldered and the antenna must be weatherproofed. Measure the distance from the end of the tee to the top of the $^{21}/_{32}$-inch tubing (at the scribe mark). Loosen the clamps on the end of the tee and on the $^{21}/_{32}$-inch tubing. Slide the tubing up the lower element until it is above the scribe mark. Slide the pipe in the end of the tee up until it touches the bottom of the bottom element. The junction of the $^{3}/_{32}$-inch and $^{1}/_{8}$-inch tubings should be at the mouth of the pipe. Carefully readjust (if necessary) the matching section until the distance from the scribe mark to the top of the tee is correct. Solder the junction of the two small tubings. Replace the $^{1}/_{2}$-inch pipe and $^{21}/_{32}$-inch tubing to their original positions. Finally, solder the $^{1}/_{8}$-inch tubing to the end cap on the stub.

The author sealed the joints in the antenna by wrapping them with vinyl electrical tape and then covering the tape with a thin coat of silicone caulking. The caulking should be spread to overlap the edges of the tape. Remove the hose clamps from the following joints: the top of the $^{21}/_{32}$-inch tubing at the bottom of the lower element, the bottom of the $^{21}/_{32}$-inch tubing, and the top of the tee. Seal these joints as described above as well as the joint where the stub enters the tee and the joint on the stub between the $^{21}/_{32}$-inch end piece and the $^{1}/_{2}$-inch pipe. The latter joints do not require a clamp since there is no stress on them.

If the center insulator was slit end-to-end, remove the hose clamps and completely seal the slit with silicone caulking. Finally, caulk the washer at the top of the decoupling sleeve. When the silicone has set, replace the hose clamps. The joint where the SO-239 mounts to the tee should not be sealed. The author sprayed his antenna with clear acrylic lacquer to help reduce corrosion.

Evaluation

The antenna performed as expected. With the top of the antenna at about 35 feet the author got a half scale reading on his ICOM 228H from a repeater 36 miles away running 25 W at 280 feet. This repeater could not be heard with a $^{5}/_{8}$-λ groundplane at 10 feet. With the antenna centered at 146 MHz, the 2:1 SWR bandwidth at the antenna with an MFJ-249 analyzer was 145.05 to 147.1 MHz. In the shack, with 78 feet of RG-213 plus one connector, the bandwidth was 144 to 147.8 MHz, according to a Bird 43 wattmeter. As predicted from the design calculations the SWR is not affected appreciably by antenna height between 12 and 35 feet at the top.

It is probable that a slight increase in gain can be achieved if four λ/4 radials (about 19 inches long) are attached to the bottom of the decoupling sleeve. The author tried this and the resonant frequency increased about 1 MHz. A rework of the matching section would have been required so the author did not pursue this refinement. Perhaps some of the experimenters who read this article can work on this.

Material Sources

Purchase a 20 foot piece of $^{1}/_{2}$-inch straight copper pipe with 0.025 inch wall thickness from a plumbing supply. The cost will be $8 to $9. Only about 12 feet are needed but this amount in a hardware store will cost more that $9.

A 20 foot piece of 1-inch copper pipe will cost about $21 at a plumbing supply. Unless several antennas will be made, the approximately 20 inches needed for the decoupling sleeve should be purchased at a hardware store or building supply for about $2.75 per foot or a total of about $4.50. The $^{1}/_{2}$-inch copper tee will cost about 75 cents from a hardware store or building supply.

The brass hobby tubing can be purchased from a hobby shop in 12 inch lengths. One length each of the following is needed: $^{3}/_{32}$-inch OD (about 40 cents), $^{1}/_{8}$-inch OD (about 50 cents), and $^{21}/_{32}$-inch OD (about $1.75).

Sheet copper (0.020 inch thick) can be purchased from a sheet metal shop. A 3 inch square scrap should cost less than $1. Sheet copper and brass (0.015 inch thick) are available at hobby shops for about $1.25 per small sheet.

Lucite or Plexiglas sheet ($^{3}/_{16}$ to $^{1}/_{4}$ inch thick) can be found at a glazier or hobby shop at close to $5 per square foot. Only a $1^{1}/_{4}$ inch diameter piece is needed, so it may be possible to pick up a scrap for practically nothing. Broken golf cart windshields are an excellent source.

Plastic stars for spacers in the transmission line can be purchased from the sewing section of department stores for about 50 cents for 25 stars (see text for description). A 10-inch length of $^{1}/_{2}$-inch PVC pipe should cost no more than $1 from a hardware store.

Three 1 inch and four $^{3}/_{4}$ inch hose clamps will cost about 75 cents apiece at a hardware store. The clamps should be marked "all stainless." Otherwise, "stainless" clamps are likely to have a steel screw, which will quickly rust.

References

[1]ELNEC, antenna analysis program by Roy Lewallen, W7EL, PO Box 6658, Beaverton, OR 97007.

[2]*The ARRL Antenna Book*, 15th Edition (Newington, CT: 1988), p 28-10.

A True Plumber's Delight for 2 Meters—An All-Copper J-Pole

By Michael P. Hood, KD8JB
2323 Jefferson Drive SE
Grand Rapids, MI 49507-3148

KD8JB was not happy with how his old J-Pole held up in the weather, so he made a much more rugged one.

A number of years ago, I built a J-Pole antenna from some scrap aluminum tubing and a 5-foot stick of TV mast.[1] The antenna tuned easily, and operated well for a few years until nature did it in. After taking it down, I found a number of problems not contemplated in the original construction, such as dissimilar metals and other types of corrosion, and deterioration of parts of the coaxial feed line. These problems are identified and explained in the accompanying sidebar.

With the shortcomings of the original J-Pole in mind, I set about planning for another, similar antenna, using parts commonly available from a single source. If you spend much time in hardware stores and home improvement places, you get a feel for what's available at a moment's notice. In my case, rigid copper tubing, fittings, and assorted hardware, came to my attention as the material of choice. See **Fig 1**. The entire assembly can be soldered together when copper is used, thus ensuring electrical integrity, and making the whole antenna weatherproof in the bargain.

I'll bet you could solder one of these together faster than using the nuts and bolts I did in my previous design. This antenna can be easily used for ARES/RACES groups that spot antennas around for emergency use, since it requires little, if any, maintenance during its lifetime.

The J-Pole will take about an hour or so out of your day to build and tune, making a great antenna for a VHF base station.

No special hardware or machined parts are used in this antenna, nor are insulating materials needed, since the antenna is always at dc ground. Best of all, even if the parts aren't on sale, the antenna can be built for less than $15. If you only build one antenna, you'll have enough tubing left over to make most of a second antenna, or perhaps to finish that small plumbing project the XYL has been hounding you about.

Materials

Copper and brass is used exclusively in this antenna. These metals get along together, so dissimilar metal corrosion is eliminated. Both metals solder well, too. **Table 1** provides a detailed parts listing for the antenna.

Construction

Cut the copper tubing to the lengths indicated in Table 1. Item 9 is a 1¼-inch nipple cut from the 20 inch length of ½-inch tubing. This leaves 18¾ inches for the $\lambda/4$-matching stub. Item 10 is a 3¼-inch long nipple cut from the 60-inch length of ¾-inch tubing. The ¾-wave element should measure 56¾ inches long. Remove burrs from the ends of the tubing after cutting, and clean the mating surfaces with sandpaper, steel wool, or emery cloth.

After cleaning, apply a very thin coat of flux to the mating elements and assemble the tubing, elbow, tee, endcaps, and stubs. Solder the assembled parts with a propane torch and rosin-core solder. Wipe off excess solder with a damp cloth, being careful not to burn yourself. The copper tubing will hold heat for a long time after you've finished soldering. After soldering, set the assembly aside to cool.

Flatten one each of the ½-inch and ¾-inch pipe clamps. Drill a hole in the flattened clamp as shown in **Fig 2**. Assemble the clamps and cut off the excess metal from the flattened clamp using the unmodified clamp as a template. Disassemble the clamps.

Assemble the ½-inch clamp around the ¼-wave element and secure with two of the screws, washers, and nuts as shown in Fig 2. Do the same with the ¾-inch clamp around the ¾-wave element. Set the clamps initially to a spot 4 inches or so above the bottom of the "J" on their respective elements. Tighten the clamps only finger tight, since you'll have to move them when tuning.

Tuning

Tuning an antenna couldn't be simpler.[2] The toughest part might be determining what type feed line you are going to use. Anything from RG-58 to open-wire line is

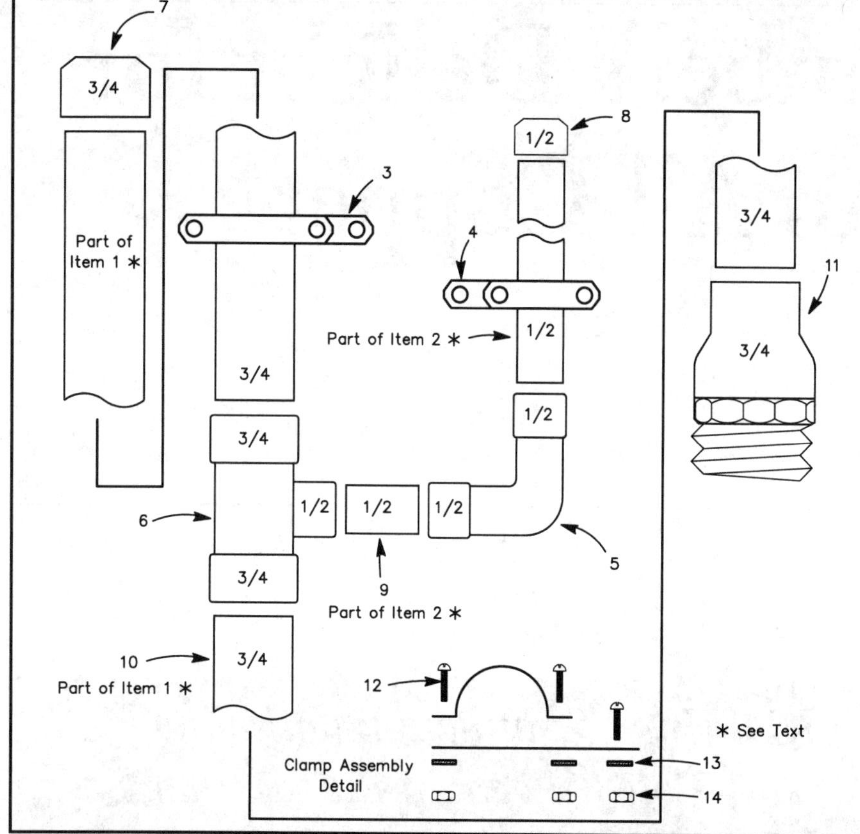

Fig 1—Exploded assembly diagram of all-copper J-Pole antenna. Item numbers refer to parts list in Table 1.

Table 1
Detailed Parts List (See Fig 1)

Item No.	*Qty*	*Part or Material Name*
1	1	3/4 inch × 10 foot length of rigid copper tubing (enough for 2 antennas, 60 inches per antenna)
2	1	1/2 inch × 10 foot length of rigid copper tubing (enough for 6 antennas, 20 inches per antenna)
3	2	3/4 inch copper pipe clamps
4	2	1/2 inch copper pipe clamps
5	1	1/2 inch copper elbow
6	1	3/4 × 1/2 inch copper tee
7	1	3/4 inch copper end cap
8	1	1/2 inch copper end cap
9	1	1/2 × 1 1/4 inch copper nipple (Make from item 2. See text)
10	1	3/4 × 3 1/4 inch copper nipple (Make from item 1. See text)
11	1	Your choice of coupling to mast fitting (3/4 × 1 inch NPT used at KD8JB)
12	6	8-32 × 1/2 inch brass machine screws (round, pan, or binder head)
13	6	no. 8 brass flat washers
14	6	8-32 brass hexnuts
15	A/R*	Rosin core solder
16	A/R*	Paste flux
17	A/R*	Fine sandpaper, steel wool, or emery cloth
18	A/R*	Solvent to clean away flux after soldering

*As required.

usable. The J-Pole can be fed directly from 50 Ω coax alone or with a 1/2-wave balun, or twin lead, or whatever. Before tuning, mount the antenna vertically, about 5 to 10 feet from the ground. A short TV mast on a tripod works well for this purpose.

When tuning VHF antennas, keep in mind that they are extremely sensitive to nearby objects—such as your body. Attach the feed line to the clamps on the antenna, and make sure all the nuts and screws are at least finger tight. If using coax, it really doesn't matter to which element (3/4-wave element or stub) you attach the coaxial center lead. I've done it both ways with no variation in operational effectiveness. Tune as follows:

1. Apply RF at the frequency you want the antenna to perform best.
2. Check the SWR.
3. Turn off the RF.
4. Move the clamps *equally* up from the original position 1/2 inch.
5. Reapply RF.
6. Check the SWR again, and turn off the RF.
   a. If the SWR went higher, move the clamps 1 inch downward, equally.
   b. If the SWR went lower, move the clamps 1/2 inch upward, equally.
7. Reapply RF.
8. Check the SWR once more, then turn off the RF. By this time you will be approaching minimum SWR (You may *never* get a 1:1 SWR. Don't sweat it, it's not important enough to worry about.)
9. Adjust the clamps a small amount in the direction of minimum SWR.
10. Repeat Steps 7 through 9 until minimum SWR is achieved.
11. Remove RF.

A point to remember about tuning the antenna is this: It will tune with any type of feed line, but the clamps will not be in the same place for all types of line. In other words, 50-Ω coaxial cable will tune further down the antenna than 600-Ω open-wire line. You have to think of this feedpoint as being similar to a delta match, which it is, except the elements run parallel and not away from each other.

Final Assembly

The final assembly of the antenna will determine its long-term survivability. Perform the following steps with care:

1. After adjusting the clamps for minimum SWR, mark the clamp positions with a pencil.
2. Remove the feed line and clamps.
3. Apply a very thin coating of flux to the inside of the clamp and the corresponding surface of the antenna element where the clamp attaches.
4. Install the clamps and tighten the clamp

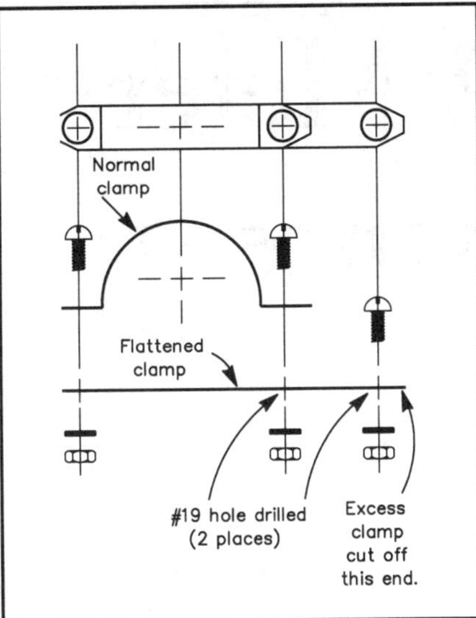

Fig 2—Detail of clamp assemblies. Both clamp assemblies are the same.

screws. Don't attach the feed line yet, and leave off the final washer and nut at the feedline attachment point.

5. Solder the clamps where they are attached to the antenna elements.
6. Apply a small amount of solder around the screw heads and nuts where they contact the clamps. Don't get solder on the screw threads!
7. Clean away excess flux with a non-corrosive solvent.

After final assembly and erecting/mounting the antenna in the desired location , attach the feed line you tuned the antenna for, and secure with the remaining washer and nut. It would be a good idea to weather-seal this joint with RTV. Otherwise, you may find yourself repairing the feed line after a couple years.

On-Air Performance

Years ago, prior to building the first J-Pole antenna for this station, I used a standard 1/4-wave ground plane vertical antenna. While there is no problem working the various repeaters around town with my 1/4 wave antenna, simplex operation left a lot to be desired, so I felt something with a little more gain was necessary. Hence the switch to the J-Pole. In on-air comparisons with the Ringo Ranger 2B, a popular antenna everywhere, a small difference in bandwidth was noted. This J-Pole's bandwidth, as built here, is slightly wider, probably from the greater element thickness, resulting in a lower Q. Actual performance differences between antennas of similar dimensions, such as those of the Ringo Ranger are negligible, although significantly better than the 1/4-wave ground-plane vertical.

Post Inspection/Evaluation of the Original J-Pole

1. ALUMINUM MOUNTING PLATE: The plate was rust-stained from contact with the non-stainless steel elements of the antenna. The alloy used for the plate was 2024 aluminum, which is quite hard and holds its finish well. However, the rust deposits affect the continuity between elements. This plate was reused after cleaning.
2. GALVANIZED PAINTED 3/4-WAVE SECTION: The painted portion of this element survived the ravages of weather very well. The small portion where the paint was removed to make contact with the aluminum plate was completely coated with rust, as were the U-bolts holding the element to the plate, and the screw (probably zinc plated) used to ensure contact between the section and the mounting plate's upper edge. The hole drilled for the stainless screw used to attach the coaxial cable shield was in as good a condition as when originally drilled owing to the liberal application of Silastic 732 RTV[4] after assembly. This section was reused as a 5-foot mast section. I would use 1 1/4-inch aluminum tubing if this antenna were rebuilt in this form.
3. NON-STAINLESS HARDWARE: All plated and unplated hardware exhibited varying degrees of oxidation (rust). Some of the galvanized hardware came through with only spot rust where the galvanized coating had either deteriorated or been scraped away. The cadmium-plated mast brackets bought at Radio Shack showed rust only where excess length on the U-bolts was removed and at the bends in the bracket plates. These brackets were reusable, but the nuts were replaced. The U-bolts holding the 3/4-wave section to the mounting plate were so badly rusted that when removal was attempted, the bolts broke off.
4. STAINLESS STEEL HARDWARE: As expected, all stainless steel came through in great shape. Even though mated with some non-stainless hardware like star washers or nuts with integral star washers, the stainless hardware was easy to disassemble. Never use other than stainless steel or brass hardware in antenna construction! It's well worth the extra cost.
5. COAXIAL CABLE: This antenna was fed with RG-8X (foam) made for LaCue Communications. Some cable was cut off for inspection when the antenna was taken down. (I do this whenever I take down an antenna to get a visual idea of how the coax is doing.) The cable was terminated at the antenna with crimp lugs, which I crimped and soldered. The shield remained in good condition throughout except for normal copper oxide near the lug. (I used RTV[4] to seal the spot where the center lead passed through the shield. The sealant worked well.) However, the foam dielectric between the shield and the terminal lug cracked in a number of places causing water to leak in and wick down the center conductor. Not good! Interestingly, the foam that had melted near the soldered terminal lug protected the center conductor well at that point.
   Lesson: When terminating coax in this fashion, the foam dielectric should be sleeved or taped to prevent cracking. In your haste to use a new antenna, always solder the lugs—never leave them just crimped. I should point out that I was responsible for the deterioration of the coaxial cable because of the way I used it. LaCue had nothing to do with what happened in this instance. I use LaCue's RG-8X cable for most of my feed lines, and when properly terminated, it works very well.

References

[1]M. P. Hood, "All-Metal 2-Meter J-Pole Antenna" (and References), *Ham Radio*, Jul 1984, pp 42-44.

[2]"A Combination 6 and 2 Meter J-Pole," *FM and Repeaters* (Newington: ARRL, 1972). (Out of print.)

[3]"Building and Using VHF Antennas," *VHF Handbook* (Newington: ARRL, 1972). (Out of print.)

[4]Silastic 732 RTV is made by the Dow Corning Company.

An Investigation Of 2-Meter HT Antenna Performance

By P. K. (Ken) Pierpont, KF4OW
204 Cedar Point Crescent
Yorktown, VA 23692

Ed Brummer, W4RTZ
108 Oyster Cove Road
Yorktown, VA 23692

Should you change the antenna on your HT? First take a look at what you really get with different antenna types and sizes. These measurements were taken under realistic conditions.

Introduction

In 1984 one of the authors (Ken, KF4OW) received his first 2-meter HT. He quickly recognized the 1-W unit didn't provide the communication desired despite its compactness, portability and convenience. An improved antenna seemed a logical first step. The first improved antenna was a λ/4 whip. It was constructed from a BNC connector with a piece of piano wire potted in place with epoxy. A thumbtack was soldered on top as a safety measure.

Next, after studying the ads and reading as much information on the subject as could be found, a commercial λ/2 telescoping antenna was purchased. A discussion with coauthor Ed Brummer, W4RTZ, led to the idea of making a few field tests to compare performance. A 5/8-λ mobile whip was borrowed to make an additional comparison with the λ/2 antenna. The first field experiments included attaching the HT to an existing wood post with rubber bands. The post was conveniently located about a mile and a half from the receiving station, and S-meter readings were used as the performance indicators. The results of these tests, performed in February, contained many anomalies and therefore were inconclusive.

In early March a second attempt was made to take an improved set of measurements. For these tests, an adjustable wood mount was built for the HT. This mount would permit rotation through 360° of azimuth angle and inclinations of the HT from antenna vertical to about 45°. Two additional popular telescoping 5/8 λ antennas, intended for HT use, were included in this second series of tests. All the antennas were tested over a range of azimuth angles and tilt angles with respect to the mouth of the operator in a normal position close to the built-in microphone.

A number of things were learned from these first two field experiments. These lessons were then employed in a final set of tests:

1. Do be aware of the difficulties in making antenna performance measurements. *The ARRL Antenna Book*[1] has some words of wisdom on this subject.

2. Be sure of the calibration of any meter used to indicate signal strength, and always take repeated reference measurements.

3. The test range must be clear of potential RF obstructions, such as trees, houses, power lines, etc.

4. Use a mounting system known to be transparent to RF at the frequencies of interest. Some plywoods, glue bonds and solid wood posts, as well as other materials like plastics, may not be truly RF-transparent. All materials proposed for use should be tested.

5. Use only a single power supply and take numerous reference data points to ensure uniform transmission levels. Transmission times should be brief to conserve battery power. The power level will be kept the same if only a fraction of the battery capacity is used. Never change battery packs during a test run.

6. Anything attached to the HT, such as a portable mike or auxiliary power cable will affect the characteristics of the emission patterns and should be avoided.

7. The presence of the operator's body will produce large effects on the emission pattern and must be controlled carefully.

A new mounting system was constructed based on these principles. RF transparent materials were used and an improved test range site was selected for the third series of tests. At the end of March, when the weather improved, a new test procedure was used to obtain what we believe are reproducible results.

Experimental Apparatus

Antenna Configurations Investigated:

The antenna configurations tested consisted of a helical-coil rubber duck supplied by the HT manufacturer, a home brew $\lambda/4$ whip antenna made of piano wire, and three commercially available telescoping antennas. One telescoping antenna was $\lambda/2$ when extended and the other two were $^5/_8\,\lambda$ when extended. All three telescoping antennas were claimed by their manufacturers to be usable when fully collapsed and therefore were also evaluated in the collapsed configuration.

Eight antennas were measured. They are shown in **Fig 1**. Because the purpose of this investigation was to evaluate performance for communication, no consideration was given to any other advantages claimed by a manufacturer.

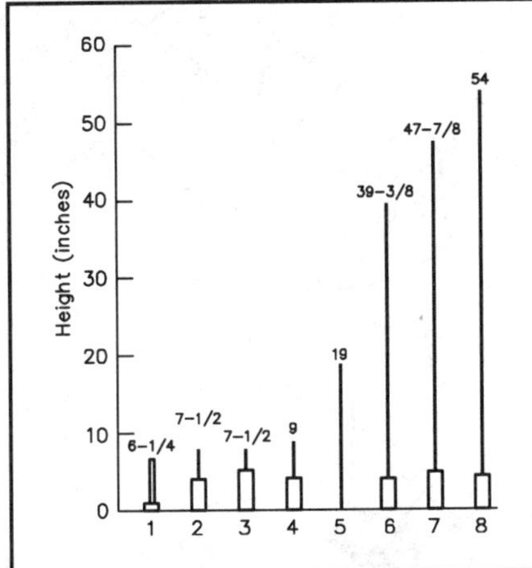

Fig 1—Description of HT antennas tested. All dimensions are measured from the top of the BNC connector. Antenna 1 is the reference rubber duck. Antennas 2, 3 and 4 are collapsed forms of telescoping antennas. Antenna 5 is a 19-inch whip, and antennas 6, 7 and 8 are telescoping antennas.

Transmitting and Receiving Equipment:

The test transmitter was a 2-meter HT capable of putting out both 1 W and 100 mW power levels with its standard battery pack. The rubber duck helical antenna supplied by the manufacturer was used as the reference antenna and is called the baseline configuration in this article. A reliable 2-meter transceiver equipped with a built-in S-meter and a 10-dB attenuator was used for received-signal measurements. Mounted on the roof at a height of about 25 feet was a homemade J-pole. It was matched to the feed line and transceiver with an SWR better than 1.1:1.

Field HT Mount:

The HT field-mount system is shown in the sketch of **Fig 2**. It was constructed of $^3/_4$-inch and 1-inch PVC pipe and clear $^1/_8$-inch plastic sheet to be entirely RF-transparent at these frequencies. All materials were tested to ensure their RF performance. The mount itself was a glued box, to which was attached the HT belt clip. A pivot made from $^1/_4$-inch PVC tubing and a second PVC tubing tilt-angle adjustment pin provides forward tilt from an initial vertical position of ϕ=0° to 45° in 15° increments. The upper-post section was 1-inch PVC pipe telescoped onto a $^3/_4$-inch section reinforced with a wood dowel inserted about a foot into a hole in the earth. Vertical adjustment was provided by a series of holes and an adjustment pin made from $^1/_4$-inch PVC tubing.

The azimuth angle was changed by rotating the entire mount. Optical alignment for direction was made with a reference compass rose laid out on the ground. Zero degrees was defined as a line connecting the receiving and transmitting stations, as determined by compass reading based on a US Coast and Geodetic Survey chart. The two primary angles of azimuth and tilt are shown in the sketch of **Fig 3**. The operator's position relative to the HT is also shown in Figs 2 and 3. In all cases, the operator head, body, arms, and hands were carefully kept in the same relative positions. Even the location of the hand and fingers affected the pattern. The importance of these factors cannot be overemphasized if reliable results are to be obtained and repeated.

Test Range:

The map of **Fig 4** shows the test range

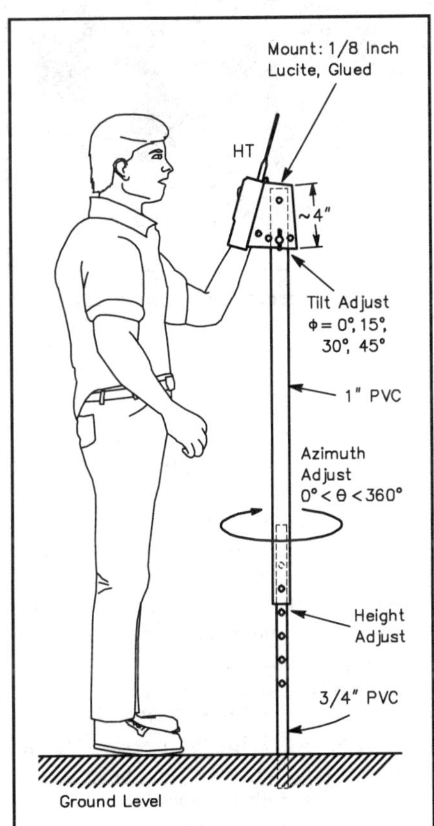

Fig 2—Field mount system.

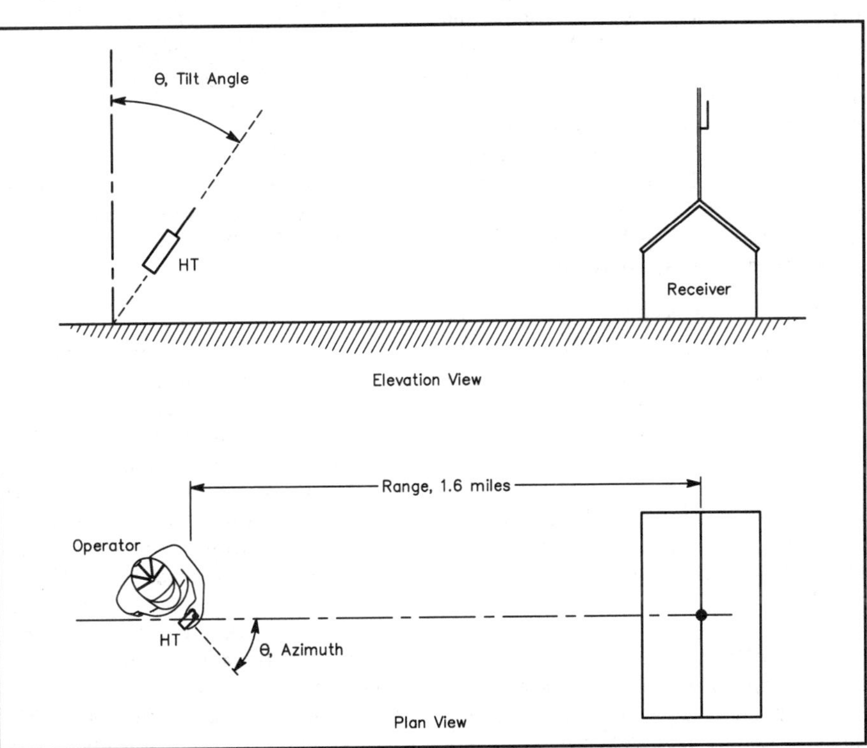

Fig 3—Definitions of azimuth and tilt angles.

used. The receiving antenna was in a clear area. The transmitting location was free of any obstructions, or RF reflectors, for at least 100 yards in any direction and more than 300 yards toward the receiving station. The nearest tree was about 150 yards to the side. The transmitting setup was located near a large pit filled with water to form a small lake, extending about 1/4 mile ahead and on both sides of the operations site.

Test Measurements and Data Reduction

All measurements were made at the receiving station by carefully determining the average value indicated by the S meter during an approximately 10 second unmodulated FM transmission. Since the distance between the receiver and transmitter was approximately 1.6 mi, the transmitter power level was kept at 100 mW. An attenuator was used to keep the received signal in the mid-scale range of the S meter. Each test run consisted of a single antenna configuration at a given angle of tilt. Data were taken for a series of azimuth angles by rotating the transmitter and operator through the 360° range. A repeat zero was always taken, and both initial and final data were taken for identical test conditions to confirm battery power level changes had not affected the results. In several cases, two test runs were made just to obtain typical data-scatter information.

A before- and after-test calibration of the S meter was made using a precision step attenuator. Thus the accuracies of all earlier calibrations were repeatedly confirmed. All data have been reduced by means of the calibration factors to a relative decibel scale using the zero-signal level as the reference value. This relative data is presented on polar coordinate graphs. Constant values (in azimuth) are therefore circles. In **Fig 5** the 5 and 10-dB circles are labeled and circles indicate 2-dB increments. Each configuration is assigned a symbol, and repeated data are represented by symbols with flags.

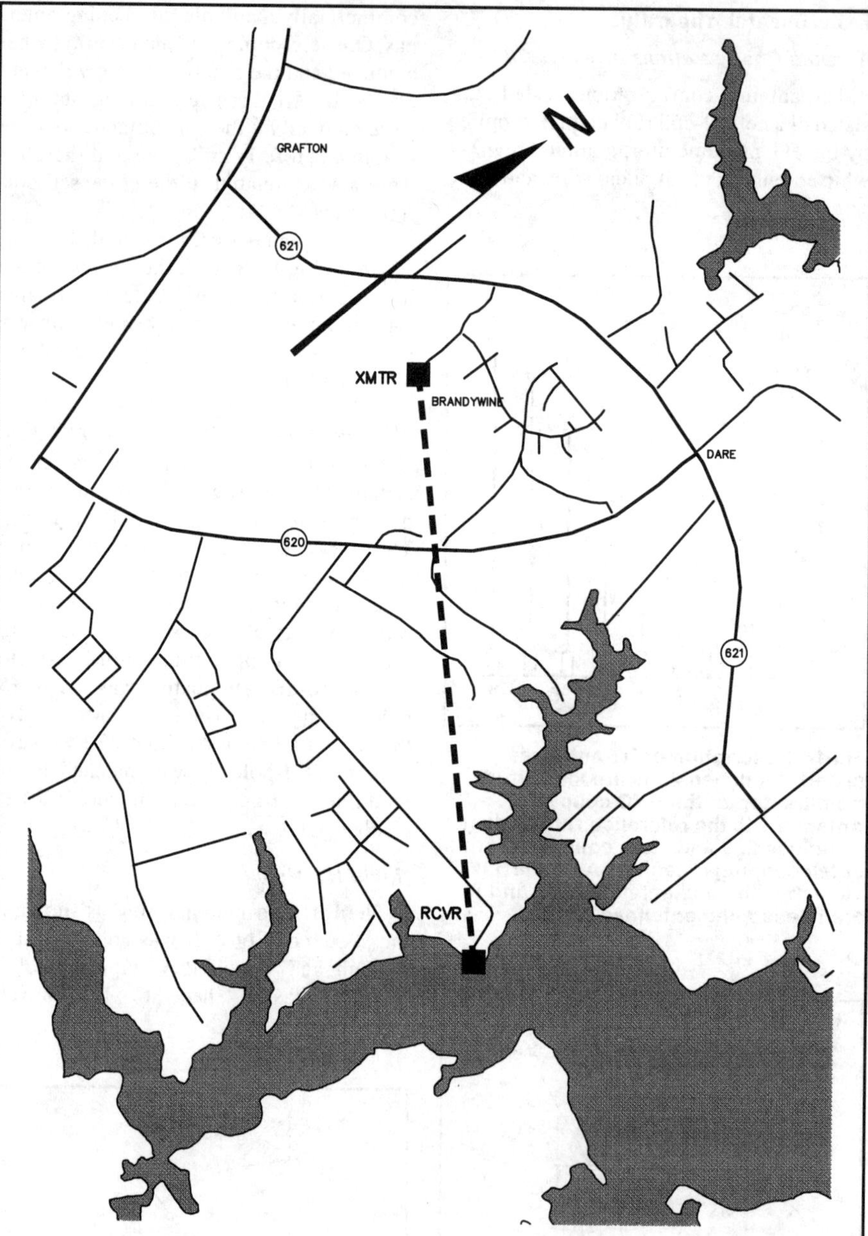

Fig 4—Test range in York County, Virginia; the distance from the transmit site to the receiver site is 1.6 mi.

Results and Discussion

The test results are divided into two basic groups. The first group consists of the short antennas, including the basic rubber duck (labeled #1) used as the baseline. The collapsed or duck versions of the telescoping antennas are labeled #2, 3, and 4. The second group consists of the long antennas, including the λ/4 whip, antenna #5, and the telescoping antennas, designated as #6, 7, and 8.

Short or "Duck" Antennas

Basic Influences: Fig 5 presents the behavior of the helical coil duck reference antenna used as the baseline for three test conditions with a tilt angle of 0° (antenna is vertical). The baseline antenna configuration data taken from different test runs illustrates clearly the gross asymmetry of the radiation pattern, showing a front-to-back ratio of more than 5 dB. Adding a stiff wire ground plane (4 wires of λ/4 spaced 90°) resulted in an almost symmetrical radiation pattern, simultaneous with a substantial decrease in forward signal level. With no ground plane, moving the operator a distance of about 5 λ laterally resulted in a near-circular radiation pattern of even greater strength than the best of the simulated hand-held configurations. This almost-circular pattern also indicates the high quality of the test setup and lends validity to the test results. The small remaining asymmetry is probably a result of the nonsymmetrical installation of the antenna on the HT itself, since it is mounted close to one side. Since this test was run with a short helical antenna, this relative level of 6 to 7 dB is probably the best attainable with this type of antenna under the most favorable antenna match conditions.

Some of the effects of tilt angle (set to 45°) are shown in **Fig 6** for the same three configurations. With the operator facing away from the receiving station (θ=180°), the signal strength is now almost equal to the case of the operator facing the receiving station, at an azimuth angle of 0°. The circu-

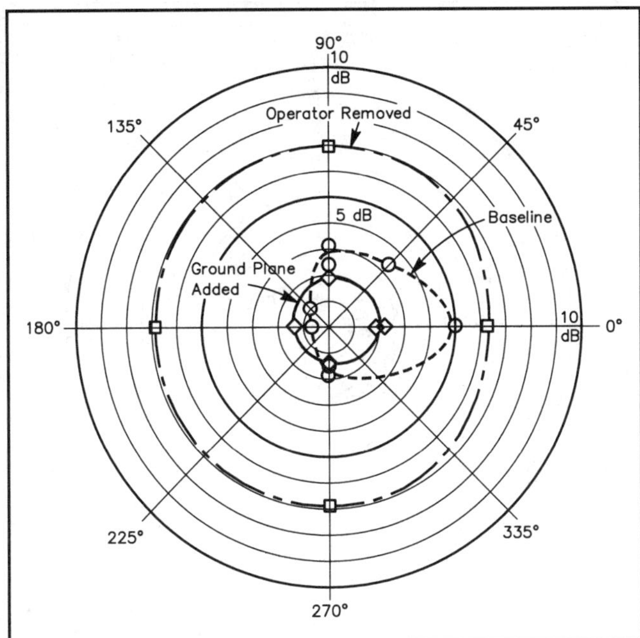

Fig 5—Effects of operator removal and ground plane addition on signal strength for antenna #1 (the baseline configuration), tilt = 0°.

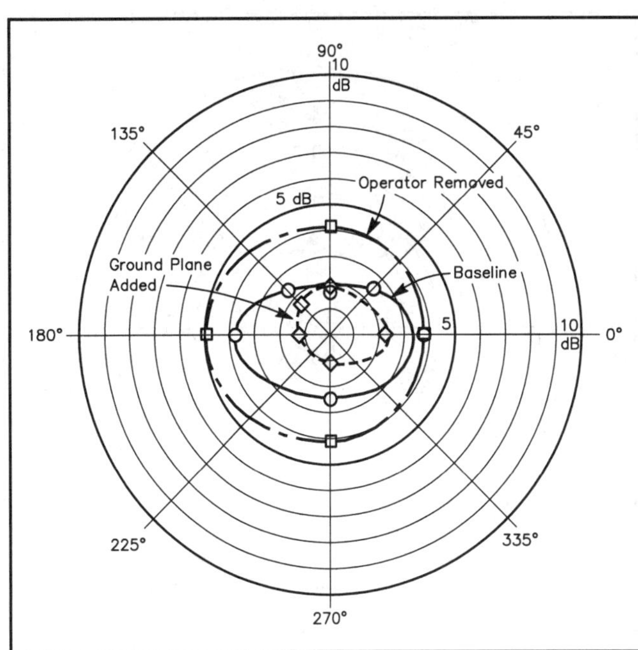

Fig 6—Effect of operator removal and ground plane addition on signal strength for antenna #1 with the tilt increased to 45°.

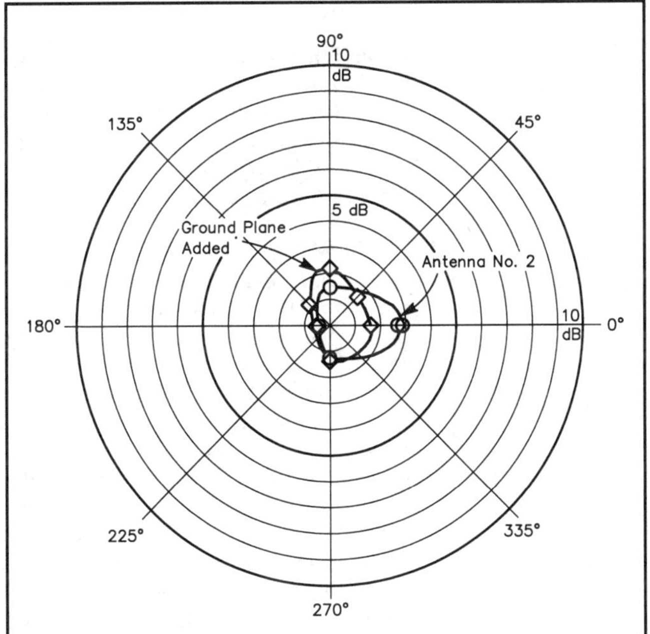

Fig 7—Effect of ground plane addition on the signal strength for antenna #2, tilt = 0°.

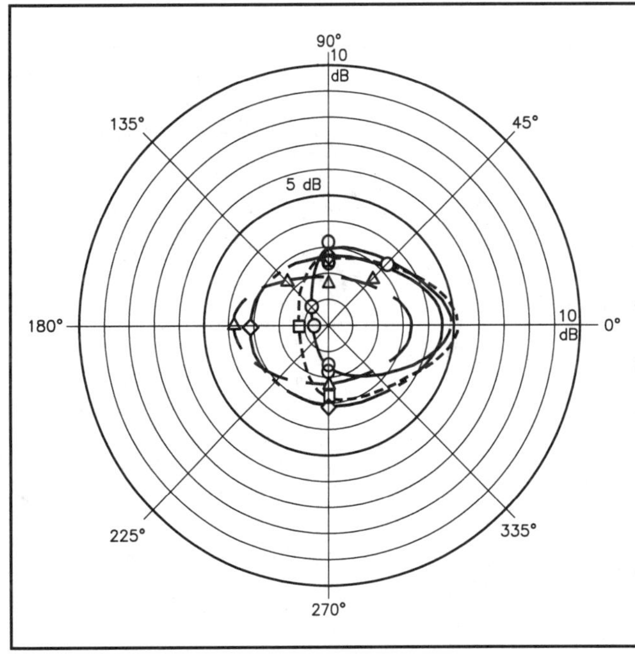

Fig 8—Effect of tilt angle on the signal strength for antenna #1, the baseline configuration.

lar pattern is distorted when the ground plane was added at this tilt angle.

When the operator was removed, and with no ground plane, a slightly distorted but still very strong signal was measured at all azimuth angles. This distortion was probably the result of the ground reflections coupled with the vertical plane lobe characteristics. These combined effects would be very difficult to express mathematically. Notice the large improvement in the signal strength, when tilted and facing away from the receiving station (see Fig 5).

Antenna Comparisons: **Fig 7** shows one of the telescoping antennas in the collapsed condition (so-called "duck" configuration) and with the ground plane added. Besides small level changes, there is a marked similarity of these normal HT configuration results compared with the helical-coil baseline antenna shown in Fig 5. Therefore, it may be inferred that this shape is characteristic of all the short antennas, whether a short helical coil, or some collapsed version of a long antenna. The level differences probably only reflect how well the antenna is matched when collapsed.

Effects of tilt angle are presented for the

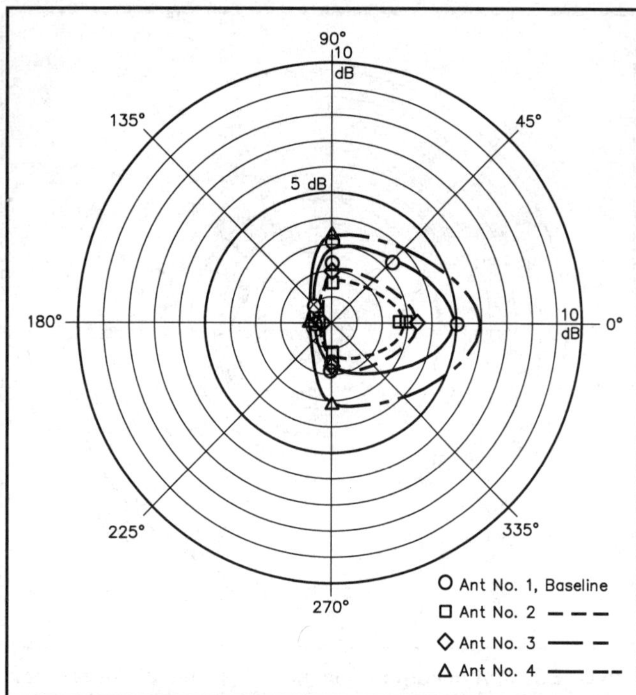

Fig 9—Comparison of the performance of four duck antennas, tilt = 0°.

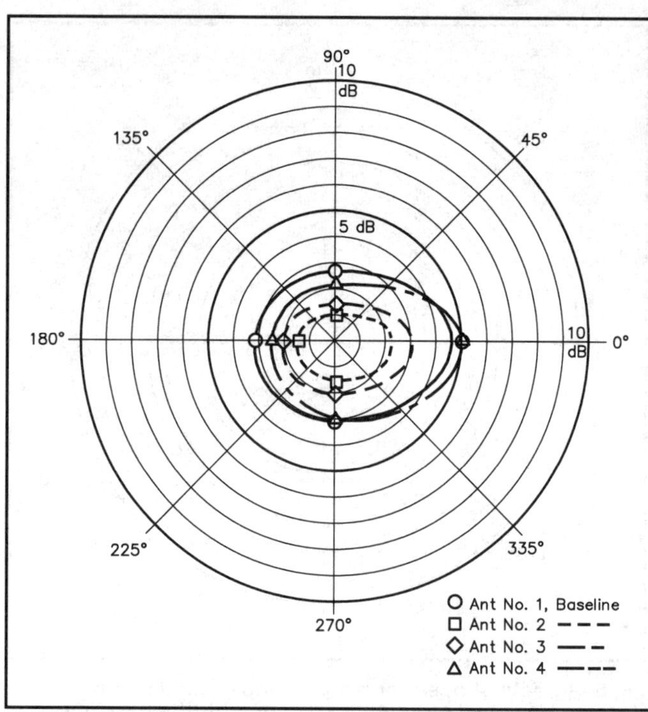

Fig 10—Comparison of the performance of four duck antennas, tilt = 30°.

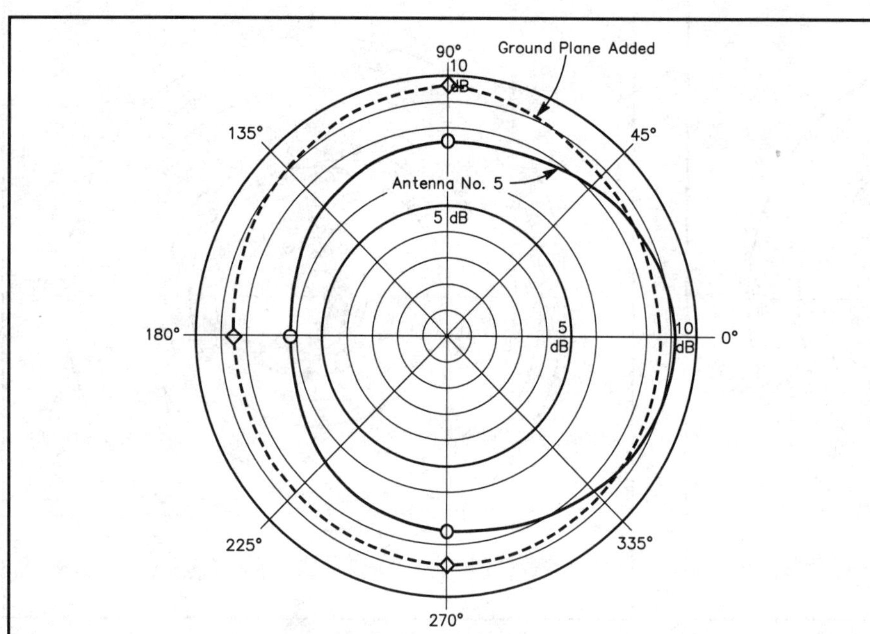

Fig 11—Effect of ground plane on signal strength with antenna #5, a λ/4 whip, tilt = 0°.

baseline duck, **Fig 8**, from the vertical (Ø=0°) to tilt angle of 45° (away from the operator). The progressive changes, especially at azimuth angle of 0° (facing the receiving station) and at 180° (facing away from the receiving station) are evident. The most rapid changes appear to occur between about 15° and 30° tilt angles.

The four "duck" configurations are compared in **Fig 9** at 0° tilt and in **Fig 10** at 30° tilt angle. The tilt angle was selected as being near the optimum, but on the high side in terms of radiation symmetry. The characteristics are all very similar except for the effect of mismatch. In all cases the forward radiation is several times greater than the rearward radiation when the antenna is vertical.

From these data it is also possible to show one design (antenna #4) has achieved a slightly higher signal level than the other designs (in the collapsed configuration) as compared with the baseline duck antenna. However, this slight difference has not been without other effects, to be pointed out later.

Long or Extended Antennas

Basic Influences: **Fig 11** shows the λ/4 whip on the HT with and without the ground plane made of four stiff wires λ/4 long, spaced 90°. Much higher signal strengths were measured for this λ/4 whip compared to those for any of the duck configurations previously shown. The signal levels are several decibels higher in all directions. For this antenna, only a moderate signal decrease results from facing away from the receiving station (θ=180°). Despite all that can be said concerning the antenna portion that does most of the radiating because of the current distribution, the upper part of the antenna is now well above the operator's head. The large, directional, operator-presence losses associated with short antennas shown earlier have now disappeared. The signal strengths measured with the λ/4 whip are in the same range as the values shown in Fig 5 for the baseline duck antenna when the operator was removed from the transmitter.

Antenna Comparisons: For antennas vertical (ϕ=0°), **Fig 12** presents a comparison of the performance of all of the long antennas referenced to the baseline duck. For this, and subsequent figures, the scale has been changed by a factor of two. If a wide paint-

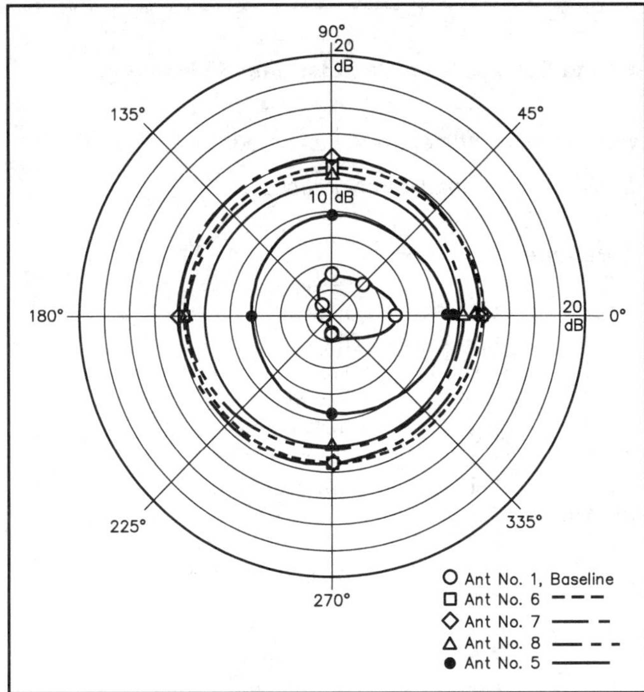

Fig 12—Comparison of the performance of the long antennas with the baseline duck and antenna #5, tilt = 0°.

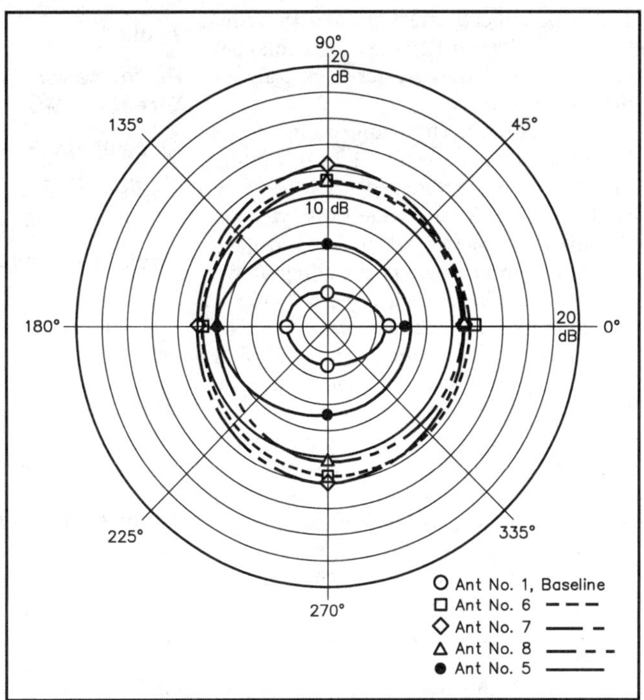

Fig 13—Comparison of the performance of the extended antennas with the baseline duck and antenna #5, tilt = 30°.

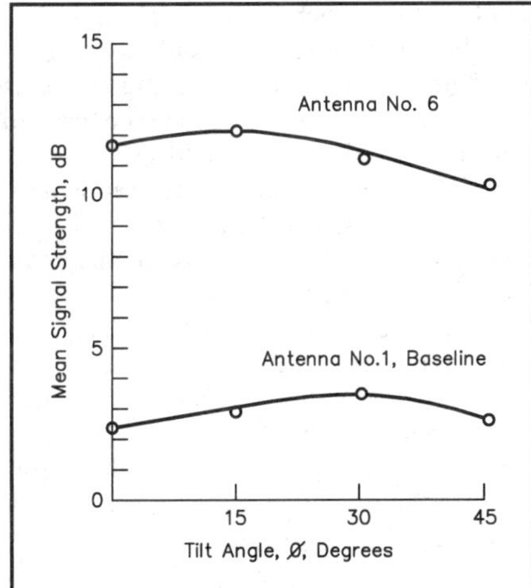

Fig 14—Variation of Mean Signal Strength, MSS, with tilt angle, antennas #1 and #6.

brush line, nearly circular in shape, were drawn, it would encompass the signal strength characteristics of all the extended versions of the telescoping antennas (whether $^1/_2\,\lambda$ or $^5/_8\,\lambda$). None of these extended antennas can be judged superior. As should be expected, they are all a little better than the $\lambda/4$ whip, antenna #5.

Fig 13 is a similar comparison of the long antennas with the baseline duck, but at a tilt angle of 30°. Once again, no one antenna is noticeably better.

Averaged Comparisons

Fig 14 is a comparison of the reference baseline duck configuration and a telescoping $\lambda/2$ antenna. The Mean Signal Strength, MSS, is plotted against tilt. MSS is an average of the measured signal strength values through the azimuth range of 0° to 360:

$$MSS = \int_0^1 (dB)\, d(\theta / 360)$$

where dB is the measured signal strength at each azimuth angle.

The results show the MSS for a short duck-type antenna will average about 25 to 30% higher if the antenna is tilted to some angle probably, not less than 20° nor more than 35°. This suggests the natural attitude for all-around best communications will not be far different from the tilt angle at which many operators normally hold their HTs. A conscious effort at tilting has already been clearly shown (see Fig 8) to be advantageous if reception behind the operator is needed.

For long antennas, tilting a few degrees appears to offer only a minor benefit. More importantly, too much tilt in this situation is bound to hurt, especially if it is more than about 15°. This is not surprising, because the combination of vertical plane signal lobes and ground reflections almost always interfere.

Finally, **Table 1** provides a comparison of the incremental value of the MSS. The incremental MSS is simply the calculated MSS referenced to the baseline configuration. Since a duck-type antenna is usually the starting point for the user, the results are shown with the duck performance as the reference.

Using this method of comparison configurations 2 and 3 (collapsed telescoping antennas) appear about 1-dB poorer than the baseline (incremental MSS –1.2 and –0.9). These same antennas, fully extended as in configurations 6 and 7, show overall improvements averaging about 10 dB (11.8 – 1.8). This amounts to an incremental MSS of about 9 dB over the baseline—a substantial increase. However configuration 8 provided a gain of only 7.5 dB (10.8 – 3.3) in its fully extended condition. This was probably due to an attempt to obtain a design in the collapsed configuration (antenna #4) that would be better than the baseline duck. For configuration 8 the net incremental MSS is about 8 dB, or about 1 dB less than the other two telescoping antennas. Of course, the difference in the ability of the receiving station to hear is barely noticeable for gains of either 8 or 9 dB, so all of the three telescoping antennas must be considered to exhibit about the same overall performance.

Configurations 9 through 14 show some other effects. The presence of the human body, configuration 9, degrades the MSS by about 4 dB (2.7–6.7 = – 4.0). Tilting this arrangement 45°, configuration 10, resulted in severe ground reflection interference. Thus, the gain of removing the operator decreased from 4.0 to a mere 1.5 dB. The addition of ground planes to short antennas always degraded the MSS. For the helical coil duck, either vertical or tilted 45° (configurations 11 and 12), the incremental decrease in MSS was greater than 1 dB, or more than 30%.

For the collapsed telescoping antenna, its MSS was already down by 1.2 dB. Adding the ground plane reduced the MSS only slightly (0.1 dB), but severely distorted the radiation pattern. Only when a ground plane was added to an antenna extending above the operator's head, such as the λ/4 whip of configuration 14, could improvements be shown. For this case, the incremental MSS is about 6, and the received signal would be four times the magnitude of the baseline antenna, configuration 1. Thus, for this simple antenna, an output of 1 W becomes effectively about 4 W. This may well make the difference in whether or not satisfactory communication can be established.

All signal strength and MSS values in Table 1 and throughout this report are in relative dB and not with respect to any absolute scale. The authors, in hindsight, regret they did not attempt to tie the measured signal strength levels to an absolute scale, such as a reference λ/2 vertical dipole.

Table 1

Performance Comparisons Derived from Calculations of Mean Signal Strength (MSS)

All configurations are with antenna vertical except as noted.

Config Number	*Antenna Number*	*MSS dB*	*Incremental MSS, dB*	*Notations*
Basic Antenna Configurations				
1	1	2.7	+0.0	Baseline
2	2	1.5	−1.2	Duck
3	3	1.8	−0.9	Duck
4	4	3.3	+0.6	Duck
5	5	7.6	+4.9	1/4-λ whip
6	6	11.6	+8.9	1/2-λ extended
7	7	11.8	+9.1	5/8-λ extended
8	8	10.8	+8.1	5/8-λ extended
Other Antenna Configurations				
9	1	6.7	+4.0	Operator removed
10	1	4.2	+1.5	No operator, $\phi = 45°$
11	1	1.7	−1.0	Ground plane
12	1	1.5	−1.2	Ground plane, $\phi = 45°$
13	2	1.4	−1.3	Ground plane
14	5	8.9	+6.2	Ground plane

Concluding Remarks

A carefully controlled experimental investigation has been performed to measure the performance of several HT antenna configurations, including three telescoping antennas. Test parameters included azimuth angle and tilt angle. The following are the principal results:

1. Test range and test mount considerations are important in order to obtain high quality data.

2. Azimuth angle is especially important when using short antennas such as ducks, or collapsed telescoping antennas.

3. Tilt angle produces large, often favorable, signal changes for all short antennas.

4. None of the telescoping antennas tested proved markedly superior in their collapsed configurations.

5. Addition of a 4-element stiff wire ground plane to a short, or duck antenna produced generally degraded signal strength patterns.

6. Operator body influences are very large, signal-degrading, and position-sensitive, particularly for short antennas.

7. Azimuth angle is relatively unimportant for long antennas extending above the operator's head, such as λ/4 or longer antennas.

8. Tilt angle is relatively unimportant for long antennas extending above the operator's head, such as λ/4 or longer.

9. None of the telescoping antennas proved markedly superior in their extended configurations; in fact, all gave similar results.

In addition, these results indicate the following recommendations for the user:

1. For station-to-station communication, always face the other station, if its location is known.

2. While operating simplex in the field with a rubber duck antenna, if one needs to blank a particular station, turn 180° away from that station. The human body will block most of the interfering signal, especially if the rubber duck is held vertical.

3. For multiple station simplex operation, or if the location of the receiving station is unknown to the operator, moderate forward tilt angles of 15° to 30° should be employed, especially if duck antennas are being used. Tilting may well make the difference between success and failure to communicate under marginal conditions;

4. An excellent, low-cost, compact, antenna is a simple λ/4 whip. One can be easily fabricated from an old portable radio telescoping antenna. Addition of a simple 4-wire λ/4 ground plane will provide performance almost as good as the best of the telescoping λ/2 or 5/8 λ antennas.

[1] *The ARRL Antenna Book*, 17th Edition (The American Radio Relay League, 1994) pp 27-43 to 27-48.

Notes

Notes

Notes

Notes

ANTENNA COMPENDIUM
VOLUME 4

PROOF OF
PURCHASE

FEEDBACK

Please use this form to give us your comments on this book and what you'd like to see in future editions.

Where did you purchase this book? ☐ From ARRL directly ☐ From an ARRL dealer

Is there a dealer who carries ARRL publications within: ☐ 5 miles ☐ 15 miles ☐ 30 miles of your location? ☐ Not sure.

License class:

☐ Novice ☐ Technician ☐ Technician with HF privileges ☐ General ☐ Advanced ☐ Extra

Name ______________________ Call sign ______________________

Address ______________________

City, State/Province, ZIP/Postal Code ______________________

Daytime Phone (   ) ______________________ Age ______________________

If licensed, how long? ______________________ ARRL member? ☐ Yes ☐ No

Other hobbies ______________________

____

Occupation ______________________

For ARRL use only	**ACV4**
Edition	1 2 3 4 5 6 7 8 9 10 11 12
Printing	2 3 4 5 6 7 8 9 10 11 12

From ______________________________________

____

____

Please affix postage. Post Office will not deliver without postage.

EDITOR, ANTENNA COMPENDIUM, VOL 4
AMERICAN RADIO RELAY LEAGUE
225 MAIN ST
NEWINGTON CT 06111-1494

please fold and tape